Marcel Berger

Geometry II

Translated from the French by M. Cole and S. Levy

With 364 Figures

Springer-Verlag Berlin Heidelberg New York
London Paris Tokyo

Marcel Berger
Institut des Hautes Etudes Scientifiques
35, route de Chartres
91440 Bures-sur-Yvette
France

Michael Cole
Academic Industrial Epistemology, 17 St. Mary's Mount
Leyburn, North Yorkshire DL8 5JB
England

Silvio Levy
Department of Mathematics, Princeton University
Princeton, NJ 08544, USA

Title of the original French edition: Géométrie (vols. 1–5)
Published by CEDIC and Fernand Nathan, Paris
© CEDIC and Fernand Nathan, Paris 1977

Mathematics Subject Classification (1980): 51-XX, 52-XX, 53-XX

ISBN 3-540-17015-4 Springer-Verlag Berlin Heidelberg New York
ISBN 0-387-17015-4 Springer-Verlag New York Berlin Heidelberg

© Springer-Verlag Berlin Heidelberg 1987
Printed in Germany

Printing and bookbinding: Druckhaus Beltz, Hemsbach/Bergstr.
2141/3140-543210

Table of contents

Volume I

Volume II

Chapter 12

Polytopes; compact convex sets

This is an important chapter; thanks to the previously accumulated material, we will be able to prove here several difficult results. Polytopes are compact polyhedra with non-empty interior; they generalize convex polygons in the plane and convex polyhedra in three-dimensional space.

The first three sections give the basic theory: faces, volume, area, duality, standard examples.

Sections 12.4, 12.5 and 12.6 are devoted to regular polytopes, a generalization of the regular polygons and polyhedra that we met in chapter 1. In spite of their apparent simplicity, the correct definition of regular polytopes requires a certain caution; thus we devote section 12.4 to regular polygons, so as to prepare the reader for a general definition. The core of section 12.5 is devoted to examples of regular polytopes. Some of these are easy: the analogs in arbitrary dimension of the regular tetrahedron, the cube and the octahedron. In contrast, demonstrating the existence of the dodecahedron and the icosahedron in dimension three is already harder, and in four dimensions two of the three polytopes we describe are very complicated.

In section 12.6 we proceed to classify regular polytopes in arbitrary dimension, and the result is perhaps surprising: not only is the case $d = 2$ exceptional, in that there exist infinitely many regular polygons, but so are the cases $d = 3$ and $d = 4$, since in higher dimension there exist only three regular polytopes (the cube, the cocube and the regular simplex), whereas we have established the existence of the icosahedron and the dodecahedron in three dimensions and of three extra regular polytopes in dimension four. There are no other regular polytopes. A heuristic remark about this is given in 12.6.8.

In section 12.7 we demonstrate Euler's formula: for every polyhedron in dimension three the number of vertices minus the number of edges plus the number of faces is equal to two. This remark is the cradle of algebraic topology.

Section 12.8 is devoted to Cauchy's theorem, which says that, contrary to the case of plane polygons, convex polyhedra in dimension three are not flexible around their edges. This result, however simple its statement, is quite difficult to demonstrate.

The last three sections contain the proof of the isoperimetric inequality: among all compact convex sets with a given area, the maximum volume is achieved by spheres. In order to get to that, one must first define the area of a compact convex set; we have already remarked in 9.12.7 how difficult it is to define a notion of area for an arbitrary compact set. Here things work smoothly because convex compact sets are very well approximated by polytopes, so one can extend by continuity the definition of area from polytopes to compact convex sets. We include two classical proofs of the isoperimetric inequality, as they are both very elegant; we also include a third proof, very recent, due to Gromov and based on an idea of H. Knothe [KT]. Gromov's proof has two advantages: it applies to the non-convex case as well, and it makes it particularly easy to show that only spheres have maximal volume, something that in other proofs tends to be a delicate business.

We have included references on historical data throughout the chapter. We also suggest the basic reference [WN] for the construction of models of polyhedra from cardboard, an endeavor which the author personally finds fun and rewarding.

In this chapter X always denotes a d-dimensional real affine space. Except
for sections 12.1 and 12.7, we also assume X to be endowed with a Euclidean
structure. In section 12.4 we assume $d = 2$, and in sections 12.7 and 12.8
we assume $d = 3$.

12.1. Definitions and examples

12.1.1. DEFINITION. *A convex polyhedron in X is a subset of X obtained as
a finite intersection of closed half-spaces* (cf. 2.7.3). *A polytope is a compact
convex polyhedron with non-empty interior* (cf. 11.2.7). *In case* $\dim X = 2$,
we talk about polygons instead of polytopes.

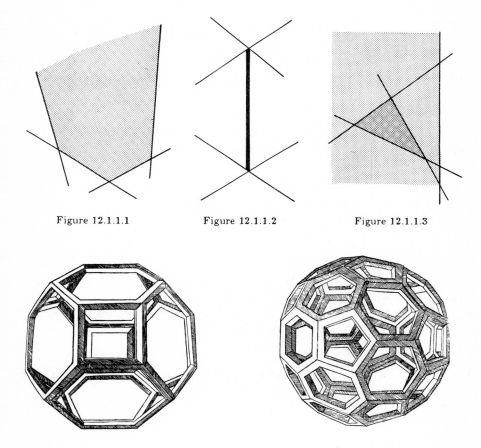

Figure 12.1.1.1   Figure 12.1.1.2   Figure 12.1.1.3

Figure 12.1.1.4
Drawn by Leonardo da Vinci for Fra Luca Pacioli's *De Divina Proportione*

Figure 12.1.1.5

Figure 12.1.1.6

(Source: [HL])

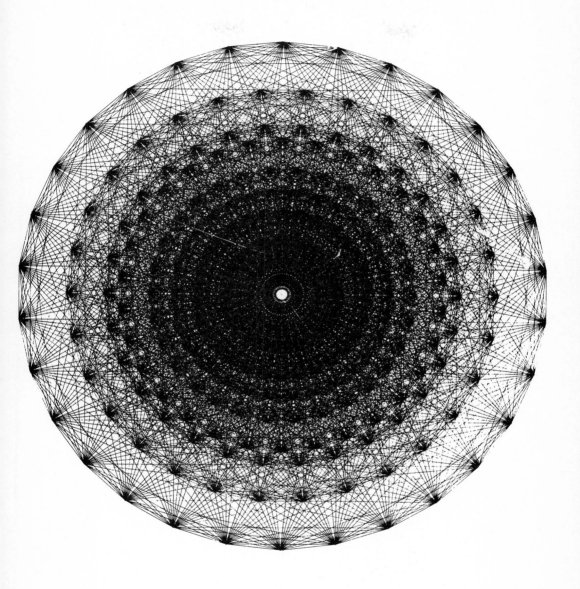

Figure 12.1.1.7 (Source: [CR4])

12.1.2. EXAMPLES

12.1.2.1. Parallelepipeds. If $(x_i)_{i=0,1,\ldots,d}$ is a simplex, the parallelepiped

$$P = \left\{ x_0 + \sum_{i=0}^{d} \lambda_i \overrightarrow{x_0 x_i} \ \Big| \ \lambda_i \in [0,1] \text{ for all } i \right\}$$

is a polytope (cf. 9.12.4.2).

12.1.2.2. Solid simplices. Similarly,

$$P = \left\{ \sum_{i=0}^{d} \lambda_i x_i \ \Big| \ \lambda_i \geq 0 \text{ and } \sum_i \lambda_i = 1 \right\}$$

is also a polytope.

12.1.2.3. A ball is compact and convex, but not a polytope if $\dim d \geq 2$.

12.1.2.4. A finite intersection of convex polyhedra is still a convex polyhedron. The intersection of a convex polyhedron with an affine subspace yields a convex polyhedron in that subspace; this is also true for polytopes if the affine subspace intersects the interior of the polytope.

12.1.2.5. Three standard polytopes should become familiar:
— the *standard cube* $\text{Cub}_d = \left\{ (x_i, \ldots, x_d) \mid |x_i| \leq 1 \text{ for all } i = 1, \ldots, d \right\}$,
— the *standard cocube* $\text{Coc}_d = \left\{ (x_1, \ldots, x_d) \mid \sum_i |x_i| \leq 1 \right\}$, and
— the *standard solid simplex*

$$\text{Simp}_d = \left\{ (x_1, \ldots, x_{d+1}) \mid \sum_i x_i = 1 \text{ and } x_i \geq 0 \text{ for all } i \right\}.$$

The cube and cocube are polytopes in $\mathbf{R}^d$, whereas Simp_d is a polytope in the hyperplane $H = \left\{ (x_1, \ldots, x_{d+1}) \mid \sum_i x_i = 1 \right\}$ of $\mathbf{R}^{d+1}$.

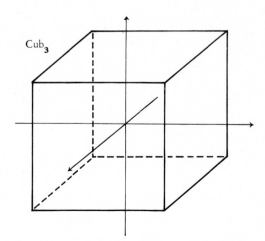

Figure 12.1.2.5.1

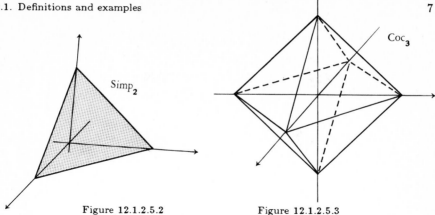

Figure 12.1.2.5.2   Figure 12.1.2.5.3

A *d*-dimensional *cube*, *cocube* or *regular simplex* is any polytope in Euclidean space similar (cf. 9.12.3) to the corresponding standard model. We remark that cubes and cocubes coincide in dimension two, but not for $d \geq 3$, as can be seen from 12.1.11.2 and 12.1.11.3, for example.

12.1.2.6. Duality. Here X is a Euclidean vector space. Let $(a_i)_{i=1,\ldots,n}$ be a finite set of points in X, and $Q = \mathcal{E}(a_1, \ldots, a_n)$ its convex hull. Then the convex polar body Q^* of Q (cf. 11.1.5) is a convex polyhedron; it is also a polytope if $0 \in \overset{\circ}{Q}$, and in that case we call Q^* the *dual* of Q. To see this, recall that, by definition,

$$Q^* = \{\, x \in X \mid (x \mid \textstyle\sum_i \lambda_i a_i) \leq 1 \text{ for any } \lambda_i \geq 0 \text{ and } \textstyle\sum_i \lambda_i = 1 \,\},$$

so $Q^* = \bigcap_i \{\, x \in X \mid (x \mid a_i) \leq 1 \,\}$ is a convex polyhedron. Now apply 11.4.8.

12.1.2.7. Example. The dual of Cub_d is Coc_d and vice versa.

We will soon verify that polytopes correspond to sets with non-empty interior that are convex hulls of finite sets (cf. 12.1.15). This and 12.1.2.6 prove the following result:

12.1.2.8. Proposition. *Let X be a Euclidean vector space. For every polytope Q containing 0 in its interior, Q^* is still a polytope with 0 in its interior, and $Q^{**} = Q$. We call Q^* the dual of Q.* $\qquad\square$

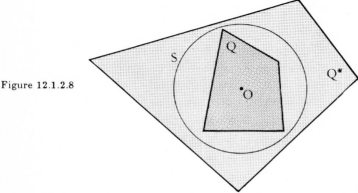

Figure 12.1.2.8

12.1.3. NOTES.

12.1.3.1. How would a (not necessarily convex) *polyhedron* be defined? One possible way is the following: a polyhedron is a finite union of compact convex polyhedra. Observe that polyhedra, thus defined, are not necessarily convex, not even simply connected. The definition just given is designed for a certain kind of geometry, cf. [HR], for example; for other definitions and other points of view, see [AW2], [ST–RA], [ZN, 229], for example.

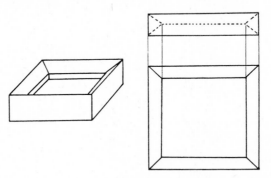

Figure 12.1.3.1

12.1.3.2. For cardboard models of polyhedra, and a great variety of them, at that, the basic reference is [WN]. It takes more than a hundred man-hours to assemble the polyhedron in figure 12.1.3.2.2, for example.

Figure 12.1.3.2.1 (Source: [WN])

Figure 12.1.3.2.2 (Source: [WN])

12.1.3.3. To represent higher-dimensional polyhedra in dimension two, there are several possible methods; some examples are illustrated in figures 12.1.1.7, 12.1.3.3, 12.5.6.1, and 12.5.6.2.1 to 12.5.6.2.4. For explanations on these figures and other possible methods, see [H–C, 145–157], [CR2, chapter XIII], and the stereographic glasses that come with [FT2].

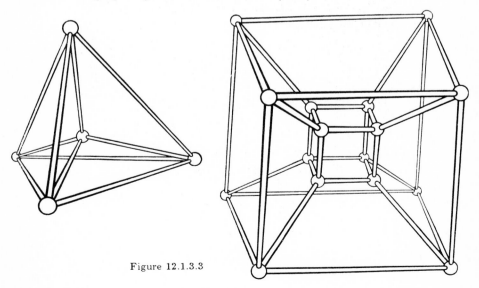

Figure 12.1.3.3

12.1.4. REMARK. In order to appreciate the next theorem, it is important to realize that there may appear superfluous half-spaces in the definition of a convex polyhedron (figure 12.1.1.3), and even if we write the polyhedron as the intersection of as few half-spaces as possible, there may still be many ways of doing so (figure 12.1.1.2, where two of the four half-spaces used in defining the convex polyhedron in question can be changed at will).

12.1.5. THEOREM (STRUCTURE OF POLYHEDRA). *Let P be a convex polyhedron with non-empty interior, defined as $P = \bigcap_{i=1}^{n} R_i$, where the R_i are closed half-spaces and we assume that n is minimal. Then*
  i) *the R_i are well-determined, up to a permutation;*
  ii) *if H_i is the hyperplane $\operatorname{Fr} R_i$ defining R_i, the intersection $H_i \cap P$ is a convex polyhedron with non-empty interior in H_i, called the i-th **face** (or side, if $d = 2$) of P, and denoted by $\operatorname{Face}_i P$;*
  iii) *the frontier of P is the union of the faces of P.*

12.1.6. CONVENTION. From now on, whenever we write a convex polyhedron with non-empty interior as the intersection of half-spaces, we assume the number of half-spaces to be minimal.

Proof. For a fixed i, let $P' = \bigcap_{j \neq i} R_j$; since we assumed n to be minimal, there exists $x \in P' \setminus R_i$. Let $a \in P$; by 11.2.4, the point $y = [a, x] \cap H_i$ lies in P', so y belongs to the interior of $H_i \cap P'$ in H_i. Since $\mathring{R}_i = R_i \setminus H_i$ and $\mathring{P} = \bigcap_i \mathring{R}_i$, we have

$$\operatorname{Fr} P = \bigcup_i \left(\operatorname{Fr} R_i \cap \left(\bigcap_{j \neq i} R_j \right) \right) = \bigcup_i (H_i \cap P) = \bigcup_i \operatorname{Face}_i P,$$

proving (iii). Since the H_i are distinct, this shows that the faces are well determined by P as subsets of X, hence so are the R_i.

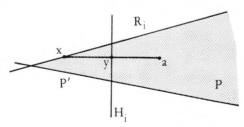

Figure 12.1.6

12.1.7. EXAMPLES. The faces of a polytope P are polytopes. If H is a hyperplane of X intersecting the interior of a polytope P, then $H \cap P$ is a polytope of H whose faces are $H \cap \operatorname{Face}_i P$.

As we can start considering the faces of the faces of P and so on, we generalize as follows:

12.1.8. DEFINITION. *Let P be a convex polyhedron with non-empty interior. A face of P is also called a $(d-1)$-face. A k-face of P is a face of a $(k+1)$-face of P $(k = 0, 1, \ldots, d-2)$. A 1-face is called an edge (or side if $d = 2$). Two k-faces are called adjacent if their intersection is a $(k-1)$-face.*

It is to be suspected that 0-faces are vertices, in the sense of 11.6.1, and this is precisely the case:

12.1.9. PROPOSITION. *Let P be a convex polyhedron with non-empty interior.*

i) *If $x \in \mathrm{Fr}\, P$, the intersection of the hyperplanes determined by faces containing x coincides with the intersection of the supporting hyperplanes to P at x. Thus the points of order α of P form the relative interior of the α-faces of P and, in particular, the vertices of P coincide with its 0-faces.*

ii) *The vertices of P coincide with the extremal points of P.*

Proof. If H is a supporting hyperplane to P, one of the half-spaces defined by H, say R, satisfies $P = P \cap R$, so 12.1.5 shows that, if H is not one of the H_i, we don't have to worry about it. This shows (i). If $x \in \mathrm{Fr}\, P$ is not extremal, there exists a segment $[y, z] \subset \mathrm{Fr}\, P$ such that $x \in]y, z[$, so x will have order greater than zero. $\square$

12.1.10. COROLLARY. *Let X be a Euclidean vector space and Q a polytope such that $0 \in \overset{\circ}{Q}$. Then the k-faces of Q are in one-to-one correspondence with the $(d-k-1)$-faces of the dual Q^*, as follows: if Y is the k-dimensional subspace that determines the face in question, the subspace Y^* determining the corresponding face of Q^* is the intersection of the polar hyperplane of the points of Y, polarity being taken with respect to the unit sphere* (cf. 10.7.11).

Proof. This follows from 12.1.9 and 11.5.3. $\square$

12.1.11. THE STANDARD EXAMPLES.

12.1.11.1. The standard simplex. The vertices of the standard simplex Simp_d are among the $d + 1$ vectors of the canonical basis $(e_i)_{i=1,\ldots,d+1}$ of $\mathbf{R}^{d+1}$, since Simp_d is the convex hull of these vectors. In fact, the vertices are exactly these $d + 1$ points, since there can be no fewer. For the same reason every $(i + 1)$-element subset of $(e_i)_{i=1,\ldots,d+1}$ has an i-face as its convex hull, and all i-faces are obtained in this way. Thus there are $\binom{d+1}{i+1}$ such i-faces, and they are all regular simplices.

12.1.11.2. The standard cube. The faces of the standard cube are the $2d$ hyperplanes $x_i = \pm 1$ $(i = 1, \ldots, d)$, since none of them is superfluous (cf. 12.1.5). But such a face is a $(d-1)$-dimensional cube, so all k-faces of Cub_d are cubes. A k-face can be explicitly expressed by fixing $d - k$ coordinates with values ± 1, and making the others range in the interval $[-1, 1]$. This shows that there are $\binom{d}{k} 2^{d-k}$ faces of dimension k. In particular, the cube has 2^d vertices and $2d$ faces; the vertices are the points $(\pm 1, \ldots, \pm 1)$.

12.1.11.3. The standard cocube. Applying 12.1.2.7, 12.1.11.2 and 12.1.10, we deduce that the faces of Coc_d are regular simplices, and that there are $\binom{d}{k+1} 2^{k+1}$ of them; in particular, Coc_d has $2d$ vertices (the points $\pm e_i$), and 2^d faces.

12.1.12. DIHEDRAL ANGLES OF A POLYTOPE. Notice that any $(d-2)$-face A of a polytope is contained in exactly 2 faces F, F'. In fact, this is, by duality (cf. 12.1.10), equivalent to saying that each edge contains exactly two vertices. The hyperplanes F, F' determine two unit vectors $\xi, \xi' \in \vec{X}$, well-defined by the condition that ξ is orthogonal to F and lies in the same side of F as F' (and similarly for ξ'). By definition, the *dihedral angle* (or *angle* if $d = 2$) of P at the $(d-2)$-face A is the angle $\overline{\xi\xi'} \in]0, \pi[$; if $d \doteq 2$, A is vertex of P.

Figure 12.1.12

Two k-faces of a polyhedron are called *adjacent* if their intersection is a $(k-1)$-face.

12.1.13. PROPOSITION. *Two arbitrary faces (of same dimension) of a polyhedron can be connected by a chain of adjacent faces.*

Proof. Assume first these are $(d-1)$-faces. Let $x, y \in P$ be arbitrary, and let Y be a plane of X containing x and y; then $Y \cap P$ is a polygon, whose faces are the intersections of the faces of P with Y. This reduces the problem to the case when P is a polygon; but then, if F is a side and $\overline{F}$ is a set of sides containing F and such that every side adjacent to an element of $\overline{F}$ is also in $\overline{F}$, we see that $\overline{F}$ is open and closed in $\mathrm{Fr}\, P$, which is a convex set. The case of faces of arbitrary dimension now follows from the previous one by recurrence. □

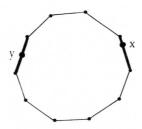

Figure 12.1.13

12.1.14. NOTE. The search for vertices of a polyhedron P defined by a certain number of explicit equations (some of which are possibly redundant) is an important problem in applied math (see 11.8.10.10). For references on algorithms to solve this problem, see [KE, 86] or [R–V, chapter V].

12.1.15. PROPOSITION. *Any polytope P has finitely many vertices, and is the convex hull of its vertices. Conversely, the convex hull of a fine number of points is a compact convex polyhedron.*

Proof. In fact, there are only finitely many k-faces ($k = 0, 1, \ldots, d - 1$), as can be seen by induction (starting from a segment, which has two faces). For the converse, we can assume that the convex hull Q of the set of points has non-empty interior, by restricting ourselves to the subspace spanned by the points (cf. 11.2.7). Vectorialize X at a point in the interior of Q and consider the polytope Q^* (cf. 12.1.2.8) for some Euclidean structure on X. By the first part of the proposition, Q^* is the convex hull of a finite number of points and, applying 12.1.2.8 again, we get that Q is a polytope from the fact that $Q = Q^{**}$.   □

12.1.16. COROLLARY. *If P is a polytope in X and $f \in A(X;Y)$ is affine, $f(P)$ is a polytope in $f(X)$.*   □

12.1.17. COROLLARY. *If P, Q are polytopes, their Minkowski sum $\lambda P + (1 - \lambda)Q$ is still one, for any λ (cf. 11.1.3).*

Proof. Set $P = \mathcal{E}\big((a_i)\big)$, $Q = \mathcal{E}\big((b_i)\big)$; we have

$$\lambda P + (1 - \lambda Q) \supset \mathcal{E}\big((\lambda a_i + (1 - \lambda)b_j)\big)$$

since $\lambda P + (1 - \lambda)Q$ is convex. On the other hand,

$$\lambda P + (1 - \lambda Q) \subset \mathcal{E}\big((\lambda a_i + (1 - \lambda)b_j)\big),$$

as can be seen by setting $p = \sum_i \lambda_i a_i$ and $q = \sum_i \mu_i b_i$ where the λ_i and μ_i are non-negative scalars satisfying $\sum_i \lambda_i = 1$ and $\sum_i \mu_i = 1$; we get

$$\lambda p + \mu q = \sum_{i.j} (\lambda_i \mu_j)\big(\lambda_i + (1 - \lambda)b_j\big).$$ □

> In sections 12.2 through 12.6 X is a Euclidean affine space.

12.2. Volume of polytopes

Now X is a Euclidean space, and we use the definition of volume and the notation introduced in 9.12. In particular, polytopes have a volume.

12.2.1. Let H be a hyperplane, $x \notin H$ a point, K a compact subset of H and $C = \mathcal{E}\big(\{x\} \cup K\big)$ the pyramid of vertex x and base K. Then

12.2.2 $$\mathcal{L}(C) = \frac{1}{d}d(x, H)\mathcal{L}_H(K).$$

The proof of 9.12.4.4 can be adapted without change.   □

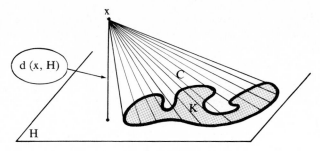

Figure 12.2.2

12.2.3. PROPOSITION. *Let $P = \bigcap_i R_i$ be a polytope (cf. 12.1.6), and $a \in \overset{\circ}{P}$ a point. Then*

$$\mathcal{L}(P) = \frac{1}{d} \sum_i d(a, H_i) \mathcal{L}_{H_i}(\mathrm{Face}_i\, P).$$

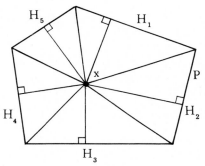

Figure 12.2.3

Proof. In fact, $P = \bigcup_i \mathcal{E}(\{a\} \cup \mathrm{Face}_i\, P)$, by 12.1.5 (iii). But

$$\mathcal{L}\big(\mathcal{E}(\{a\} \cup \mathrm{Face}_i\, P) \cap \mathcal{E}(\{a\} \cup \mathrm{Face}_j\, P)\big) = \mathcal{L}\big(\mathcal{E}(\{a\} \cup (\mathrm{Face}_i\, P \cap \mathrm{Face}_j\, P))\big) = 0,$$

since $\mathrm{Face}_i\, P \cup \mathrm{Face}_j\, P$ is contained in a codimension-two subspace. Applying 12.2.2 and adding up we get the result. $\qquad\square$

12.2.4. NOTE. The reader should verify that 12.2.3 is valid for any point $a \in X$ if we replace the "geometric" distance $d(a, \cdot)$ by the "algebraic" distance $d'(a, \cdot)$ defined by

$$d'(a, H_i) = \begin{cases} d(a, H_i) & \text{if } a \in R_i \\ -d(a, H_i) & \text{if } a \notin R_i \end{cases}.$$

12.2.5. ELEMENTARY VOLUME OF POLYTOPES. The reader can argue, not without reason, that using Lebesgue measure to define the volume of polytopes is overkill, as they are such elementary objects (especially plane polygons and triangles). Indeed, there is an alternative definition, based on the following fact, which can be proved by elementary means:

12.2.5.1. Theorem. *Given a Euclidean space X and a cube $C \subset X$ of edge 1, there exists a unique function Φ from the set of polytopes in X into* **R** *satisfying the following three axioms:*

EV1) *For any translation t and any polytope P, we have $\Phi\big(t(P)\big) = \Phi(P)$;*

EV2) *For any two polytopes P and Q such that $\overset{\circ}{P} \cap \overset{\circ}{Q} = \emptyset$ and $P \cup Q$ is a polytope, we have $\Phi(P \cup Q) = \Phi(P) + \Phi(Q)$.*

EV3) $\Phi(C) = 1.$ □

The proof is elementary but long; see [HR, chapter 2]. It consists first in proving that Φ must necessarily be given by the formula in 12.2.3 for arbitrary a (using the convention of 12.2.4), and then that the formula indeed gives a function satisfying the three axioms. It follows that $\Phi\big(f(P)\big) = \Phi(P)$ for every $f \in \mathrm{Is}(X)$, so that, in fact, Φ does not depend on the cube C (which is reassuring). The function Φ is called the *elementary volume* of polytopes.

12.2.5.2. The reader will find in [HR] many interesting facts on the elementary volume, and also on $(d - 1)$-dimensional measures (cf. 9.12.7). In spite of its apparent simplicity, the elementary volume gives rise to a number of questions, including the famous Hilbert–Dehn problem. The motivation for this problem is the following: in the proof of 12.2.5.1, one uses analytical arguments, including the continuity of Φ and, in order to show that the volume of a simplex S is given by $1/d$ the altitude times the volume of the base, a limit argument involving approximations of S by inscribed, prismatic polytopes (figure 12.2.5). In the two-dimensional case this is not necessary, since a parallelogram is the union of two isometric triangles, so that we can construct Φ in an even more elementary fashion if we strengthen (VE1) to apply to isometries instead of translations only (cf. [HD, volume 1, 292]). The question put by Hilbert is whether two polytopes of same volume, in a space X of dimension ≥ 3, can always be decomposed into polytopes isometric to one another. Dehn was the first to find counterexamples in $\mathbf{R}^3$, and he even gave an explicit criterion for which such a decomposition is possible. The unit cube and the regular simplex (cf. 12.1.2.5) of volume 1 form the simplest counterexample, cf. [HR, 49–52].

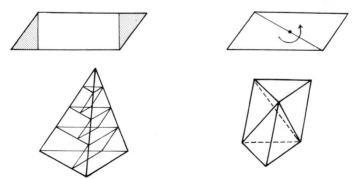

Figure 12.2.5

12.3. Area of polytopes

Let $P = \bigcap_i R_i$ be a polytope in the Euclidean space X. Its faces $\mathrm{Face}_i\, P$ are polytopes in the Euclidean spaces $H_i \subset X$, with volume $\mathcal{L}_{H_i}(\mathrm{Face}_i\, P)$. Thus the frontier $\mathrm{Fr}\, P$ of P, equal to $\bigcup_i \mathrm{Face}_i\, P$ by 12.1.5 (iii), has a natural $(d-1)$-dimensional volume (cf. 9.12.7) equal to $\sum_i \mathcal{L}_{H_i}(\mathrm{Face}_i\, P)$, which we call the *area* of P, by analogy with the everyday case $d = 3$ (and so as to avoid a heavier notation involving $d - 1$). All of this justifies the following

12.3.1. DEFINITION. *The area of the polytope $P = \bigcap_i R_i$, denoted by $\mathcal{A}(P)$, is the positive real number*

$$\mathcal{A}(P) = \sum_i \mathcal{L}_{H_i}(\mathrm{Face}_i\, P).$$

If $\dim d = 2$, *we call $\mathcal{A}(P)$ the perimeter of P, instead of area.*

12.3.2. EXAMPLE. If f is a similarity of X of ratio μ, we have, for every polytope P:

$$\mathcal{A}\big(f(P)\big) = \mu^{d-1}\mathcal{A}(P).$$

This follows from the definition and from 9.12.3.

The next formula can be interpreted as saying that, in some sense, the area of a polytope is the average, over all hyperplane directions, of the volumes of the projections of the polytope onto those hyperplanes. In order to formalize this observation, let K be a compact set, H a hyperplane and $p : X \to H$ the associated orthogonal projection. Then $\mathcal{L}_H\big(p(K)\big)$ depends only on $\vec{H}$ and not on H. If ξ is a unit vector in X, that is, if $\xi \in S = S(0,1)$, the unit sphere in $\vec{X}$, we denote by $\mathcal{L}\big(p_\xi(K)\big)$ the volume in H of the projection $p(K)$ of K on any hyperplane H such that $\xi \in (\vec{H})^\perp$. We then have, using the notation of 9.12.4.5 and denoting by σ the canonical measure on S (cf. 18.3.7):

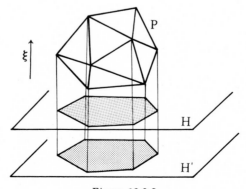

Figure 12.3.3

12.3.3. THEOREM (CAUCHY'S FORMULA). *For any* $d \geq 2$, *the area of a polytope* P *is given by*

$$A(P) = \frac{1}{\beta(d-1)} \int_{\xi \in S} \mathcal{L}\big(p_\xi(P)\big) \sigma.$$

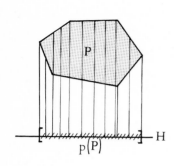
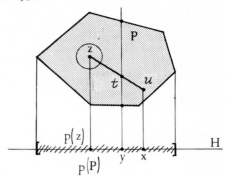

Figure 12.3.3.1

Proof.

12.3.3.1. First step. Fix $\xi \in S$, subject to the sole condition that $\xi \notin \vec{H}_i$, where the H_i are the hyperplanes containing the faces of $P = \bigcap_i R_i$. (Observe that the set of such vectors $\xi \in S$ has full measure in S, that is, its complement has measure zero.) We will show that

$$\sum_i \mathcal{L}\big(p_\xi(\text{Face}_i\, P)\big) = 2\mathcal{L}\big(p_\xi(P)\big);$$

this boils down to showing that, up to a set of measure zero, every point in $p(P)$ is the image under p of exactly two points of $\text{Fr}\, P$ (see figure 12.3.3.1).

We know that $\mathcal{L}\big(p(P)\big) = \mathcal{L}\big(\overset{\circ}{\widehat{p(P)}}\big)$; we must show that, for $y \in \overset{\circ}{\widehat{p(P)}}$, we have $p^{-1}(y) \cap \overset{\circ}{P} \neq \emptyset$, for then $p^{-1}(y)$ will be a segment and its frontier will have two points. Choose $z \in \overset{\circ}{P}$ arbitrarily. If $p(z) = y$, we're done; otherwise there exists $x \in p(P)$ such that $y \in [p(z), x[$, and, for any $u \in p^{-1}(x)$, we conclude that $t = p^{-1}(u) \cap [z, u] \in P$ by 11.2.4.

12.3.3.2. Second step. For each H_i, fix $u_i \in S$ so that $u_i \in (\vec{H}_i)^\perp$. By 9.12.4 we have

$$\mathcal{L}\big(p_\xi(\text{Face}_i\, P)\big) = |(\xi \mid u_i)| \mathcal{L}_{H_i}(\text{Face}_i\, P),$$

so that 11.3.1 gives

$$\int_{\xi \in S} \mathcal{L}\big(p_\xi(P)\big)\sigma = \frac{1}{2} \int_{\xi \in S} \left(\sum_i \mathcal{L}_{H_i}(\text{Face}_i\, P)\right) |(\xi \mid u_i)|\sigma$$

$$= \frac{1}{2} \sum_i \left(\mathcal{L}_{H_i}(\text{Face}_i\, P) \int_{\xi \in S} |(\xi \mid u_i)|\sigma\right).$$

The proof is concluded by applying the formula $\int_{\xi \in S} |(\xi \mid u)| \sigma = 2\beta(d-1)$, which holds for any unit vector u. This formula is classical and can be proved using calculus (see 12.12.19), but also by symmetry considerations: the integral is invariant under isometries, so it must be a scalar $k(d)$ independent of ξ. Its value follows by applying 12.10.4.1. $\qquad\square$

Formula 12.3.3 is pleasant in itself, but it will also be used in 12.10 to define the area of an arbitrary compact convex set.

12.3.4. NOTE. A remarkable problem, whose solution is due to Minkowski, consists in finding a polytope whose faces have a prescribed volume and direction; see the excellent references [LU], [AW2] and [PV1].

12.3.5. THE STEINER–MINKOWSKI FORMULA FOR POLYTOPES. Let P be a polytope and λ a positive real number. Recall the notation

$$B(P, \lambda) = \{ x \in X \mid d(x, P) \leq \lambda \};$$

we are interested in studying the volume $\mathcal{L}\big(B(P, \lambda)\big)$ and its growth with λ. Figure 12.3.5.1 shows that, for $d = 2$, we can decompose $B(P, \lambda)$ into three parts: P itself, plus a union of rectangles of altitude λ constructed on the sides of P, and finally a union of circular sectors of radius λ. The second component has area

$$\sum_i \lambda \mathcal{L}_{H_i}(\text{Face}_i\, P) = \lambda \mathcal{A}(P)$$

by definition of $\mathcal{A}(P)$. As for the third, when we bring together all the circular sectors we observe that they form a whole disc, so its area is $\pi\lambda^2$ (cf. 9.12.4.4). All the unions are disjoint up to sets of measure zero, so we can write

$$\mathcal{L}\big(B(P, \lambda)\big) = \mathcal{L}(P) + \mathcal{A}(P)\lambda + \pi\lambda^2.$$

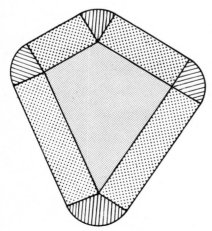

Figure 12.3.5.1

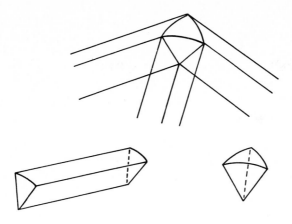

Figure 12.3.5.2

For $d = 3$, we find that $B(P, \lambda)$ is naturally decomposable into four subsets: P again, then right parallelepipeds of height λ built on each face, then orthogonal products of circular sectors and edges, then spheric sectors whose union is the ball. The third component gives terms in λ^2. These heuristic observations are generalized and formalized as follows:

12.3.6. PROPOSITION. *Every d-dimensional polytope P has associated scalars $\mathcal{L}_i(P) \in \mathbf{R}_+^*$ ($i = 0, 1, \ldots, d$) such that, for any $\lambda \in \mathbf{R}_+^*$, the volume of $B(P, \lambda)$ is given by*

$$\mathcal{L}\big(B(P, \lambda)\big) = \sum_{i=0}^{d} \mathcal{L}_i(P) \lambda^i.$$

As special cases we have $\mathcal{L}_0(P) = \mathcal{L}(P)$, $\mathcal{L}_1(P) = \mathcal{A}(P)$, $\mathcal{L}_{,}(P) = \beta(P)$.
Proof.

12.3.6.1. Take $y \notin P$ and let $x \in P$ be the unique point (cf. 11.1.7.1) such that $d(x, y) = d(y, P)$. It follows from 9.2.2 and the proof of 11.6.2 that, for every $k \geq 0$, we have $d(x + k\overrightarrow{xy}, P) = kd(x, y)$. Denote by CN_x the normal cone at $x \in \operatorname{Fr} P$, by CN'_x the translate of CN_x in $\vec{X}$ with vertex at the origin, and set $S_x = CN_x \cap S(0, 1)$ (cf. 11.6.2). What the reasoning above shows is that

$$B(P, \lambda) \setminus \overset{\circ}{P} = \bigcup_{x \in \operatorname{Fr} P} \bigcup_{\xi \in S_x} [x, x + \lambda \xi].$$

12.3.6.2. For $x \in \operatorname{Fr} P$, denote by ω_x the order of x and set

$$\Omega_i = \{ x \in \operatorname{Fr} P \mid \omega_x = i \} \qquad (i = 0, 1, \ldots, d - 1),$$

that is, Ω_i is the union of the relative interiors of the i-faces of P (cf. 12.1.9). Set, for $i = 0, 1, \ldots, d - 1$,

$$B_i(\lambda) = \bigcup_{x \in \Omega_i} \bigcup_{\xi \in S_x} [x, x + \lambda \xi];$$

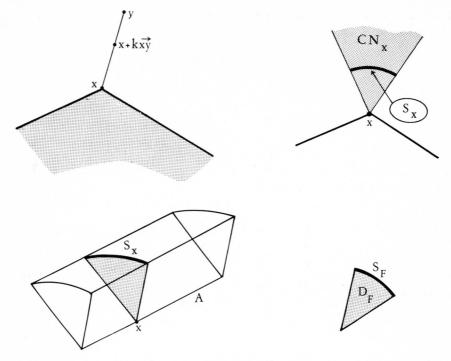

Figure 12.3.6

then $B(P, \lambda) \setminus \mathring{P} = \bigcup_{i=0}^{d-1} B_i(\lambda)$, so we have obtained a partition of $B(P, \lambda) \setminus \mathring{P}$. We can then write

$$\mathcal{L}(B(P, \lambda)) = \mathcal{L}(P) + \sum_{i=0}^{d-1} \mathcal{L}(B_i(\lambda)).$$

12.3.6.3. The essential remark is that, as x ranges over the relative interior $F^\bullet$ of a given face F, the cone CN'_x, hence also S_x, remains fixed. This is because CN_x is determined by the hyperplanes containing the faces to which x belongs. We can thus set $S_F = S_x$, for any $x \in F$, and $D_F = \bigcup_{\xi \in S_x}[0, \xi]$.

For F a given i-face, we can write the following equality (in fact, an isometry, where the right-hand side is a direct orthogonal product):

$$\bigcup_{x \in F^\bullet} \bigcup_{\xi \in S_x} [x, x + \lambda \xi] = F^\bullet \times \big([0, \lambda]S_F\big).$$

Applying the two examples in 9.12.3 we get

$$\mathcal{L}\big(F^\bullet \times ([0, \lambda]S_F)\big) = \mathcal{L}(F^\bullet)\mathcal{L}([0, \lambda]S_F) = \mathcal{L}(F)\mathcal{L}(D_F)\lambda^{d-1},$$

where the volumes $\mathcal{L}(F)$ and $\mathcal{L}(D_F)$ are of course i- and $(d-i)$-dimensional, respectively.

12.3.6.4. Denote by Φ_i the set of i-faces of P; by the reasoning above we have $\mathcal{L}\big(B_i(\lambda)\big) = \mathcal{L}_{d-i}(P)\lambda^{d-i}$, where

$$\mathcal{L}_{d-1}(P) = \sum_{F \in \Phi_i} \mathcal{L}(D_F).$$

In particular, $i = d - 1$ gives

$$\mathcal{L}_1(P) = \sum_{F \in \Phi_{i-1}} \mathcal{L}(F) = \mathcal{A}(P)$$

because $D_F = [0,1]$; for $i = 0$, the remarks above and 11.6.2 show that $\bigcup_{x \in \Phi_0} S_x = S$, or $\bigcup_{F \in \Phi_0} D_F = B(0,1)$, whence $\mathcal{L}_d(P) = \mathcal{L}\big(B(0,1)\big) = \beta(d)$, since F is a point and $\mathcal{L}(F) = 1$. □

The next proposition will be necessary in 12.10.6:

12.3.7. PROPOSITION. *If $a \in X$ and $r > 0$ are given, the functions $\mathcal{L}_i(\cdot)$ are bounded over the set of polytopes contained in $B(a,r)$.*

Proof. Fix $\lambda > 0$. We have $B(P,\lambda) \subset B(a, \lambda + r)$ and $\mathcal{L}\big(B(P,\lambda)\big) \le \beta(d) \cdot (r + \lambda)^d$; since all terms of $\mathcal{L}\big(B(P,\lambda)\big) = \sum_i \mathcal{L}_i(P)\lambda^i \le \beta(d)(r + \lambda)^d$ are positive, we have

$$\mathcal{L}_i(P) \le \beta(d)\frac{(r + \lambda)^d}{\lambda^i}$$

for every i. □

12.3.8. REMARK. We can now justify the "painting" idea expressed in 9.12.7. If we paint P with a very thin coating of thickness λ, we can neglect the terms of order higher than one in λ, and the amount of paint necessary is $\mathcal{L}\big(B(P,\lambda)\big) = \mathcal{L}(P) = \mathcal{A}(P)\lambda$, which is proportional to the area. We shall expand on this point in 12.10.7.

12.4. Regular polygons

In this section X has dimension two.

This rather elementary section is designed to pave the way for the definition of polytopes in arbitrary dimension.

12.4.1. DEFINITION. *A polygon is called regular if all its sides have the same length and all its angles are the same.*

Both conditions are necessary (figure 12.4.1). By 10.5.2 the common value of the angles of an n-sided polygon is $\dfrac{n-2}{n}\pi$.

 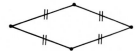

Figure 12.4.1

Castel del Monte, Apulia—aereal view from the southeast

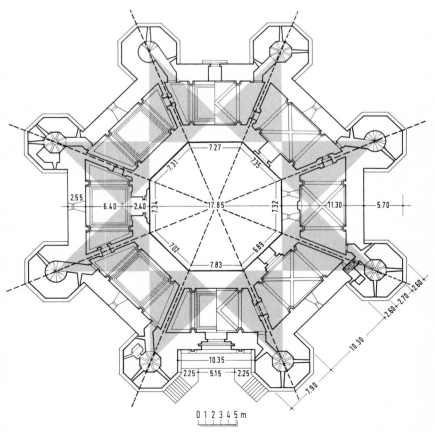

Castel del Monte—ground plan with highlighted eight-pointed star

Source: [GZ]

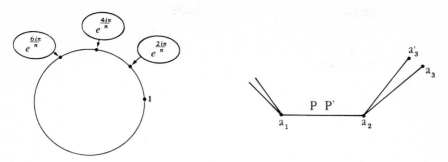

Figure 12.4.2

12.4.2. PROPOSITION. *For every integer $n \geq 3$ there exists a regular n-sided polygon. Two polygons with the same number of sides are similar.*

Proof. To prove existence, take the points $e^{2ik\pi/n} \in \mathbf{C} = \mathbf{R}^2$, for $k = 0, 1, \ldots,$ $n - 1$. To prove uniqueness, observe that by 9.6.2 we can suppose that, up to similarity, $P = \mathcal{E}(a_i)$ and $P' = \mathcal{E}(a_i')$ have a side in common, say $a_1 = a_1'$ and $a_2 = a_2'$, and also that P and P' lie on the same side of $\langle a_1, a_2 \rangle$. Then the fact that sides and angles are the same shows that $a_3 = a_3'$, and, by induction, $P = P'$. $\qquad\square$

12.4.3. FORMULA. By 12.4.2, every regular polygon can be inscribed in a circle, whose center is called the *center* of the polygon. If the radius of the circle is r and l is the length of each side of the polygon, we have

$$l = 2r \sin \frac{\pi}{n}.$$

In order to extend the notion of a regular polygon to arbitrary dimension, we introduce the stabilizer $G(P) = \mathrm{Is}_P(X)$ of a polygon (cf. 9.8.1).

12.4.4. PROPOSITION. *In order for the polygon P to be regular it is necessary and sufficient that $G(P)$ act transitively on pairs (x, F) consisting of a vertex x and a side $F \ni x$ of P.*

Proof. To prove necessity it is enough to work with the case of $\{e^{2ik\pi/n}\}$. This polygon is invariant under the group of rotations $\mathbf{Z}_n$, which acts transitively on vertices, so we can just consider the case of two pairs (x, F) and (x, F') sharing a vertex. But these two can certainly be mapped to one another: if $F \neq F'$, just reflect through the line $\langle 0, x \rangle$.

Sufficiency is proved by observing that both angles and sides have to be all equal because $G(P)$ acts transitively on them. $\qquad\square$

12.4.5. By 12.1.12, the number of pairs (x, F) with $x \in F$ is equal to $2n$, so by 9.6.2 we have $\#G(P) = 2n$. This group is the dihedral group $\mathcal{D}_{2n}$, the semidirect product $\mathbf{Z}_n \times \mathbf{Z}_2$, studied in 1.8.3.4.

12.4.6. GEOMETRIC CONSTRUCTIONS. For $n = 6$ and $n = 10$, see figures 12.4.6.1 and 12.4.6.2; these also take care of the cases $n = 3$ and $n = 5$,

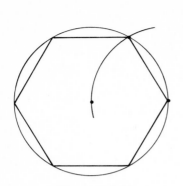
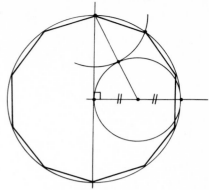

Figure 12.4.6.1   Figure 12.4.6.2

of course. Knowing the case $n = 5$ is necessary is you want to build paper dodecahedra (cf. 12.5.5) or soccer balls (figure 12.1.1.5).

If one can construct n-sided regular polygons, one can construct $2n$-sided ones, since bisectors can be found with ruler and compass. Constructing an n-sided regular polygon with ruler and compass gives rise to an interesting algebraic problem, which was solved by Gauss; the only number of sides for which it is possible to do that are those of the form $2^k n_1 \cdots n_k$, where the n_i are distinct, of the form $2^{2^{\alpha_i}} + 1$ and primes. Numbers of the form $2^{2^{\alpha_i}} + 1$ are called *Fermat numbers*. Which Fermat numbers are prime is an arithmetic problem as yet unsolved; one knows that $F_0 = 3$, $F_1 = 5$, $F_2 = 17$, $F_3 = 257$ and $F_4 = 65537$ are prime, but those are the only ones known. We also know several non-prime Fermat numbers; one conjecture is that there are only finitely many Fermat numbers. See [HA–WR, 14–15].

The cases of figures 12.4.6.1 and 12.4.6.2 are F_0 and F_1. For a construction of regular polygons with 17 or more sides, and for a proof of Gauss's result, see [LB1, 110–153], or the more recent [SW, chapter 17].

12.5. Regular polytopes: definition and examples

Despite their simple appearance, three-dimensional regular polytopes (also called *regular polyhedra*) are not easily defined. The conditions have to be quite strong, as the reader will realize if he tries to state them himself, or by consulting 12.12.7 and the additional references [FT2], [CR2], [HD, volume 2]. The right definition is suggested by 12.4.4:

12.5.1. DEFINITION. *A flag of a d-dimensional polytope P is a d-tuple $(F_0, F_1, \ldots, F_{d-1})$ consisting of i-faces F_i of P such that $F_i \subset F_{i+1}$ for $i = 0, 1, \ldots, d-2$. A polytope P is called regular if its stabilizer $G(P) = \mathrm{Is}_P(X)$ acts transitively on the flags of P.*

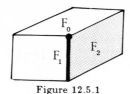

Figure 12.5.1

12.5.2. CONSEQUENCES. Let P be a regular polytope.

12.5.2.1. The center of mass O of the vertices of P, called the *center* of P, is fixed under $G(P)$; since $G(P)$ acts transitively on vertices, this implies that all the vertices belong to a sphere centered at O, called the sphere *circumscribed* around P.

12.5.2.2. Any i-face of P, for $i = 2, \ldots, d - 1$, is an i-dimensional regular polytope.

12.5.2.3. Let F and F' be adjacent faces of P (cf. 12.1.12), and H the hyperplane of P spanned by $F \cap F'$ and centered at O, that is, $H = \langle O, F \cap F' \rangle$. Then $\sigma_H(F) = F'$, that is, the reflection σ_H maps F onto F'.

12.5.2.4. Let X be the vectorialization of P with respect to O. The dual P^* (cf. 12.1.2.7) of P is also a regular polytope, and we have $G(P^*) = G(P)$. In fact, the group $G(P)$ fixes O and preserves polarity, and we use 12.5.1 and 12.1.10.

12.5.2.5. The group $G(P)$ acts simply transitively on flags. In fact, any flag, extended by P itself, determines by induction a unique associated orthonormal frame (figure 12.5.2). In particular, the order of $G(P)$ is equal to the number of flags of P.

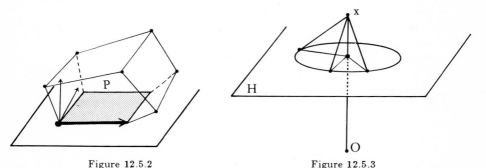

Figure 12.5.2   Figure 12.5.3

12.5.3. THE LINK OF A REGULAR POLYTOPE

12.5.3.1. Let P a regular polytope, O its center, and x a vertex of P. For A an edge (cf. 12.1.8) of P containing x, let y be the other endpoint of A. Since $G(P)$ acts transitively on flags, and in particular on pairs (x, A) with $A \ni x$, all points y as above lie on the same hyperplane H, orthogonal to $\overrightarrow{xO}$. The intersection $P \cap H$ is a polytope, whose vertices are obviously the points y, and whose $(i-1)$-faces are the intersections of H with the i-faces of P containing x. This shows that $P \cap H$ is a regular polytope: since $G(P)$ acts transitively on the flags of P, it acts transitively on those whose first element is x, and $G(P \cap H)$ acts transitively on the flags of $P \cap H$.

12.5.3.2. Definition. *The regular polytope thus constructed is called the link of x in P, and denoted by* $\operatorname{link}_x P$, *or simply by* $\operatorname{link} P$ (*since all links of a polytope are isometric*).

12.5.3.3. From 1.5.5 we get that the order of $G(P)$ is equal to the order of $G(\operatorname{link} P)$ times the number of vertices of P.

12.5.4. THE EASY EXAMPLES

12.5.4.1. The regular simplex. The flags of Simp_d are in one-to-one correspondence with the d-tuples $(e_{i_1}, \ldots, e_{i_d})$ of $(e_i)_{i=1,\ldots,d+1}$; but for any two d-tuples $(e_{i_1}, \ldots, e_{i_d})$ and $(e_{j_1}, \ldots, e_{j_d})$, there exists $f \in O(d+1)$ taking one onto the other, and taking the remaining vertex $e_{i_{d+1}}$ into the corresponding $e_{j_{d+1}}$ (cf. 8.2.7). Thus $f \in G(\operatorname{Simp}_d)$; this also shows that $G(\operatorname{Simp}_d) = S_{d+1}$, the symmetric group on $d+1$ elements. Further, $\operatorname{link}(\operatorname{Simp}_d)$ is a regular simplex $\operatorname{Simp}_{d-1}$.

12.5.4.2. The cube. The reflections through coordinate hyperplane generate a group of isometries of Cub_d that acts transitively on the vertices $(\pm 1, \ldots \pm 1)$ (cf. 12.1.11.2). There remains to see that $G(\operatorname{Cub}_d)$ acts transitively on flags containing a given vertex; to do that, consider the regular cube $C = [0, 1]^d \subset \mathbf{R}^d$, which is, of course, similar to Cub_d. Set $(0, \ldots, 0)$; the edges of C containing x are the segments $[0, e_i]$ $(i = 1, \ldots, d)$, where $(e_i)_{i=1,\ldots,d}$ is the canonical basis of $\mathbf{R}^d$. Thus the flags containing x are in one-to-one correspondence with the flags of $\operatorname{Simp}_{d-1}$; since $G(\operatorname{Simp}_{d-1}) \subset G(C)$, we conclude that Cub_d is regular, and that $\operatorname{link}(\operatorname{Cub}_d) = \operatorname{Simp}_{d-1}$.

link_x is the set of points marked $\bullet$

Simp_4
$\operatorname{link}_x = \operatorname{Simp}_3$

Cub_3
$\operatorname{link}_x = \operatorname{Simp}_2$

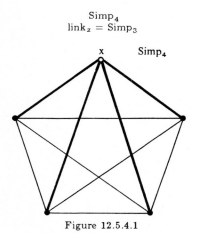

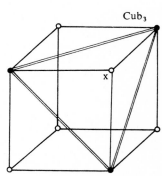

Figure 12.5.4.1   Figure 12.5.4.2

To determine the group $G(\text{Cub}_d)$, we first notice that its order is $2^d d!$, from 12.5.3.3 and 12.5.4.1. But we know a subgroup of $G(\text{Cub}_d)$ with $2^d d!$ elements, namely, the semi-direct product of the $d!$ coordinate permutations and the group of order 2^d generated by reflections through coordinate hyperplanes.

12.5.4.3. The cocube. By duality, we deduce from 12.1.2.7, 12.1.10 and 12.5.2.4 that Coc_d is a regular polyhedron and that its symmetry group is the same as that of Cub_d. We can determine $\text{link}(\text{Coc}_d)$ by duality, too, or directly: the edges containing the vertex e_1, say, are the segments $[e_1, \pm e_i]$, for $i = 2, \ldots, d$, showing that $\text{link}(\text{Coc}_d) = \text{Coc}_{d-1}$.

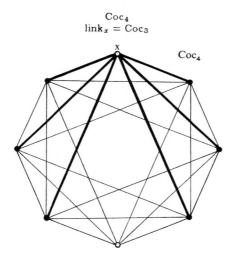

link_x is the set of points marked $\bullet$

Coc_4
$\text{link}_x = \text{Coc}_3$

x

Coc_4

Figure 12.5.4.3

12.5.5. THE DIFFICULT EXAMPLES IN DIMENSION THREE. We have remarked in 1.8.3.4 that the existence of the dodecahedron and of the icosahedron is not obvious; we give two proofs for it. Notice that the regular dodecahedron and icosahedron are dual to each other, so that, if one exists, so does the other.

12.5.5.1. The geometric method. The idea is that, if we do have a regular dodecahedron, we can inscribe a cube inside it in a natural way; see figure 12.5.5.1. Conversely, we will show that it is possible to start from a cube and assemble regular pentagons around it, so as to form a dodecahedron.

12.5.5.2. Lemma. *Let F, F' and F'' be identical regular pentagons. There is a unique way, up to isometry, to glue the three along common edges A, A' and A'', so that they share a vertex x (in other words, the solid obtained is rigid). The lines $\langle x, y \rangle$, $\langle x', y' \rangle$ and $\langle x'', y'' \rangle$ in figure 12.5.5.2 are pairwise orthogonal.*

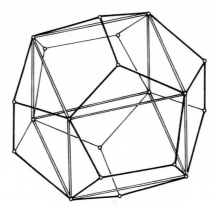

Figure 12.5.5.1

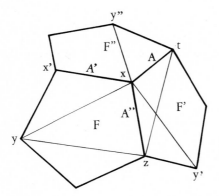

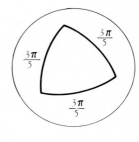

Figure 12.5.5.2

Proof. The easiest proof uses spherical trigonometry. That the three pentagons can be glued together is equivalent to the existence of a spherical triangle whose sides are all three equal to the angle of a regular pentagon, which is equal to $3\pi/5$; since $3 \times 2\pi/5 < 2\pi$, the spherical triangle exists by 18.6.10. Uniqueness up to isometries follows from 18.6.13.10. To show that $\langle x, y \rangle$ and $\langle x', y' \rangle$ are orthogonal, we can proceed as follows: in the figure, $\langle y, z \rangle$ and $\langle z, t \rangle$ are orthogonal because $\langle y, z \rangle$ is parallel to the edge A', which is orthogonal to $\langle z, t \rangle$ (since $xz = xt$ and $x'z = x't$, for example). But the reflection σ_H through the plane equidistant from x and z takes the triple (y, z, t) into (y, x, y'), completing the proof.   □

Here it might be argued that existence is proven, since the gluing process and the reflection σ_H can be repeatedly applied until we're done, as in the building of a cardboard dodecahedron (first glue together F, F', F'', then glue $F''' = \sigma_H(F'')$ to F and F', and so on); but this is not enough because one must show that at the end of the construction things fit together nicely, which is not at all evident. To sew up the argument, we start from a cube C

with vertex x, and install on the three edges of C containing x the three regular pentagons F, F', F'', according to figure 12.5.5.1; this works because, by 12.5.5.2, $\langle x, y \rangle$, $\langle x', y' \rangle$ and $\langle x'', y'' \rangle$ are perpendicular and have the same length. Applying the symmetries of C that switch each pair of opposite faces, we construct a polyhedron whose twelve faces are regular pentagons. A dodecahedron constructed in this way is uniquely determined by giving a face and specifying in which half-space determined by that face the polyhedron lies, by the rigidity result of 12.5.5.2: this implies that the polyhedron is regular.

12.5.5.3. Algebraic method. We give an explicit method for the construction of the icosahedron. The idea is to locate its twelve vertices on the edges of an octahedron (that is, a three-dimensional cocube), so that the faces are equilateral triangles; this is obviously possible by continuity. Unfortunately, it is not obvious that a polytope formed by twenty equal equilateral triangles glued five by five is regular, because gluing five equilateral triangles around a common vertex x, as we did in 12.5.5.2, does not yield a unique result, but rather an eminently *flexible* construct, in the same way that a plane polygon with $n > 3$ sides of fixed length is deformable. We will return to such questions in 12.8.6.

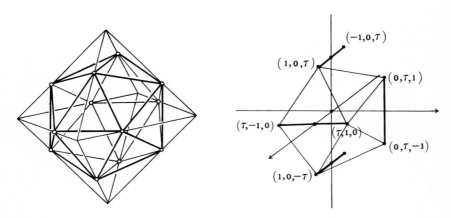

Figure 12.5.5.3   Figure 12.5.5.4

The vertices of our icosahedron, following the octahedron idea, will be of the form $(0, \pm\tau, \pm 1)$, $(\pm 1, 0, \pm\tau)$, $(\pm\tau, \pm 1, 0)$, where τ is to be determined. Starting from $(\tau, 1, 0)$, its distances to the five other vertices $(\tau, -1, 0)$, $(0, \tau, \pm 1)$ and $(1, 0, \pm\tau)$, have the values 2 and $\sqrt{2\tau^2 - 2\tau + 2}$, so that we must have $\tau^2 - \tau + 1 = 0$ and $\tau > 0$, that is,

12.5.5.4
$$\tau = \frac{\sqrt{5} + 1}{2}.$$

With this value of τ, the point $(\tau, 1, 0)$ is equidistant from the five vertices above. This is true of all vertices, if only because there is an isometry group acting transitively on our set of twelve vertices, namely, the group generated by a cyclic permutation of the axes and by reflections through the coordinate hyperplanes.

We now remark that the five points $(\tau, -1, 0)$, $(0, \tau, \pm 1)$ and $(1, 0, \pm \tau)$ lie on the plane with equation $\tau x + y - \tau = 0$. This implies that the solid formed by gluing the five equilateral triangles around each vertex (fig. 12.5.5.4) is rigid; consequently, a rotation of order five around the axis containing $(\tau, 1, 0)$ and orthogonal to the plane $\tau x + y - \tau = 0$ preserves the icosahedron. Adjoining this rotation to the previously mentioned group, we obtain a group that acts transitively on flags.

12.5.5.5. The coordinates of a dodecahedron constructed on the cube with vertices $(\pm 1, \pm 1, \pm 1)$ by the method of 12.5.5.1 are $(0, \pm \tau^{-1}, \pm \tau)$, $(\pm \tau, 0, \pm \tau^{-1})$ and $(\pm \tau^{-1}, \pm \tau, 0)$ (where τ is as in 12.5.5.4). This can be easily checked by the reader, using 12.5.5.1, for example, or duality.

12.5.5.6. The icosahedron group. By 12.5.3.3 and 12.4.5, this group has order $12 \times 10 = 120$; but it is not isomorphic to the symmetric group $\mathcal{S}_5$. To see that, observe that the thirty edges of the regular dodecahedron determine fifteen lines joining their midpoint to the center of the dodecahedron. By 12.5.5.1, they are grouped into three triples of pairwise orthogonal lines. It is clear that the group G^+ of orientation-preserving isometries of the dodec-ahedron acts faithfully on the set of those five triples; since G^+ has order 60, we have $G^+ \cong \mathcal{A}_5$. On the other hand, G does not act faithfully, because it contains the reflection through the origin; thus, G is not isomorphic to $\mathcal{S}_5$, but to the direct product $\mathcal{A}_5 \times \mathbf{Z}_2$.

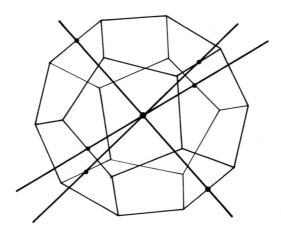

Figure 12.5.5.6

12.5.5.7. Notes. Regular dodecahedra and icosahedra, or objects possessing their symmetry group, are unusual in nature. There are extremely few crystals in that class; among living creatures, this kind of symmetry is somewhat more common (cf. figure 12.5.5.7 and [WL, 75]).

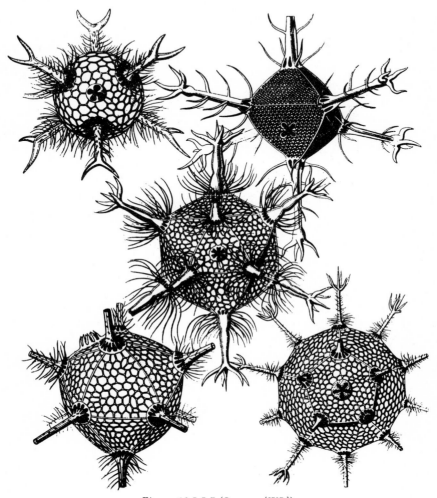

Figure 12.5.5.7 (Source: [WL])

As for cultural constructs, a steatite dodecahedron has been found, made by the Etruscans no later than 500 BC; there is also a pair of icosahedral dice from the Ptolemaic dynasty in the British Museum. The Greeks speculated about the five regular polyhedra; see Plato's *Timaeus* and precise references in [CR2, 13], [FT2, 120–121], as well as figure 12.5.5.8.

Kepler also purported to have found a relation between the five regular polyhedra and the orbits of the planets in the solar system (figure 12.5.5.9).

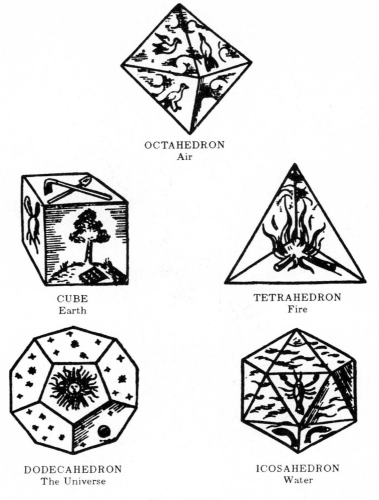

OCTAHEDRON
Air

CUBE
Earth

TETRAHEDRON
Fire

DODECAHEDRON
The Universe

ICOSAHEDRON
Water

Figure 12.5.5.8

12.5.6. THE HARD EXAMPLES IN DIMENSION FOUR. Here we merely describe these three polytopes, which were first discovered through very laborious geometric considerations (see [CR2, 13 and 141]); a less painful way of arriving at them will be given in 12.6.7, in the course of a classification of all possible regular polytopes (cf. table in 12.6.7.3). Their naming will be justified in 12.6.1.

12.5.6.1. The standard $\{3, 4, 3\}$ polytope. We define this polytope P by giving its 24 vertices: $(\pm 2, 0, 0, 0)$, $(0, \pm 2, 0, 0)$, $(0, 0, \pm 2, 0)$, $(0, 0, 0, \pm 2)$, $(\pm 1, \pm 1, \pm 1, \pm 1)$, that is, the union of the vertices of Cub_4 and those of a homothetic image of Coc_4 of ratio 2. Let us show these points define a regular polytope.

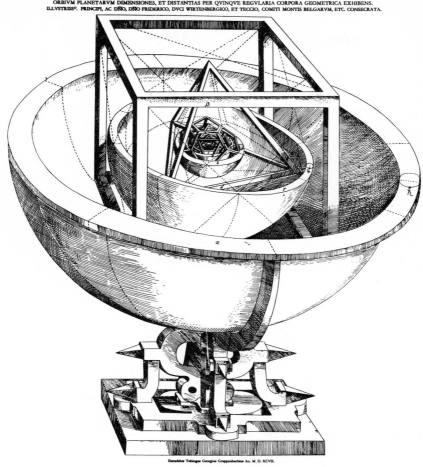

TABELLA III.
ORBIVM PLANETARVM DIMENSIONES, ET DISTANTIAS PER QVINQVE REGVLARIA CORPORA GEOMETRICA EXHIBENS.
ILLVSTRISS°. PRINCIPI, AC DÑO, DÑO FRIDERICO, DVCI WIRTENBERGICO, ET TECCIO, COMITI MONTIS BELGARVM, ETC. CONSECRATA.

Figure 12.5.5.9

The vertex $x = (2, 0, 0, 0)$ lies at a distance 2 from the eight vertices with coordinates $(1, \pm 1, \pm 1, \pm 1)$; thus the set of faces of P containing x is in one-to-one correspondence with the faces of the cube defined by these eight vertices. In particular, the group $G(P)$ acts transitively on flags containing x, and all that is left to show is that $G(P)$ acts transitively on vertices. Since the group of the cube Cub_2 acts transitively on the vertices of Cub_4 and on the vertices of $2\,\mathrm{Coc}_4$, it suffices to exhibit an isometry of P taking $(2, 0, 0, 0)$ and $(1, -1, -1, -1)$ to each other. A good candidate is the reflection through the hyperplane equidistant from the two points, which, by 8.2.10, takes a point (x, y, z, t) into

$$\left(\frac{x - y - z - t}{2}, \frac{-x + y - z - t}{2}, \frac{-x - y + z - t}{2}, \frac{-x - y - z + t}{2} \right).$$

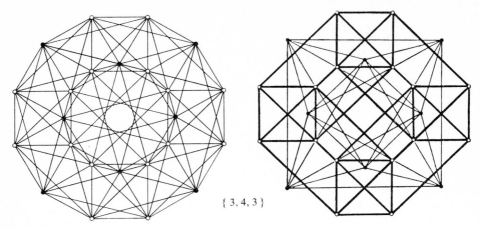

$\{3, 4, 3\}$

Figure 12.5.6.1 (Source: [CR2])

This reflection is easily seen to fix the set of vertices of P. The same considerations also show that $\operatorname{link} P = \mathrm{Cub}_3$; thus the group $G(P)$ has order $48 \times 24 = 1152$, by 12.5.4.2 and 12.5.3.3.

12.5.6.2. The standard $\{3, 3, 5\}$ polytope. This polytope and its dual, called $\{5, 3, 3\}$, are the most complicated of all regular polytopes. The 120 vertices of $\{3, 3, 5\}$ are

— the 24 vertices of the standard $\{3, 4, 3\}$ polytope, and
— all points obtained from $(\pm\tau, \pm 1, \pm\tau^{-1}, 0)$ by an even permutation of coordinates, where τ is as in 12.5.5.4.

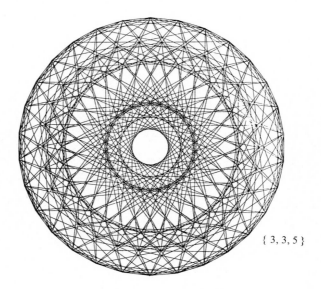

$\{3, 3, 5\}$

Figure 12.5.6.2.1 (Source: [CR2])

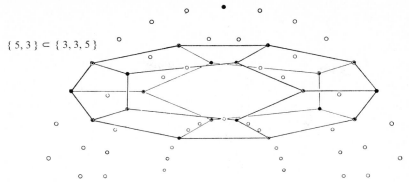

$$\{\,5,3\,\}\subset\{\,3,3,5\,\}$$

Figure 12.5.6.2.2

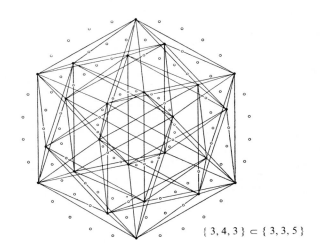

$$\{\,3,4,3\,\}\subset\{\,3,3,5\,\}$$

Figure 12.5.6.2.3

Source: [CR2]

One shows that this polytope Q is regular by using the same method of 12.5.6.1. The edges containing $x = (2,0,0,0)$ are the twelve segments connecting x to the points $(\tau, \pm 1, \pm\tau^{-1}, 0)$, $(\tau, 0, \pm 1, \pm\tau^{-1})$ and $(\tau, \pm\tau^{-1}, 0, \pm 1)$. These points form a regular icosahedron (cf. 12.5.5.3), whose symmetry group, extended by the identity on the $(1,0,0,0)$-axis, acts transitively on the flags of Q containing x. This group also preserves Q, as seen by considering each subset of vertices with the same first coordinate: if that coordinate is τ or $-\tau$, the subset is preserved by construction. If it is $+1$, the points are $(1, \pm 1, \pm 1, \pm 1)$, $(1, \pm\tau, 0, \pm\tau^{-1})$, $(1, 0, \pm\tau^{-1}, \pm\tau)$ and $(1, \pm\tau^{-1}, \pm\tau, 0)$, which form a dodecahedron, left fixed by the symmetries of the icosahedron. Finally, the points with first coordinate zero are

$$(0, \pm 2, 0, 0), \quad (0, 0, \pm 2, 0), \quad (0, 0, 0, \pm 2),$$
$$(0, \pm\tau, 1, \pm\tau^{-1}), \quad (0, 1, \pm\tau^{-1}, \pm\tau), \quad (0, \pm\tau^{-1}, \pm\tau, 1).$$

$$\{\,5,3,3\,\}$$

Figure 12.5.6.2.4 (Source: [CR2])

But these are the midpoints of the edges of an icosahedron $2\tau^{-1}$ times larger than the standard icosahedron 12.5.5.3, and are again preserved by the our group.

There remains to see that $G(Q)$ acts transitively on vertices; this is done as in 12.5.6.1. By 12.5.5.6, the group of Q has order $120 \times 120 = 14400$.

12.6. Classification of regular polytopes

We will show that there the only possible regular polytopes are the ones already described; this result is due to Schläfli and dates from 1850, approximately. The proof is based on two ideas: the symbol of a regular polytope, and formula 12.6.5, which allows an induction argument.

12.6.1. DEFINITION. *The symbol of a d-dimensional regular polytope P, denoted by $\{r_1(P), r_2(P), \ldots, r_{d-1}(P)\}$, is the sequence of $d-1$ integers defined by induction on d as follows: $r_1(P)$ is the number of sides of the two-faces of P (cf. 12.5.2.2 and 12.4.2), and $\{r_2(P), \ldots, r_{d-1}(P)\}$ is the symbol of* link P *(cf. 12.5.3.2).*

12.6.2. EXAMPLES.

12.6.2.1. Observe that all the r_i are greater than or equal to 3.

12.6.2.2. The symbol of a face of P is $\{r_1(P), \ldots, r_{d-2}(P)\}$, by 12.5.3.1.

12.6.2.3. The symbol of Simp_d is $\{3, \ldots, 3\}$. The symbol of Cub_d is $\{4, 3, \ldots, 3\}$, and that of Coc_d is $\{3, \ldots, 3, 4\}$. For the other polytopes described: icosahedron, $\{3, 5\}$; dodecahedron, $\{5, 3\}$; example 12.5.6.1, $\{3, 4, 3\}$; and example 12.5.6.2, $\{3, 3, 5\}$. The reader can check this by recalling what the link of each polyhedron is (12.5.4.1, 12.5.4.2, 12.5.4.3, 12.5.6.1, 12.5.6.2) and working by induction.

12.6.2.4. The symbol of the dual P^* of P (cf. 12.5.2.4) is

$$\{r_{d-1}(P), \ldots, r_1(P)\},$$

as follows from 12.6.1, 12.1.10 and 12.6.2.2. For example, the dual of the standard $\{3, 3, 5\}$ polytope is a regular polytope with symbol $\{5, 3, 3\}$.

12.6.3. A FUNDAMENTAL RELATION.

12.6.4. NOTATION. Let P be a regular polytope. We denote by l the length of the edges of P and by r the radius of the sphere circumscribed around P (cf. 12.5.2.1). We also set $\rho(P) = l^2/4r^2$.

12.6.5. LEMMA. If P has symbol $\{r_1, \ldots, r_{d-1}\}$, the following relation holds between $\rho(P)$ and $\rho(\mathrm{link}\, P)$:

$$\rho(P) = 1 - \frac{\cos^2 \pi/r_1}{\rho(\mathrm{link}\, P)}.$$

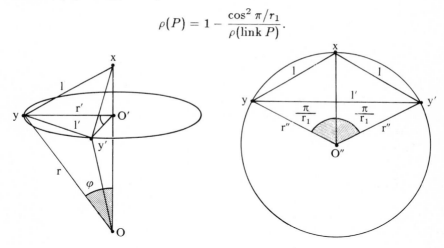

Figure 12.6.5

Proof. Let $x \in P$ be a vertex, O the center of P and O' the center of $\mathrm{link}_x P$. Let r (resp. r') be the radius of the sphere circumscribed around P (resp. $\mathrm{link}_x P$). Let y, y' be the endpoints of an edge of $\mathrm{link}_x P$, and $l' = yy'$ the length of such an edge. The points y, x, y' are consecutive vertices (by 12.5.3.1) of a 2-face of P, which is an r_1-sided regular polygon; letting r'' be the radius of a circle of center O'' circumscribed around that 2-face, we have, by 12.4.3:

$$l = yx = xy' = 2r'' \sin \frac{\pi}{r_1}, \qquad l' = 2r'' \frac{2\pi}{r_1},$$

whence $l' = 2l \cos(\pi/r_1)$. By definition,

$$\rho(P) = \frac{l^2}{4r^2}, \qquad \rho(\text{link } P) = \rho(\text{link}_x P) = \frac{l'^2}{4r'^2}.$$

Denoting by 2ϕ the angle of the triangle $\{y, O, x\}$ at O, we have $r' = l \cos \phi$, $l = 2r \sin \phi$; throwing everything together, we get

$$\rho(P) = \sin^2 \phi, \quad \rho(\text{link } P) = \frac{4l^2 \cos^2 \frac{\pi}{r_1}}{4l^2 \cos^2 \phi} = \frac{\cos^2 \frac{\pi}{r_1}}{\cos^2 \phi}. \qquad \square$$

12.6.6. CONSEQUENCE. The number $\rho(P)$ depends only on the $(d-1)$-tuple $\{r_1, \ldots, r_{d-1}\}$, and shall be denoted by $\rho(r_1, \ldots, r_{d-1})$. For any regular polytope with symbol $\{r_1, \ldots, r_{d-1}\}$, the relation

$$\rho(r_1, \ldots, r_{d-1}) = 1 - \frac{\cos^2(\pi/r_1)}{\rho(r_2, \ldots, r_{d-1})}$$

holds. This follows from 12.6.5, applied to link P, link(link P), and so on.

12.6.7. THEOREM (SCHLÄFLI, 1850). *The only possible symbols for a regular polytope are given by the following list:*

    $d = 2$: $\{n\}$, *where* $n \geq 3$ *is an arbitrary integer;*
    $d = 3$: $\{3, 3\}$, $\{3, 4\}$, $\{4, 3\}$, $\{3, 5\}$, $\{5, 3\}$;
    $d = 4$: $\{3, 3, 4\}$, $\{4, 3, 3\}$, $\{3, 4, 3\}$, $\{3, 3, 5\}$, $\{5, 3, 3\}$;
    $d \geq 5$: $\{3, \ldots, 3\}$, $\{3, \ldots, 3, 4\}$, $\{4, 3, \ldots, 3\}$.

For each symbol in this list, there exists a regular polytope with that symbol. Two regular polytopes with the same symbol are similar.

Proof.

12.6.7.1. Existence follows from 12.4.2, 12.6.2.3 and 12.6.2.4.

12.6.7.2. Uniqueness up to similarity is proven by induction. Let P and P' be regular polytopes with same symbol; by the induction assumption and 12.6.2.2, the faces of P and P' are similar. After applying the appropriate similarity, we can assume them isometric, and then we can make a face of P coincide with one of P', in such a way that P, P' lie in the same half-space determined by that face F.

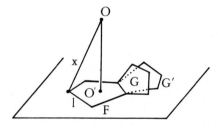

Figure 12.6.7.2

But then P and P' have the same center O, because O necessarily lies on the perpendicular to the hyperplane of F going through the center O' of F, and its distance r to a vertex of F is also known, being given by

$$\rho(P) = \rho(P') = \frac{l^2}{4r^2}$$

(cf. 12.6.4 and 12.6.6), where l is the length of an edge of F. Now let A be a $(d-2)$-face of F, and let G and G' be faces of P and P' adjacent to F along A (cf. 12.1.12). If $H = \langle O, A \rangle$ is the hyperplane going through O and A, it follows from 12.5.2.3 that $\sigma_H(F) = G$ and $\sigma_H(F) = G'$, so $G = G'$. Using induction and 12.1.13 we get $P = P'$.

12.6.7.3. The list of possible symbols is established by induction, considering all cases allowed by the fundamental relation 12.6.6. Since $r_1 \geq 3$ (cf. 12.6.2.1), no value of $\rho(r_1, \ldots, r_{d-1})$ is possible when $\rho(r_2, \ldots, r_{d-1}) \leq 1/4$. Thus there exist only a finite number of cases.

This leads to table 12.6.7.4, where the only things to be proven are the entries $(d+1)/2d$ and $1/d$ in the last column. This is done by recurrence. $\square$

12.6.7.4. Table of symbols.

$d = 2$		$d = 3$		$d = 4$		$d \geq 5$	
symbol	$\rho(\cdot)$	symbol	$\rho(\cdot)$	symbol	$\rho(\cdot)$	symbol	$\rho(\cdot)$
						$\{3,3,\ldots,3\}$	$\dfrac{d-1}{2d} \Rightarrow$
				$\{3,3,3\}$	$\dfrac{5}{8} \Rightarrow$		
		$\{3,3\}$	$\dfrac{2}{3} \Rightarrow$			$\{4,3,\ldots,3\}$	$\dfrac{1}{d} \bullet$
				$\{4,3,3\}$	$\dfrac{1}{4} \bullet$		
$\{3\}$	$\dfrac{3}{4} \Rightarrow$			$\{5,3,3\}$	$\dfrac{7-3\sqrt{5}}{16} \bullet$		
		$\{4,3\}$	$\dfrac{1}{3} \Rightarrow$	$\{3,4,3\}$	$\dfrac{1}{4} \bullet$		
		$\{5,3\}$	$\dfrac{3-\sqrt{5}}{6} \bullet$				
$\{4\}$	$\dfrac{1}{2} \Rightarrow$	$\{3,4\}$	$\dfrac{1}{2} \Rightarrow$	$\{3,3,4\}$	$\dfrac{1}{2} \Rightarrow$	$\{3,\ldots,3,4\}$	$\dfrac{1}{2} \Rightarrow$
$\{5\}$	$\dfrac{5-\sqrt{5}}{8} \Rightarrow$	$\{3,5\}$	$\dfrac{5-\sqrt{5}}{10} \Rightarrow$	$\{3,3,5\}$	$\dfrac{3-\sqrt{5}}{8} \bullet$		
$\{n \geq 6\}$	$\sin^2 \dfrac{\pi}{n} \bullet$						

Quantities marked $\bullet$ are $\leq \frac{1}{4}$

12.6.8. A HEURISTIC REMARK. We verify that there are more regular polytopes in low dimension than in high: there are but three general series, but in dimension three and four there are exceptional cases, and in dimension two there are infinitely many polygons. This fact illustrates one of the halves of the following principle, due to Thom: rich structures are more numerous in low dimension, and poor structures are more numerous in high dimension. Here are more examples of this contrast, some of which may be familiar to the reader:

rich structures:
— simple Lie groups ([SE3, the whole book]),
— division algebras ([KH, 249]),
— quadratic fields ([SU, 167] and [B–S, 342–355]),
— the orthogonal group is not simple in dimension four (cf. 8.9.10),
— compact manifolds of constant negative curvature can be deformed only in dimension two ([MW1], [MW2]),
— the symmetric group S_n is solvable for $n \leq 4$ ([SW, chapter 13]);

poor structures:
— topological vector spaces (in finite dimension they are all homeomorphic),
— finite fields are all commutative and easy to classify ([SE2, chapter 1]),
— singularities of differentiable maps become more complicated as the dimension increases ([MR]).

12.6.9. A CULTURAL REMARK. In spite of their purely geometric character and their specialized appearance, regular polytopes are present in important mathematical structures, via their isometry groups. These fall into the class of groups "generated by reflections" (in this case hyperplane reflections), and all finite groups of this type have been classified. They play a fundamental role in the study of Lie groups (the Weyl group) and of algebraic groups. For a general theory, see [BI4]; see also [CR2, chapter 1]. From this general theory one could deduce a classification of regular polytopes. See also 1.8.7.

12.6.10. NOTES.

12.6.10.1. The most complete discussion of regular polytopes, including a rigorous historical study, is to be found in [CR2].

12.6.10.2. A very good reference for the study of $\{3, 4, 3\}$ and $\{3, 3, 5\}$ and of their groups, carried out with the help of quaternions (not surprising in dimensions three and four), is the book [VL].

12.6.10.3. There is a connection between the icosahedron and the general quintic equation, established via Galois theory (since the group of the icosahedron is isomorphic to A_4, cf. 12.5.5.6). See [KN1], a classical book devoted to this subject, or [SW, 148–149], a more modern reference.

12.6.10.4. Formula 12.6.6 also leads to a classification of tilings of Euclidean space by regular polytopes. We know of tilings of the plane by equilateral triangles, squares and regular hexagons, and also tilings of arbitrary-dimensional spaces by cubes. Are there any others? The answer is that regular polytopes with symbol $\{r_1, \ldots, r_{d-1}\}$ tile the d-dimensional Euclidean space if and only if there exists a regular polytope with symbol $\{r_2, \ldots, r_d\}$ such that

$$\rho(r_1, r_2, \ldots, r_{d-1}, r_d) = 0;$$

the polytope with symbol $\{r_2, \ldots, r_d\}$ corresponds to the "link" of the tiling. By studying table 12.6.7.4, one easily sees that, apart from the already mentioned cases, with symbols $\{3,6\}$, $\{4,4\}$, $\{6,3\}$ and $\{4,3,\ldots,3,4\}$, the only remaining tilings occur in four dimensions, with symbols $\{3,4,3,3\}$ and $\{3,3,4,3\}$ (one is a tiling by cocubes, the other by $\{3,4,3\}$ polytopes). These two are dual to each other, as are the tilings by triangles and hexagons (figure 12.6.10.4); tilings by cubes are self-dual (the duality is realized by taking the center of each tile).

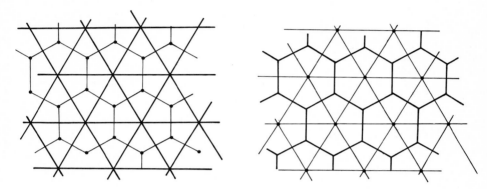

Figure 12.6.10.4

12.6.10.5. There exist (non-convex) regular *star polytopes*; they too have been classified ([CR2, chapter 14]). Figures 12.6.10.5.1 and 12.6.10.5.2 show the four examples in dimension three, all having the symmetry group of the icosahedron:

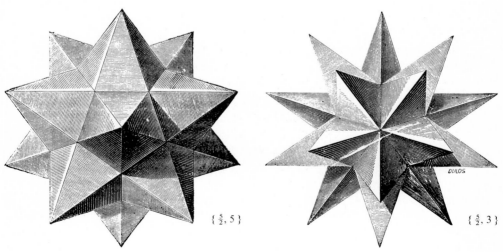

$\{\frac{5}{2}, 5\}$ $\{\frac{5}{2}, 3\}$

Figure 12.6.10.5.1 (Source: [R–C])

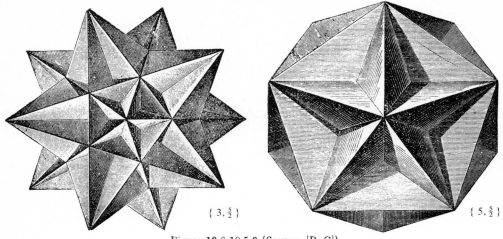

$\{3, \frac{5}{2}\}$ $\{5, \frac{5}{2}\}$

Figure 12.6.10.5.2 (Source: [R–C])

$\{\frac{5}{2}, 3, 3\}$

Figure 12.6.10.5.3 (Source: [CR2])

In dimension four, there exist ten regular star polyhedra, two of which are represented in figures 12.6.10.5.3 and 12.6.10.5.4. All ten have the symmetry group of the $\{3,3,5\}$ regular polytope (with 14400 elements). Star polyhedra can also be given Schläfli symbols (shown in the figures), whose entries are no longer integers, but rational numbers. The classification in dimension two (regular star polygons) can be undertaken by the reader; in dimension five and above, there are no star polytopes.

$$\{\tfrac{5}{2}, 3, 5\} , \{\tfrac{5}{2}, 5, \tfrac{5}{2}\}$$
$$\{3, \tfrac{5}{2}, 5\} , \{3, 3, \tfrac{5}{2}\}$$

Figure 12.6.10.5.4 (Source: [CR2])

12.7. Euler's formula

In this section (except at the end) X is a three-dimensional affine space, which will be given a Euclidean structure in 12.7.3.1 and 12.7.3.2 for the purposes of a proof.

12.7.1. NOTATION. For P a polytope, set:

$$\Xi = \text{set of edges of } P, \qquad\qquad\qquad \alpha = \#\Xi;$$
$$\Sigma_i = \text{set of vertices of } P \text{ belonging to } i \text{ faces}, \qquad \sigma_i = \#\Sigma_i;$$
$$\Sigma = \text{set of all vertices of } P, \qquad\qquad\qquad \sigma = \#\Sigma;$$
$$\Phi_i = \text{set of faces of } P \text{ containing to } i \text{ edges}, \qquad \phi_i = \#\Phi_i;$$
$$\Phi = \text{set of all faces of } P, \qquad\qquad\qquad \phi = \#\Phi.$$

By 12.1.12, for example, the values of i to be considered are all greater than or equal to 3. Clearly,

12.7.2 $$\sigma = \sum_i \sigma_i, \qquad \phi = \sum_i \phi_i, \qquad 2\alpha = \sum_i i\sigma_i = \sum_i i\phi_i.$$

After calculating σ, α and ϕ for figures 1.8.4.2 to 1.8.4.6 and 12.1.1.4 and 12.1.1.5, the reader will find each time that $\sigma - \alpha + \phi = 2$. This is a general phenomenon:

12.7.3. THEOREM (EULER'S FORMULA). *For every three-dimensional polytope,* $\sigma - \alpha + \phi = 2$.

We give two proofs of this formula; the first is "rounder" and more pleasant, but has the drawback of requiring Girard's formula involving the area of a spherical triangle. The second imitates the first, but only uses the fact that the sum of the angles of a triangle is equal to π. See also 12.7.5.3.

12.7.3.1. First proof. Take an arbitrary $O \in \overset{\circ}{P}$ and let $S = S(0, 1)$ be the unit sphere of X_O. Project the boundary of P onto S via the map $p : \operatorname{Fr} P \ni x \mapsto \dfrac{x}{\|x\|} \in S$. The vertices of P become points of S, the edges arcs of great circles and the faces spherical polygons (for spherical objects, see chapter 18). Since P is convex and $O \in \overset{\circ}{P}$, hyperplanes containing O and an edge do cut P into two pieces (cf. 12.1.5). It follows that the spherical polygons obtained are all convex and that the number of those with i faces is ϕ_i, that every edge is shared by two distinct faces and there are α of them, that the number of vertices belonging to i faces is σ_i and that each edge contains two vertices (cf. 12.1.12).

The idea now is to add up the angles of all the convex polygons in two different ways: first the angles around each vertex, then the angles of each polygon. Since $p(\operatorname{Fr} P) = S$ and the convex polygons do not overlap, the sum of angles around a vertex is 2π and the total sum of all angles is

$$\sum_{s \in \Sigma} 2\pi = 2\pi\sigma.$$

On the other hand, the sum of the angles of a polygon $F \in \Phi_i$ is, by 18.3.8.5, equal to area $F + (i - 2)\pi$, so we obtain the following value for the sum of all the angles:

$$\sum_i \sum_{F \in \Phi_i} \big(\text{area}\, F + (i-2)\pi \big) = \sum_{F \in \Phi} \text{area}\, F + \sum_i (i-2)\pi\phi_i$$

$$= 4\pi + \pi \sum_i i\phi_i - 2\pi \sum_i \phi_i = 4\pi + 2\alpha\pi - 2\pi\phi,$$

using 12.7.2 and again the fact that $p(\text{Fr}\, P) = S$. Comparing the two sums we get $\sigma - \alpha + \phi = 2$.

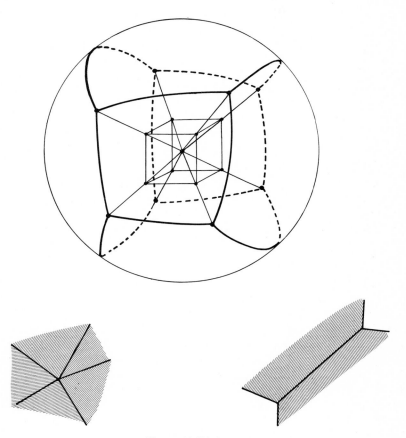

Figure 12.7.3.1

12.7.3.2. Second proof. Here we work on a plane; the idea is to remove a face of P and apply stereographic projection from a point close to the removed face. After that it's just a matter of counting objects like before, except that we have to distinguish between interior and boundary vertices.

Set $P = \bigcap_{i=1}^n R_i$ (cf. 12.1.6), $F = F_1$, $a \in \bigcap_{i \geq 2} \overset{\circ}{R}_i \setminus P$ and let Y be a plane parallel to F and separated from a by F. Projecting from a onto Y,

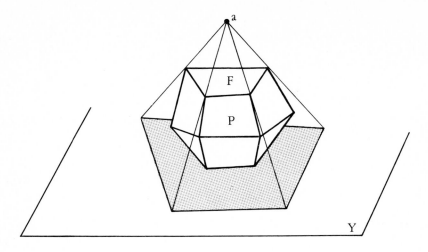

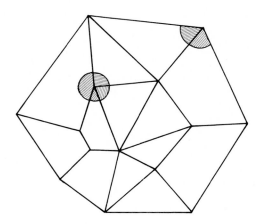

Figure 12.7.3.2

via the map $p : x \mapsto \langle a, x \rangle \cap Y$, we define a bijection between the vertices, edges and faces of P (other than F) and a set of points, segments and convex polygons of Y satisfying the following properties: the polygons add up to a large convex polygon $p(F_1)$ and each meets the next only along an edge and its vertices. Let σ' be the number of vertices; of these there are some boundary vertices, as many as the number of sides of $p(F_1)$. Thus, from 10.5.2, we get the following value for the sum of the angles of all the polygons:

$$2\pi(\sigma' - k) + (k - 2)\pi = \pi(2\sigma' - k - 2),$$

since the sum of the angles at an interior vertex is 2π, and at a boundary vertex it is equal to the correspondent angle of $p(F_i)$.

Let ϕ_i' be the number of polygons with i sides, and let ϕ' be the total number of polygons; again from 10.5.2, we get the another estimate for the total sum of angles:

$$\sum_i (i-2)\pi\phi_i' = \pi\sum_i i\phi_i' - 2\phi'.$$

Here $\sum_i i\phi_i$ is not equal to the number α' of sides, but only to $2\phi'-k$, because of the boundary. Then we get

$$2\sigma' - k - 2 = 2\alpha' - k - 2\phi',$$

or $\sigma' - \alpha' - \phi' = 1$. We conclude by observing that $\sigma = \sigma'$, $\alpha = \alpha'$ and $\phi = \phi' + 1$.

12.7.4. APPLICATION. Euler's formula's primary use will be in the proof of Cauchy's theorem (12.8.6), but we devote this section to showing how one can derive the list of three-dimensional regular polytopes from it.

12.7.4.1. Proposition. *Let P be a polytope in three-dimensional affine space satisfying the condition that each vertex belongs to a constant number r of faces and each face contains a constant number s of vertices. Then the pair $\{r, s\}$ must be one of the following: $\{3,3\}$, $\{3,4\}$, $\{3,5\}$, $\{4,3\}$ and $\{5,3\}$.*

Proof. We deduce from 12.7.2 that $2a = r\phi = s\sigma$, so 12.7.3 gives

$$\frac{1}{r} + \frac{1}{s} = \frac{1}{2} + \frac{1}{\alpha},$$

where $r, s \geq 3$ and $\alpha > 0$. The list is then obtained by inspection. $\square$

12.7.4.2. Note. This proof also determines the numbers σ, α and ϕ in each case. Notice that the existence of polyhedra with the property of the statement is trivial—a picture is sufficient proof, and one does not have to resort to 12.5.5. Combinatorially, then, there exists exactly one polytope of each of the five types:

$\{r, s\}$	$\{3,3\}$	$\{3,4\}$	$\{4,3\}$	$\{3,5\}$	$\{5,3\}$
σ	4	8	6	20	12
α	6	12	12	30	30
ϕ	4	6	8	12	20

12.7.5. NOTES

12.7.5.1. The six-pointed shaddock. The proof given in 12.7.3.1 shows that Euler's formula is still valid for star-shaped compact polygons (11.1.2.4), not only convex ones. We take the opportunity to point out the possible fallacy of assuming that such a non-convex polyhedron P, star-shaped at $O \in \overset{\circ}{P}$, can be made convex by sliding each vertex along the line that connects it to O. This is in general not possible; see in [DY] a counterexample (the

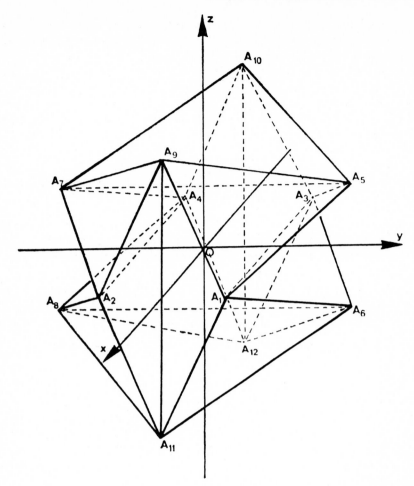

Figure 12.7.5.1 (Source: [DY])

six-pointed shaddock of figure 12.7.5.1) and a discussion of the relevance of
this fact to algebraic geometry.

12.7.5.2. The proof found in 12.7.3.1, and even the one in 12.7.3.2, are merely
particular cases of the proof of the Gauss–Bonnet theorem for compact two-
dimensional Riemannian manifolds; in 12.7.3.1 the manifold is the sphere
of constant curvature 1, and in 12.7.3.2 a disc of curvature zero. See, for
example, [KG1, 143] or [SB, 290]. See also 18.3.8.6 and 19.5.4.

12.7.5.3. Proofs 12.7.3.1 and 12.7.3.2 may have left the reader unsatisfied, in
that one should not have to invoke Riemannian geometry to prove an result in
affine geometry. In fact, the result holds in the even more general "simplicial"
setting, which only has to do with the incidence relations between vertices,
edges and faces. In this setting the convexity condition must be replaced by

the assumption that the polyhedron is simply connected, otherwise the result is false (see below). Simple connectedness is most often defined nowadays as the property that the fundamental group is zero, that is, that every closed path can be continuously deformed into a constant path (a point). Previously the condition was, for a two-dimensional polyhedron, that any simple closed curve would split the polyhedron into two connected components; see in [VN, 27–30] a proof of Euler's formula with this definition. Check also that the original and the modern definition are equivalent. All this belongs to the realm of algebraic topology, and is somewhat outside the scope of this book.

12.7.5.4. For an arbitrary polyhedron, in the sense of 12.1.3, Euler's formula is not true: for example, gluing γ polyhedra like the ones in figure 12.1.1.2 we obtain figure 12.7.5.4 and the following table:

γ	σ	α	ϕ	$\sigma - \alpha + \phi$
1	12	24	12	0
2	20	44	22	-2
γ	$12\gamma - 4(\gamma - 1)$	$24\gamma - 4(\gamma - 1)$	$12\gamma - 2(\gamma - 1)$	$2(1 - \gamma)$

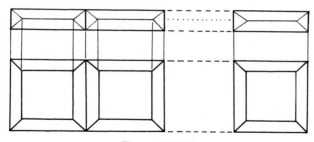

Figure 12.7.5.4

This example, together with the convex case, gives the values $2(1 - \gamma)$ for $\sigma - \alpha + \phi$, where $\gamma \geq 0$ is an integer. It can be shown that $\sigma - \alpha + \phi$ can take no other value for any polyhedron (or, equivalently, for any surface embedded in $\mathbf{R}^3$), and that the number γ is an invariant under continuous deformations; in other words, every polyhedron is homeomorphic to either a convex polyhedron ($\gamma = 0$) or exactly one polyhedron of the type shown in figure 12.7.5.4. The invariant γ is called the *genus* or *number of holes* of the polyhedron, and $\sigma - \alpha + \phi$ its *Euler characteristic*. For a complete discussion of this question see [SE–TH, chapter 6].

12.7.5.5. You may be asking yourself what happens for a convex polyhedron P of arbitrary dimension d; the answer is, if we denote by ϕ_i the number of i-faces of P $(i = 0, 1, \ldots, d - 1)$:

$$\sum_{i=0}^{d-1}(-1)^i \phi_i = 1 + (-1)^{d-1}.$$

You can verify this for $d = 2$, and also for $d = 4$ and P a regular polytope. For a proof, see [GG2, 102–103], and [CR2, chapter IX].

12.8. Cauchy's theorem

> Here X is a three-dimensional Euclidean affine space.

This is a perfect example of a theorem that is easy to state but very difficult to prove. If you have manufactured a convex polyhedron out of cardboard, you will have noticed that, in the beginning, sets of four or more faces sharing a common vertex tend to be flexible around the edges, but once the whole polyhedron has been assembled it is not flexible anymore; if you press too hard on it, you either deform its faces or tear it open. This is a consequence of the following stronger result:

12.8.1. THEOREM (CAUCHY). *Let P, P' be two convex polytopes in X and $f : \operatorname{Fr} P \to \operatorname{Fr} P'$ a bijection taking vertices into vertices, edges into edges and faces into faces. Assume that, for each face F of P, the restriction $f|_F : F \to f(F)$ is an isometry. Then there exists an isometry $\overline{f}$ of X such that $\overline{f}(P) = P'$ and $\overline{f}|_{\operatorname{Fr} P} = f$; in particular, P and P' are isometric.*

12.8.2. COROLLARY. *A convex polytope P is not flexible, that is, if $P(t)$, $t \in [0, 1]$ is a family of (not necessarily convex) polyhedra with $P(0) = P$ and $f_t : \operatorname{Fr}(P(0)) \to \operatorname{Fr}(P(t))$ $(t \in [0, 1])$ is a family of bijections such that $f_t|_F$ is an isometry between F and $f_t(F)$ for every face F of $P(0)$, that $f_0 = \operatorname{Id}_P$ and that*

$$f : \operatorname{Fr} P \times [0, 1] \ni (x, t) \mapsto f_t(x) \in X$$

is continuous, then there exists a family $\overline{f}_t \in \operatorname{Isom}(X)$ $(t \in [0, 1])$, such that $\overline{f}_t|_{\operatorname{Fr}(P(0))} = f_t$ and $\overline{f}_t(P(0)) = P(t)$ for any $t \in [0, 1]$.

Proof. The set of points $t \in [0, 1]$ for which there exists an isometry $\overline{f}_t$ extending f_t and satisfying $\overline{f}_t(P(0)) = P(t)$ is open and closed in $[0, 1]$: closed by continuity and open because, if $P(t)$ is a polytope, $P(t')$ is still a polytope for t' close enough to t, and we can apply 12.8.1. □

12.8.3. COROLLARY. *If P is a polytope such that (i) each face of P is a regular polygon having a fixed number of faces, and (ii) each vertex of P belongs to the same number of faces, then P is regular.*

Proof. By 12.7.4 and 12.6.7, there exists a regular polytope P' satisfying the same conditions as P, so we can construct a bijection $f : \operatorname{Fr} P \to \operatorname{Fr} P'$ satisfying the assumptions of 12.8.1. This gives an isometry between P and P'. □

12.8.4. NOTES.

12.8.4.1. Theorem 12.8.1 can be false if P is a polytope but P' is a non-convex polyhedron. For example, in figure 12.8.4, P' can be derived from P by reflecting the four faces on the right through a plane:

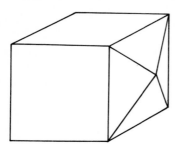 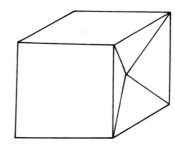

Figure 12.8.4

But notice that P is not flexible; we cannot go from P to P' continuously and isometrically.

12.8.4.2. Ever since Cauchy's original paper in 1813, it had been an open problem whether there existed (necessarily non-convex) flexible polyhedra. The question was recently answered in the affirmative by Robert Connelly; see [CL2], where he exhibits a simply connected flexible polyhedron. See also the earlier results in [GL] and [CL1], and the preprint [CL3] for a practical construction. See also [BAN] and [LB3].

12.8.4.3. Theorem 12.8.1 is false in dimension two; a polygon with more than three sides can easily be deformed with constant side lengths (see 10.8.3, even!). In dimension greater than three, on the other hand, the theorem is true. The proof is by induction, based on the following idea: theorem 12.8.1 is still valid for convex spherical polyhedra in the three-dimensional sphere S^3 (instead of polyhedra in Euclidean affine space); the proof below can be carried over with no changes. Now let P and P' be as in 12.8.1, in dimension four; for every $x \in P$, consider the link $\text{link}_x P$, which is the intersection of P with a sufficiently small sphere centered at x. The assumption of 12.8.1 guarantees that $\text{link}_x P$ and $\text{link}_{f(x)} P'$, after a suitable normalization, are spherical polyhedra in S^3 satisfying the same assumption, so they are isometric. Thinking now in terms of P and P', this means that f preserves dihedral angles, and this, by lemma 12.8.5.1, gives an isometry between the two polyhedra. Of course, this method does not work in dimension three, since a polygon of S^2 with more than three sides is eminently deformable.

12.8.5. AN ELEMENTARY LEMMA. If P is a polytope and A is an edge of P, we denote by $\delta_A(P)$ the dihedral angle of P at A (cf. 12.1.12).

12.8.5.1. Lemma. *If the bijection f of 12.8.1 also preserves dihedral angles, that is, if $\delta_{f(A)}(P') = \delta_A(P)$ for any edge A of P, it can be extended into an isometry $\overline{f} : X \to X$ such that $\overline{f}(P) = P'$.*

Proof. The proof mimics that of 12.4.2. Let F be a face of P; we can assume that $f(F) = F$, and also that P and P' lie on the same half-space determined by F. Let A be an edge of F and let G, G' be the faces of P, P' adjacent to F at A; since $\delta_A(P) = \delta_A(P')$, saying that G and G' are isometric and lie on the same side of F is to say that $G = G'$. Now 12.1.13 brings home the bacon.

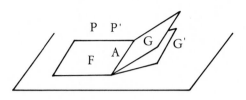

Figure 12.8.5

12.8.6. PROOF OF CAUCHY'S THEOREM.

12.8.6.1. By 12.8.5.1 all we have to do is show that f preserves dihedral angles. Observe that, according to the criterion for equality of spherical triangles (cf. 18.6.13.10) this would be easy if each vertex of P belonged to three faces; but once we have four or more faces, we get deformable spherical polygons.

The real proof is based on a crucial lemma, concocted by Cauchy for this purpose, and saying that when a spherical polygon is deformed with sides fixed, the angles that increase must to a certain extent alternate with those that decrease (18.7.16). We can count all alternations in two independent ways (first over each face, then over each vertex), and compare the two sums to obtain a contradiction with Euler's formula. The proof is much easier if $\delta_{f(A)}(P') \neq \delta_A(P)$ for every edge A of P.

12.8.6.2. Let's introduce the following notation, for P, P' and f as in the statement of 12.8.1: Ξ is the set of edges of P and $\epsilon : \Xi \to \{-1, 0, 1\}$ is the function defined by $\epsilon(A) = -1$ (resp. 0, 1) if $\delta_A(P)$ is smaller than (resp. equal to, greater than) $\delta_{f(A)}(P')$. We also say that two edges A and B of P are *adjacent* if they belong to the same face of P and $A \cap B$ is a vertex of P.

12.8.6.3. First case: $\epsilon(A) \neq 0$ for any $A \in \Xi$. We say that the pair $\{A, B\}$ of adjacent edges has a sign flip if $\epsilon(A)\epsilon(B) = -1$, and we let ν be the total number of pairs of adjacent edges having a sign flip.

Let x be a vertex of S, and π (resp. π') the convex spherical polygon obtained by intersecting P (resp. P') with a small enough sphere centered at x (resp. $x' = f(x)$). By assumption, once π and π' are transferred to the unit sphere, there is a bijection between the two that preserves side lengths.

Cauchy's lemma (cf. 18.7.16) says exactly that there are at least four sign flips among the pairs of adjacent edges originating at a fixed vertex x; thus, borrowing the notation introduced in 12.7, we get $\nu \geq 4\sigma$. Now count

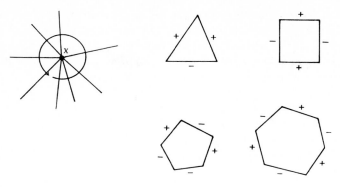

Figure 12.8.6.3

sign flips for pairs of edges around each i-sided face; there are at most two for $i = 3$, at most 4 for $i = 4$ or 5, and so on. We get

$$\nu \le 2\phi_3 + 4\phi_4 + 4\phi_5 + 6\phi_6 + 6\phi_7 + \cdots$$

Using 12.7.2, we obtain

$$4\alpha - 4\phi = 6\phi_3 + 8\phi_4 + 10\phi_5 + 12\phi_6 + \cdots - (4\phi_3 + 4\phi_4 + 4\phi_5 + 4\phi_6 + \cdots)$$
$$= 2\phi_3 + 4\phi_4 + 6\phi_5 + 8\phi_6 + \cdots \ge \nu \ge 4\sigma,$$

whence $\sigma - \alpha + \phi \le 0$, contradicting 12.7.3.

12.8.6.4. Second case: the $\epsilon(A)$ are arbitrary. Let the *ghost edges* of P be the edges $A \in \Xi$ such that $\epsilon(A) = 0$, and *live edges* the others. Let Ξ' be the set of live edges, and α' its cardinal. Remove from $\operatorname{Fr} P$ the set of live edges, to obtain the topological space $U = \operatorname{Fr} P \setminus \bigcup_{A \in \Xi'} A$, and call a *live face* of P the closure of any connected component of U. Let Φ' be the set of live faces and ϕ' its cardinality. Finally, let the *live vertices* of P be the vertices of P belonging to at least one live edge, let Σ' be the set of live vertices and $\sigma' = \#\Sigma'$. Notice that Cauchy's lemma implies that every live vertex belongs to at least two live edges, that is, there is no live edge with a free endpoint.

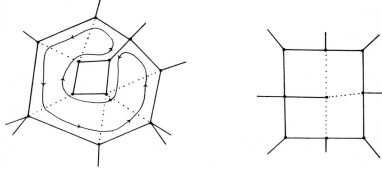

Figure 12.8.6.4

12.8.6.5. We claim that $\sigma' - \alpha' + \phi' \geq 2$. In fact, we can restore the
ghost edges one by one, with the precaution of always attaching them to a
live vertex or a vertex that has already been restored (in this process we
may temporarily leave edges hanging). This gives a sequence of intermediate
values σ_t, α_t and ϕ_t $(t = 1, \ldots, \alpha - \alpha')$ for the cardinalities. At each stage t,
we have $\alpha_{t+1} = \alpha_t + 1$, and either $\sigma_{t+1} = \sigma_t$ and $\phi_{t+1} = \phi_t$ or $\phi_t + 1$, or else
$\sigma_{t+1} = \sigma_t + 1$ and then $\phi_{t+1} = \phi_t$. Thus the function $\sigma_t - \alpha_t + \phi_t$ is always
non-decreasing, and, since it has the value 2 for $t = \alpha - \alpha'$ by 12.7.3, the
claim is proved. For a similar formula, see [LU, chapter 13], [ST–RA, §33]
and [AW2, 78, formula (3)].

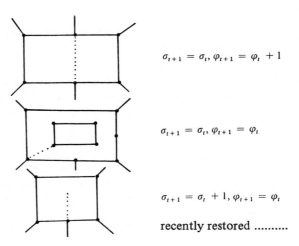

$$\sigma_{t+1} = \sigma_t,\ \varphi_{t+1} = \varphi_t + 1$$

$$\sigma_{t+1} = \sigma_t,\ \varphi_{t+1} = \varphi_t$$

$$\sigma_{t+1} = \sigma_t + 1,\ \varphi_{t+1} = \varphi_t$$

recently restored ....

Figure 12.8.6.5

12.8.6.6. We can now proceed as in the first case, defining adjacent live
edges, sign flips and the number ν. The only difference is that a live edge
does not necessarily belong to two live faces, but to one or two; this is taken
care of by counting the number of sides of a live face as the number of live
sides plus the number of live edges that are represented twice in the boundary
of the face (see figure 12.8.6.4). With this convention, let Φ_i' the set of i-sided
live faces and set $\phi_i' = \#\Phi_i'$. Applying Cauchy's lemma to the σ' live vertices,
we have

$$2\alpha' = \sum_i i\phi_i', \qquad \phi' = \sum_i \phi_i', \qquad \nu \geq 4\phi';$$

counting the number of flips for each live face, we get

$$\nu \leq 2\phi_3' + 4\phi_4' + 4\phi_5' + 6\phi_6' + 6\phi_7' + \cdots,$$

whence we get, as in 12.6.3, $4\alpha' - 4\phi' \geq \nu \geq 4\sigma'$, contradicting 12.8.6.5.
This shows our "live polyhedron" is impossible, that is, all edges are ghost
edges, and $\delta_A(P) = \delta_{f(A)}(P')$ for all $A \in \Xi$. Now 12.8.5.1 concludes the
proof.

> From now till the end of the chapter X is a Euclidean affine space.

12.9. Approximation of compact convex sets by polytopes

In this section we approximate compact convex sets by polytopes in the Hausdorff metric (cf. 9.11). This will be fundamental at several points: to define the volume of convex sets in an elementary way (12.9.3.2), to show that this volume varies continuously with the set (12.9.2.4) and that the frontier of convex sets has measure zero (12.9.3.4), and to define the area of a convex set (12.10.2).

12.9.1. Convex sets and the Hausdorff distance. Recall the notation $\mathcal{K}$ introduced in 9.11, and let $\mathcal{E} : \mathcal{K} \to \mathcal{K}$ be the map taking a set $K \in \mathcal{K}$ to $\mathcal{E}(K)$.

12.9.1.1. Proposition. *The map $\mathcal{E} : \mathcal{K} \to \mathcal{K}$ is Lipschitz with constant 1.*

Proof. From definition 9.11.1, it suffices to show that if $F, G \in \mathcal{K}$ satisfy $F \subset B(G, \rho)$ $(\rho \geq 0)$, we have $\mathcal{E}(F) \subset B(\mathcal{E}(G), \rho)$. This is a consequence of 11.8.7.6. $\square$

12.9.1.2. Corollary. *Set $C = \{ F \in \mathcal{K} \mid F$ is convex $\}$; then C is closed in $\mathcal{K}$. In particular, C is complete and, for $a \in X$ and $r \geq 0$, the set*

$$C_{a.r} = \mathcal{K}_{a.r} \cap C$$

is compact. $\square$

The next lemma gives a good feeling for the notion of limit among convex sets:

12.9.1.3. Lemma. *Let A, C and D be compact convex sets such that $\overset{\circ}{C} \neq \emptyset$, $D \subset C \subset A$, $\mathrm{Fr}\, A \cap C = \emptyset$ and $\mathrm{Fr}\, C \cap D = \emptyset$. Then there exists $\eta > 0$ such that every convex set S such that $\delta(C, S) \leq \eta$ satisfies $D \subset S \subset A$.*

Proof. Showing that $S \subset A$ is easy, just take $\eta = d(\mathrm{Fr}\, A, C)$. To get $S \supset D$, set $\eta = d(\mathrm{Fr}\, C, D)$ and assume that for some S such that $\delta(S, C) \leq \eta$ there is a point $x \in D \setminus S$. Take y such that $d(x, y) = d(x, S)$; we know from 11.1.7.2 that $S \subset H$, where H is one of the half-spaces defined by the hyperplane orthogonal to $\langle x, y \rangle$ at y. Since $x \in \overset{\circ}{C}$, we can take the point z of $\langle x, y \rangle \cap \mathrm{Fr}\, C$ that does not lie on H. This point satisfies

$$d(z, S) = d(z, y) > d(z, x) \geq \eta$$

by the definition of η; but, since $z \in C$, this contradicts the assumption $C \subset B(S, \eta)$. $\square$

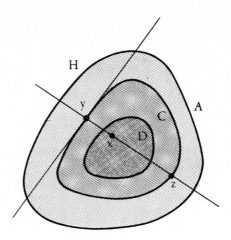

Figure 12.9.1

12.9.2. APPROXIMATION LEMMAS. We *denote* by $\mathcal{P}$ the set of all compact convex polyhedra, by $\mathcal{P}^\bullet$ the set of all polytopes, and by $\mathcal{C}^\bullet$ the set $\{\, C \in \mathcal{C} \mid \dim C = \dim X \,\}$ (or such that $\overset{\circ}{C} \neq \emptyset$, cf. 11.2.7). If confusion can arise, we write $\mathcal{P}(X)$, $\mathcal{P}^\bullet(X)$, $\mathcal{C}(X)$ and $\mathcal{C}^\bullet(X)$.

12.9.2.1. Lemma. *For every $\epsilon > 0$ and every $C \in \mathcal{C}^\bullet$ there exists $P \in \mathcal{P}^\bullet$ such that $P \subset C \subset B(P, \epsilon)$ (in particular, $\delta(P, C) \leq \epsilon$).*

Proof. Since C is compact, it can be covered by n balls $B(a_i, \epsilon)$, $a_i \in C$. Take for P the convex hull $\mathcal{E}(a_1, \ldots, a_n)$, cf. 12.1.15. □

12.9.2.2. Corollary. $\mathcal{P}$ *is dense in* $\mathcal{C}$.

Proof. We have just seen that $\mathcal{P}^\bullet$ is dense in $\mathcal{C}^\bullet$. Now take $C \in \mathcal{C} \setminus \mathcal{C}^\bullet$, and let $Y = \langle C \rangle$ be the subspace spanned by C. In Y, the convex set C has non-empty interior (cf. 11.2.7), so it can be approximated by $P \in \mathcal{P}(Y)$. Then $P \times I_\epsilon$, where I_ϵ is a ball of radius ϵ in the orthogonal $Y^\perp$ to Y, will approximate C in $\mathcal{C}$. □

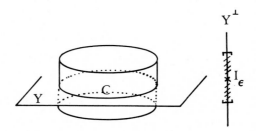

Figure 12.9.2.2

In what follows it will be useful to frame C both inside and outside by polytopes, not just inside as in the previous lemma, where $B(P, \epsilon)$ is not a polytope.

12.9.2.3. Lemma. *For every $C \in \mathcal{C}^{\bullet}$, $a \in \overset{\circ}{C}$ and $\eta > 1$, there exists $P \in \mathcal{P}^{\bullet}$ such that $P \subset C \subset H_{a.\eta}(P)$, $\mathrm{Fr}(C) \cap P = \emptyset$ and $C \cap \mathrm{Fr}(H_{a.\eta}P) = \emptyset$.*

Proof. Choose $r > 0$ such that $B(a, r) \subset \overset{\circ}{C}$, and ϵ such that $0 < \epsilon < r(\eta - 1)$; by 12.9.2.1, we can also choose $P \in \mathcal{P}^{\bullet}$ such that $P \subset C \subset B(P, \epsilon)$. By 12.9.1.3, P contains $B(a, r)$ if ϵ is small enough. By construction, the distance between a face F of P and $H_{a.\eta}(F)$ is greater than or equal to $(\eta - 1)r > \epsilon$; it follows that

$$H_{a.\eta}(P) \supset B(P, \epsilon) \supset C.$$

As to the conditions on the frontiers, they can be achieved by applying to P a homothety of ratio smaller than but sufficiently close to 1. □

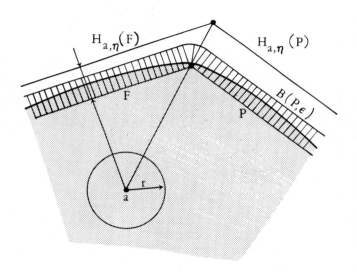

Figure 12.9.2.3

12.9.2.4. Corollary. *The frontier of any convex set C has measure zero* (cf. 9.12.5).

Proof. If $\overset{\circ}{C}$ is empty, C lies in a proper subspace and has measure zero itself. If $\overset{\circ}{C} \neq \emptyset$, apply 12.9.2.3; the boundary of C is contained in $P' \setminus P$, where $P' = H_{a.\eta}P$. Since, by 9.12.3,

$$\mathcal{L}(P' \setminus P) = \mathcal{L}(P)(\eta^d - 1) \leq \mathcal{L}(C)(\eta^d - 1),$$

the volume between the two polyhedra tends to zero as η tends to 1.

12.9.3. VOLUME OF COMPACT CONVEX SETS. Same notation as in 9.12.

12.9.3.1. Proposition. *For $C \in \mathcal{C}$ we have*

$$\mathcal{L}(C) = \sup \big\{ \mathcal{L}(P) \mid P \in \mathcal{P}, \; P \subset C \big\} = \inf \big\{ \mathcal{L}(P) \mid P \in \mathcal{P}, \; P \supset C \big\}.$$

Proof. If $\mathring{C} = \emptyset$, we pass to the case $\mathring{C} \neq \emptyset$ as in 12.9.2.2. If $\mathring{C} \neq \emptyset$ we apply 12.9.2.3. □

12.9.3.2. By lemma 12.9.2.3 we can use 12.9.3.1 as the definition of $\mathcal{L}(\cdot)$ on convex sets, starting with the elementary definition of volume given in 12.2.5. Thus we can avoid integration theory altogether, if we so desire.

12.9.3.3. Proposition. *The volume function $\mathcal{L} : \mathcal{C}^{\bullet} \to \mathbf{R}$ is strictly increasing, that is, if D and C are distinct elements of $\mathcal{C}$, with $D \subset C$, then $\mathcal{L}(D) < \mathcal{L}(C)$.*

This result is obviously false in $\mathcal{K}$.

Proof. In fact, if $C \supset D$, $C \neq D$, lemma 11.2.4 shows that $\mathring{C} \setminus D \neq \emptyset$. Let $\epsilon > 0$ and $x \in \mathring{C} \setminus D$ be such that $B(x, \epsilon) \subset C \setminus D$; we have

$$\mathcal{L}(C) \geq \mathcal{L}(D) + \mathcal{L}\big(B(x, \epsilon)\big) > \mathcal{L}(D).$$

□

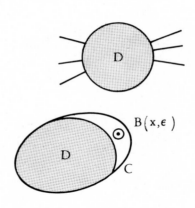

Figure 12.9.3

Recall (cf. 9.12.5) that the volume function is not continuous on $\mathcal{K}$. On $\mathcal{C}$ we have the following

12.9.3.4. Proposition. *The volume function $\mathcal{L} : \mathcal{C} \to \mathbf{R}$ is continuous.*

Proof. We work as in 9.12.6. On $X \setminus \mathrm{Fr}\, C$, we have $\lim_{n \to \infty} \chi_{C_n} = \chi_C$ if $\lim_{n \to \infty} C_n = C$. But $\mathrm{Fr}\, C$ has measure zero, so $\lim_{n \to \infty} \int_X \chi_{C_n} \mu = \int_X \chi_C \mu$ on the whole of X.

An elementary proof can also be given: if $C = \lim_{n \to \infty} C_n$ has empty interior, it can be boxed in as in the proof of 12.9.2.2. Now assume that $\mathring{C} \neq \emptyset$. By 12.9.2.3, we can find polytopes P, P' such that $P \subset C \subset P'$ with $\mathrm{Fr}\, C \cap P = \emptyset$ and $\mathrm{Fr}\, P' \cap C = \emptyset$, and that and $\mathcal{L}(P') - \mathcal{L}(P) \leq \epsilon$. Applying 12.9.1.3, we see that for small enough η the inequality $\delta(C, D) \leq \eta$ implies $P \subset D \subset D'$, whence $\big| \mathcal{L}(D) - \mathcal{L}(C) \big| \leq \mathcal{L}(P') - \mathcal{L}(P) \leq \epsilon$.

12.10. Area of compact convex sets

We have seen in 9.12.7 that defining k-dimensional volume, for $k < d$, is tricky. In the case of convex sets, it is natural to expect to find the $(d-1)$-dimensional volume (here called area) of their boundary by approximating them by polytopes and taking the limit of the area of the approximating sets, which has been defined in 12.3 and has reasonable properties. This approach is possible, but we shall instead take Cauchy's formula (12.3.3) as the definition of area. We must first verify a lemma.

12.10.1. LEMMA. *With the same notation as in 12.3, the function*

$$\mathcal{L}(p(\cdot)) : S \times \mathcal{C} \ni (\xi, C) \mapsto \mathcal{L}(p_\xi(C)) \in \mathbf{R}$$

is continuous.

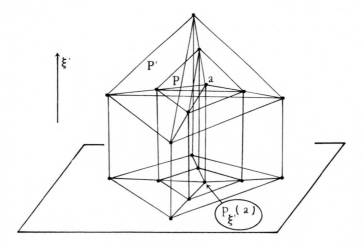

Figure 12.10.1

Proof. Fix $\xi \in S$, $C \in \mathcal{C}^\bullet$, $a \in \overset{\circ}{C}$ and $\epsilon > 0$. For every $\eta > 1$ there exists (cf. 12.9.2.3) some $P \in \mathcal{P}^\bullet$ satisfying $P \subset C \subset P' = H_{a.\eta}P$, $\mathcal{L}(P') - \mathcal{L}(P) \leq \epsilon/3$, Fr $C \cap P = \emptyset$ and Fr $P' \cap C = \emptyset$. Since $H_{p\xi'(a).\eta} \circ p_{\xi'} = p_{\xi'} \circ H_{a.\eta}$ for every $\xi' \in S$, we have

$$\mathcal{L}(p_{\xi'}(P')) = \eta^{d-1}\mathcal{L}(p_{\xi'}(P));$$

since there exists r with $B(a, r) \supset C$, we have (cf. 9.12.4.5):

$$\left|\mathcal{L}(p_{\xi'}(P')) - \mathcal{L}(p_{\xi'}(P))\right| \leq (\eta^{d-1} - 1)r^{d-1}\beta(d-1)$$

for every $\xi' \in S$. We can thus fix η, hence also P, such that

$$\left|\mathcal{L}(p_{\xi'}(P')) - \mathcal{L}(p_{\xi'}(P))\right| \leq \epsilon/3$$

for any $\xi' \in S$. But the formula in 12.3.3.2 shows that the map $S \ni \xi' \mapsto \mathcal{L}(p_{\xi'}(P)) \in \mathbf{R}$ is continuous in ξ', so that there exists ς such that $\|\xi' - \xi\| \le \varsigma$ implies

$$\left|\mathcal{L}(p_{\xi'}(P)) - \mathcal{L}(p_\xi(P))\right| \le \epsilon/3.$$

By 12.9.1.3 we can choose θ small enough that $\delta(D, C) \le 0$ implies $P \subset D \subset P'$; putting everything together we get

$$\left|\mathcal{L}(p_{\xi'}(D)) - \mathcal{L}(p_\xi(C))\right| \le \left|\mathcal{L}(p_{\xi'}(D)) - \mathcal{L}(p_{\xi'}(P))\right|$$
$$+ \left|\mathcal{L}(p_{\xi'}(P)) - \mathcal{L}(p_\xi(P))\right| + \left|\mathcal{L}(p_\xi(P)) - \mathcal{L}(p_\xi(C))\right|$$
$$\le \left|\mathcal{L}(p_{\xi'}(P')) - \mathcal{L}(p_{\xi'}(P))\right| + \epsilon/3 + \left|\mathcal{L}(p_\xi(P')) - \mathcal{L}(p_\xi(P))\right| \le \epsilon.$$

12.10.2. DEFINITION AND PROPOSITION. *For every $C \in \mathcal{C}^\bullet$ the integral*

$$\mathcal{A}(C) = \left(\beta(d-1)\right)^{-1} \int_{\xi \in S} \mathcal{L}(p_\xi(C)) \sigma$$

exists, and is called the area *of C (or* length *if $d = 2$). This definition of area is compatible with the one given for polytopes. The map $\mathcal{A} : \mathcal{C}^\bullet \to \mathbf{R}$ is continuous, strictly increasing, and invariant under isometries.*

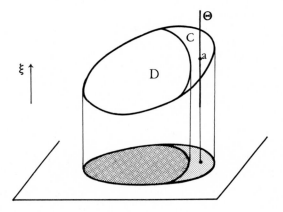

Figure 12.10.2

Proof. Existence and continuity derive from 12.10.1 and integration theory. That this area is compatible with that of polytopes follows form 12.3.3. Finally, take $D \subset C$, $D \ne C$, and choose a ball $B(a, \epsilon) \subset C \setminus D$, with $\epsilon > 0$, as in 12.9.3.3. By 11.4.1 there exists a line Θ containing a and not intersecting D; take $\xi \in \vec{\Theta} \cap S$. By continuity, any line through a whose direction η is close enough to ξ still doesn't intersect D. From 12.9.3.3 we get $\mathcal{L}(p_\eta(D)) < \mathcal{L}(p_\eta(C))$ for any such η, and finally $\mathcal{A}(D) < \mathcal{A}(C)$. $\square$

12.10.3. COROLLARY. *For every $C \in \mathcal{C}^\bullet$ we have*

$$\mathcal{A}(C) = \sup\left\{\, \mathcal{A}(P) \mid P \in \mathcal{P}^\bullet, P \subset C \,\right\} = \inf\left\{\, \mathcal{A}(P) \mid P \in \mathcal{P}^\bullet, P \supset C \,\right\}.$$

$\square$

12.10.4. NOTES.

12.10.4.1. The area of the sphere $C = S(a, r)$ is $\mathcal{A}(C) = r^{d-1}\alpha(d)$. In fact, $p_\xi(S(a, r))$ is a ball of radius r for every ξ, so that

$$\mathcal{L}\big(p_\xi(S(a, r))\big) = r^{d-1}\beta(d-1),$$

whence

$$\mathcal{A}\big(S(a, r)\big) = r^{d-1}\int_S \sigma = r^{d-1}\alpha(d).$$

12.10.4.2. It is natural to try to equate this length of a convex set C in the plane with the length of the curve $\mathrm{Fr}\, C$. Rigorously speaking, $\mathrm{Fr}\, C$ is not yet a curve, but 11.3.4 allows us to parametrize it continuously. Then the perimeter of a polygon inscribed in $\mathrm{Fr}\, C$ is precisely a sum as in definition 9.9.1, which proves the equality of the two notions (cf. 12.11.6).

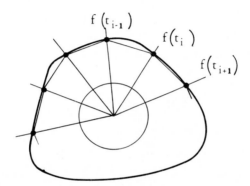

Figure 12.10.4.2

12.10.5. APPLICATION: CURVES OF CONSTANT WIDTH. A curve (or convex set) of *constant width* is a plane convex set C such that the segment $p_\xi(C)$ has the same length for any $\xi \in S$, or, equivalently, such that the two supporting lines of C in the direction of ξ (cf. 11.5.6) are the same distance apart for any ξ. Discs are not the only sets with this property; other examples are the Reuleaux triangle and, more generally, the evolute of any appropriately chosen curve with an odd number of cusps (of which there are many). But one thing we deduce from 12.10.2 and 12.10.4 is that the length of a curve of constant width is π times its width.

Solids of constant width have been the object of much study, but there subsist a number of open problems on the subject. For example, a result of Blaschke–Lebesgue says that the Reuleaux triangle is the unique plane convex set of constant width that has minimal area, but no analogue is known so far in higher dimension (cf. [SL1], [SL2]). References for solids of constant are: [EN, chapter 7], [VE, 156], [BLA1, 150], [FA], [FI], [B–F, paragraph 15] and the delightful book [I–B, chapter 7]. A recent and extensive report can be found in [CN–GR].

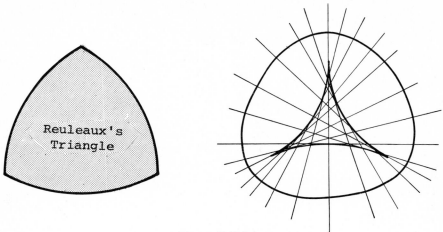

Figure 12.10.5.1

The Reuleaux triangle is an element in many mechanical devices, for example, movie projectors (see [I–B, 72]) and the Wankel engine. See 9.14.34.6 and [LF–AR, 433–435].

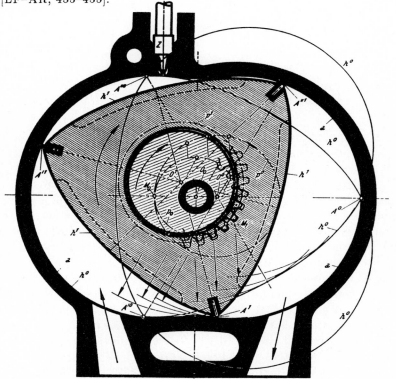

Figure 12.10.5.2 (Source: [GK])

We shall now extend formula 12.3.6 to arbitrary convex sets, by approximating them with polytopes.

12.10.6. THEOREM (STEINER–MINKOWSKI). *To every d-dimensional convex set $C \in \mathcal{C}^\bullet$ we can associate scalars $\mathcal{L}_i(C)$ $(i = 0, 1, \ldots, d)$ such that, for any $\lambda \in \mathbf{R}_+^*$ we have*

$$\mathcal{L}(B(C, \lambda)) = \sum_{i=0}^{d} \mathcal{L}_i(C)\lambda^i.$$

The functions $\mathcal{L}_i : \mathcal{C}^\bullet \to \mathbf{R}$ are continuous; in addition, we have $\mathcal{L}_0(C) = \mathcal{L}(C)$, $\mathcal{L}_1(C) = \mathcal{A}(C)$ and $\mathcal{L}_d(C) = \beta(d)$ for every C.

Proof. Write C as $\lim_{n \to \infty} P_n$, where $P_n \in \mathcal{P}^\bullet$ for all n. Since each P_n is bounded, 12.3.7 says that $\mathcal{L}_i(P_n)$ is bounded for all i; we can thus suppose, after passing to a subsequence if necessary, that there exist k_i $(i = 0, 1, \ldots, d)$ such that $k_i = \lim_{n \to \infty} \mathcal{L}_i(P_n)$. Then

$$\lim_{n \to \infty} \mathcal{L}(B(P_n, \lambda)) = \sum_{i=0}^{d} k_i \lambda^i.$$

But

$$\lim_{n \to \infty} \mathcal{L}(B(P_n, \lambda)) = \mathcal{L}(\lim_{n \to \infty} B(P_n, \lambda))$$

by 12.9.3.4, and, by 9.11.7,

$$\lim_{n \to \infty} B(P_n, \lambda) = B(\lim_{n \to \infty} P_n, \lambda) = B(C, \lambda),$$

whence $\mathcal{L}(B(C, \lambda)) = \sum_{i=0}^{d} k_i \lambda^i$.

Thus $\mathcal{L}(B(C, \lambda))$ is a polynomial in λ, and, since its values depend only on C and not on the approximating sequence, the same is true of the $k_i = \mathcal{L}_i(C)$. Further, the map $\mathcal{C}^\bullet \times \mathbf{R}_+ \ni (C, \lambda) \mapsto \mathcal{L}(B(C, \lambda)) \in \mathbf{R}$ is continuous, so the coefficients of the polynomial giving $\mathcal{L}(B(C, \lambda))$ must also be continuous (see, for example, the formula of the "successive differences" in [CH1, 6.3.6]). Cf. 11.1.3.3 and 12.9.3.4.

Finally, the values of $\mathcal{L}_0$, $\mathcal{L}_1$ and $\mathcal{L}_d$ follow from 12.3.6 and from continuity. $\qquad\square$

We now verify (cf. 12.3.8) that painting a convex set is a way to measure its area:

12.10.7. COROLLARY. *For every convex set $C \in \mathcal{C}^\bullet$,*

$$\mathcal{A}(C) = \lim_{\lambda \to 0} \frac{\mathcal{L}(B(C, \lambda)) - \mathcal{L}(C)}{\lambda}.$$

12.10.8. EXAMPLE. Take $C = B(0, 1) \subset \mathbf{R}^d$. Then

$$B(C, \lambda) = B(0, \lambda + 1), \qquad \lambda(B(C, \lambda)) = (\lambda + 1)^d \beta(d),$$

$$\mathcal{A}(C) = \lim_{\lambda \to 0} \frac{\beta(d)(\lambda + 1)^d - \beta(d)}{\lambda} = d\beta(d).$$

Since $\mathcal{A}(C) = \alpha(d)$ (cf. 12.10.4), we obtain the relation $\alpha(d) = d\beta(d)$.

12.10.9. NOTES.

12.10.9.1. The numbers $\mathcal{L}_i(\cdot)$ are interesting isometry invariants of convex compact sets. In dimension three, we already know what $\mathcal{L}_0(C)$, $\mathcal{L}_1(C)$ and $\mathcal{L}_3(C)$ are. The invariant $\mathcal{L}_2(C)$ can be called the *total mean curvature of C* because one can prove that when $\mathrm{Fr}\, C$ is a C^2 submanifold of X this number is equal to $\int_{\mathrm{Fr}\, C} \mu\sigma$, where σ is the canonical measure on $\mathrm{Fr}\, C$ and $\mu : \mathrm{Fr}\, C \to \mathbf{R}$ is its mean curvature. See, for example, [B–F, 63, formula (12)].

12.10.9.2. Another well-behaved class of sets is that of C^2 submanifolds of $V \subset X$; it can be shown (see [B–G, 6.9]) that for such sets $\mathcal{L}\big(B(V, \lambda)\big)$ is still a polynomial, but only for small enough values of λ. For example, for the circle $S = S(0, 1) \subset \mathbf{R}^2$, we have

$$\mathcal{L}\big(B(S, \lambda)\big) = \begin{cases} 4\pi\lambda & \text{for } \mathcal{L} \in [0, 1] \\ \mathcal{L}\big(B(S, \lambda)\big) = \pi(\lambda + 1)^2 & \text{for } \lambda \geq 1 \end{cases}.$$

Also, for λ small enough, only every other power of λ is represented because the other terms cancel out. Another significant difference is that here the term in λ^d (again for λ small enough) is $\chi(V)\beta(d)\lambda^d$, where $\chi(V)$ is the Euler characteristic of the manifold V, whereas for convex sets it is just $\beta(d)\lambda^d$. For example, for a torus with γ holes (figure 12.10.9.2.2) we have

$$\mathcal{L}\big(B(V, \lambda)\big) = 2\mathrm{area}(V)\lambda + \frac{8\pi(1 - \gamma)}{3}\lambda^3,$$

for small enough values of λ (see [B–G, 7.5.7] and compare with 12.7.5.4).

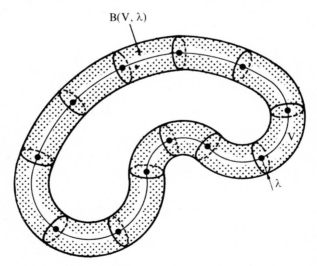

Figure 12.10.9.2.1 (Source: [B–G])

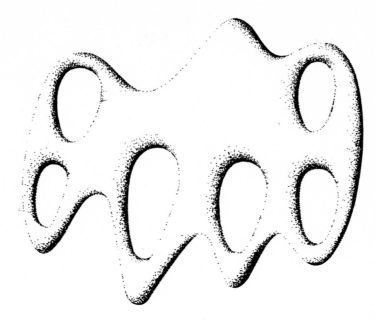

Figure 12.10.9.2.2

12.10.9.3. On the other hand, if K is just a compact set and no other restrictions are imposed, $\mathcal{L}\big(B(K, \lambda)\big)$ will be a wild function, even for small values of λ. Even the limit $\displaystyle \lim_{\lambda \to 0} \frac{\mathcal{L}\big(B(K, \lambda)\big) - \mathcal{L}(\lambda)}{\lambda}$ does not exist in general (cf. 12.12.9); the best we can do is define the *upper* and *lower Minkowski area* of K as

$$\mathcal{M}^+(K) = \limsup_{\lambda \to 0} \frac{\mathcal{L}\big(B(K, \lambda)\big) - \mathcal{L}(K)}{\lambda},$$

$$\mathcal{M}^-(K) = \liminf_{\lambda \to 0} \frac{\mathcal{L}\big(B(K, \lambda)\big) - \mathcal{L}(K)}{\lambda}.$$

$\mathcal{M}^+(K)$ $\mathcal{M}^-(K)$ See 12.11.5.2, [FR, 273] and [HR, 185] for properties of these areas.

12.10.10. AREAS AND THE STEINER SYMMETRIZATION. Same notation as in 9.13 and 11.1.3. Our objective is to show that Steiner symmetrization does not increase the area of convex sets.

12.10.10.1. Lemma. *Let A, $B \in \mathcal{C}$ be convex sets, and H a hyperplane. For every $\lambda \in [0, 1]$ we have*

$$\mathrm{st}_H\big(\lambda A + (1 - \lambda)B\big) \supset \lambda\, \mathrm{st}_H(A) + (1 - \lambda)\, \mathrm{st}_H(B).$$

If X is a vector space, we also have, for λ, $\mu \geq 0$:

$$\mathrm{st}_H\big(\lambda A + \mu B\big) \supset \lambda\, \mathrm{st}_H(A) + \mu\, \mathrm{st}_H(B).$$

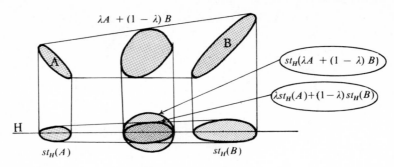

Figure 12.10.10.1.1

Figure 12.10.10.1.1 shows that equality does not hold in general.

Proof. Let $x = \lambda y + (1 - \lambda)z$, where $y \in \mathrm{st}_H(A)$ and $z \in \mathrm{st}_H(B)$. Call D_x, D_y and D_z the lines orthogonal to H and passing through x, y and z, set $T_y = \mathrm{st}_H(A) \cap D_y$ and $T_z = \mathrm{st}_H(B) \cap D_z$, and let T'_y (resp. T'_z) be the segment of $D_y \cap A$, (resp. $D_z \cap B$) which gives rise to T_y (resp. T_z). We have

$$\lambda A + (1 - \lambda) \supset \lambda T'_y + (1 - \lambda)T'_z,$$

and this, by 9.13.1, shows that

$$\mathrm{st}_H\left(\lambda A + (1 - \lambda)B\right) \supset \lambda T_y + (1 - \lambda)T_z \ni x.$$

The vector case can be reduced to the affine case by a homothety of ratio $(\lambda + \mu)^{-1}$.

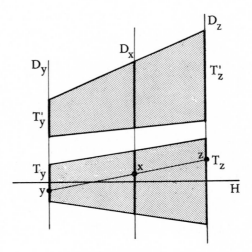

Figure 12.10.10.1.2

12.10.10.2. Proposition. *For every $C \in \mathcal{C}^\bullet$ and every hyperplane H we have*

$$\mathcal{A}\left(\mathrm{st}_H(C)\right) \le \mathcal{A}(C).$$

Proof. Vectorialize X at $0 \in H$ and let $S \in B(0,1)$ be the unit ball of X. From 11.1.3.2, 12.10.7, 12.10.10.1 and 9.13.4 we get

$$\mathcal{L}\big(B(C,\lambda)\big) = \mathcal{L}(C) + \mathcal{A}(C)\lambda + o(\lambda)$$
$$\mathcal{L}\big(\mathrm{st}_H\big(B(C,\lambda)\big)\big) = \mathcal{L}\big(\mathrm{st}_H\big(C + \lambda S\big)\big)$$
$$\geq \mathcal{L}\big(\mathrm{st}_H(C) + \lambda\,\mathrm{st}_H(S)\big)$$
$$= \mathcal{L}\big(\mathrm{st}_H(C) + \lambda S\big) = \mathcal{L}\big(B(\mathrm{st}_H(C)),\lambda\big)$$
$$= \mathcal{L}\big(\mathrm{st}_H(C)\big) + \mathcal{A}\big(\mathrm{st}_H(C)\big)\lambda + o(\lambda)$$
$$= \mathcal{L}(C) + \mathcal{A}\big(\mathrm{st}_H(C)\big)\lambda + o(\lambda).$$

Now, since $\lim\limits_{\lambda \to 0} \dfrac{o(\lambda)}{\lambda} = 0$, we conclude that $\mathcal{A}(C) \geq \mathcal{A}\big(\mathrm{st}_H(C)\big)$. □

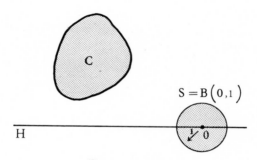

Figure 12.10.10.2

It is reasonable to ask when equality holds in 12.10.10.2. It certainly does when $\mathrm{st}_H(C)$ is equal to C or obtained from it by a translation. The converse is true:

12.10.10.3. Theorem. *If the convex set $C \in \mathcal{C}^\bullet$ and the hyperplane H satisfy*

$$\mathcal{A}\big(\mathrm{st}_H(C)\big) = \mathcal{A}(C),$$

there exists a hyperplane H' parallel to H such that $\mathrm{st}_{H'}(C) = C$.

Proof. The proof of 12.10.10.2 cannot be used here, since it includes a passage to the limit. We use instead the explicit expression for $\mathcal{A}(C)$ given in 12.10.11.1, the notation being the same. By definition, the symmetrization $\mathrm{st}_H(C)$ of a set C defined by the two functions f, g is defined by the new functions $(f-g)/2, (g-f)/2$. We then have

$$\mathcal{A}(C) = \int_{\mathrm{Fr}\,D} (f-g)\sigma + \int_D \big(\sqrt{1 + \|f'\|^2} + \sqrt{1 + \|g'\|^2}\big)\mu,$$

$$\mathcal{A}\big(\mathrm{st}_H(C)\big) = \int_{\mathrm{Fr}\,D} (f-g)\sigma + \int_D \sqrt{4 + \|f' - g'\|^2}\,\mu.$$

for 8.1.2.4 we get

$$\sqrt{1 + \|\xi\|^2} + \sqrt{1 + \|\eta\|^2} \geq \sqrt{4 + \|\xi - \eta\|^2}$$

for any two vectors ξ, η in a Euclidean space, and equality only holds if $\eta = -\xi$. This shows that $f' = -g'$ almost everywhere, so $f + g$ is constant a.e., hence constant (since it is continuous). This is the desired conclusion.

$\square$

12.10.11. EXPLICIT CALCULATION OF THE AREA. Consider $C \in \mathcal{C}^\bullet$ and a hyperplane H of X, and write X as a product $X = H \times \mathbf{R}$. Call $p : X \to H$ the projection on H, let μ be the Lebesgue measure on H, and set $D = p(C)$. We have the following result:

12.10.11.1. Theorem.   *There exist a measure σ on $\mathrm{Fr}\, D$ and functions $f, g : D \to \mathbf{R}$ such that:*

   i) *f and g are almost everywhere differentiable on D, and the functions $\sqrt{1 + \|f'\|^2}$ and $\sqrt{1 + \|g'\|^2}$ are μ-integrable (the norm being the one defined on $(\vec{H})^*$ in 8.1.8.2);*

  ii) *we have $C = \{ (x, t) \in H \times \mathbf{R} \mid x \in D \text{ and } f(x) \geq t \geq g(x) \}$;*

 iii) *we have $\mathcal{A}(C) = \int_{\mathrm{Fr}\, D} (f - g)\sigma + \int_D (\sqrt{1 + \|f'\|^2} + \sqrt{1 + \|g'\|^2})\mu.$*

Proof. The proof is very technical and we only include its highlights. The functions f and g are defined on $\overset{\circ}{D}$ in the way shown in figure 12.3.3 or 12.10.11.1 and proof 12.3.3.1, and their extension to D is well-defined thanks to compactness and convexity. Differentiability almost everywhere derives from 11.8.10.5 and 11.8.3.

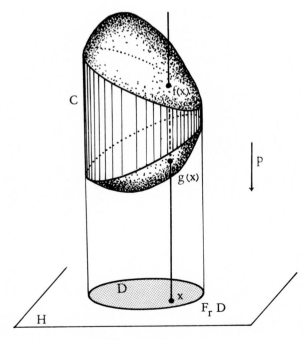

Figure 12.10.11.1

The theorem is first proved for polytopes, by using 9.12.4.9, where the measure σ on $\mathrm{Fr}\, D$ is the union of the Lebesgue measures induced on the faces of D. We conclude by applying 12.9.2.3 and convergence follows from integration theory.   □

12.10.11.2. Corollary. *Assume that $C \in \mathcal{C}^\bullet(X)$ is such that $\mathrm{Fr}\, C$ can be partitioned into finitely many differentiable submanifolds of X of dimension $0, 1, \ldots, d-1$, where $d = \dim X$. Then each $(d-1)$-dimensional submanifold in that partition has finite volume (in the sense of [B–G, 6.5]) and $\mathcal{A}(C)$ is the sum of volumes of those $(d-1)$-dimensional submanifolds.*

Proof. This is a direct consequence of 12.10.1.1 and the generalization of formula 6.4.2.3 of [B–G] to arbitrary dimension. If fact, for the graph

$$g : (x_1, \ldots, x_n) \mapsto \big(x_1, \ldots, x_n, f(x_1, \ldots, x_n)\big)$$

we have

$$|g^*\omega| = \left\| \frac{\partial g}{\partial x_1} \wedge \cdots \wedge \frac{\partial g}{\partial x_n} \right\| |dx_1 \wedge \cdots \wedge dx_n|.$$

But 8.11.11 shows that the coordinates of $\dfrac{\partial g}{\partial x_1} \wedge \cdots \wedge \dfrac{\partial g}{\partial x_n}$ are

$$\left(-\frac{\partial f}{\partial x_1}, \ldots, -\frac{\partial f}{\partial x_n}, 1 \right).$$ □

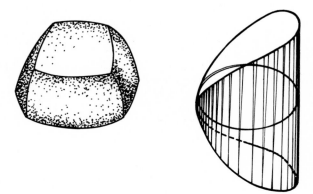

Figure 12.10.11.2

This corollary allows the calculation of the area of convex compact sets found in the physical world by using, for example, [B–G, 6.6.2]. See the formulary in 12.12.20.

12.11. The isoperimetric inequality

This is a simply stated and fundamental result, whose long and interesting history can be looked up in [PR]. Other references are [BLA1, 79–82],
[OS1], [OS2] and [BU–ZA]. Historical interest lies in that, starting with of
Steiner's original idea, it took many attempts for a completely rigorous proof
to evolve.

12.11.1. THEOREM. *For every convex subset C of a d-dimensional affine
space X such that $\overset{\circ}{C} \neq \emptyset$ we have*

$$\mathcal{A}(C) \geq d\big(\beta(d)\big)^{1/d}\mathcal{L}(C)^{(d-1)/d}, \qquad or \qquad \frac{\mathcal{A}(C)}{\alpha(d)} \geq \left(\frac{\mathcal{L}(C)}{\beta(d)}\right)^{(d-1)/d}.$$

Equality holds if and only if C is a ball.

In other words, among all convex sets (with non-empty interior) having a
fixed volume, balls are the ones with the smallest area; or, conversely, among
convex sets having a fixed area, balls are the ones with the largest volume.
And they are the only ones.

The equivalence between the two inequalities of the statement follows
from 9.12.4.8 or 12.10.8. We shall give three demonstrations of the isoperimetric inequality, the first and third of which show that equality only holds
for balls. The third proof applies to any domain, not only convex ones.

12.11.2. FIRST PROOF. The first proof is based on Steiner symmetrization
and the Kugelungsatz. Given a convex compact set $C \in \mathcal{C}^\bullet$, set

$$\mathcal{F} = \big\{\, F \in \mathcal{C}^\bullet \mid \mathcal{L}(F) = \mathcal{L}(C) \text{ and } \mathcal{A}(F) \leq \mathcal{A}(C) \,\big\}.$$

This set is closed in $\mathcal{C}$. In fact, if $A = \lim_{n \to \infty} A_n$, with $A \in \mathcal{C}$ and $A_n \in \mathcal{F}$
for all n, we have $\mathcal{L}(A_n) = \mathcal{L}(C) \neq 0$, so $A_n \in \mathcal{C}^\bullet$. By 12.9.3.4, this gives
$\mathcal{L}(A) \neq 0$, so $A \in \mathcal{C}^\bullet$. Then, by 12.10.2, we have $\mathcal{A}(A) = \lim_{n \to \infty} \mathcal{A}(A_n) \leq
\mathcal{A}(C)$, so $A \in \mathcal{F}$.

Furthermore, $\mathcal{F}$ is invariant under Steiner symmetrization, by 9.13.4 and
12.10.10.2. We now apply 9.13.6 to find a ball $B(a, r) \in \mathcal{F}$, where $a \in X$ and
$r > 0$. Thus

$$\mathcal{A}(C) \geq \mathcal{A}\big(B(a, r)\big) = \alpha(d)r^{d-1}, \qquad \mathcal{L}(C) = \mathcal{L}\big(B(a, r)\big) = \beta(d)r^d.$$

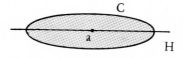

Figure 12.11.2

If equality is achieved, 12.10.10.3 implies that, for any hyperplane direction ξ hyperplanes there exists H such that $\mathrm{st}_H(C) = C$. This implies $\sigma_H(C) = C$, by the definition of st_H; but then all these hyperplanes H must go through a fixed point a of X (cf. 9.8.6). Since C is invariant under the orthogonal group $O(X_a)$, it must be a ball. $\square$

12.11.3. SECOND PROOF. This one is based on the theorem of Brunn–Minkowski and the fact that the area is a derivative (Steiner–Minkowski). Take $C \in \mathcal{C}^\bullet$; vectorialize X at an arbitrary point 0, and consider the ball $S = B(0,1)$. We have $B(C,\lambda) = C + \lambda S$ (cf. 11.1.3.2). From 11.8.8.5 and the binomial formula we have

$$\mathcal{L}\big(B(C,\lambda)\big) = \mathcal{L}(C + \lambda S) \geq \big((\mathcal{L}(C))^{1/d} + \lambda(\beta(d))^{1/d}\big)^d$$
$$\geq \mathcal{L}(C) + d\mathcal{L}(C)^{(d-1)/d}\big(\beta(d)\big)^{1/d}\lambda,$$

which implies 12.11.1 by 12.10.7.

12.11.4. THIRD PROOF. This proof, due to Gromov and based on [KT], is very recent. It works for domains C of X whose frontier is a differentiable hypersurface (or whenever we can approximate the frontier by such a hypersurface). In particular, C does not have to be convex. This generality is achieved at some cost; the proof requires differential calculus, involving as it does an inspired application of Stokes' formula. We denote by $\mathcal{L}(C)$ the Lebesgue measure of C and by $\mathcal{A}(C)$ the canonical measure of the hypersurface $H = \partial C$.

Vectorialize X at a point O and consider an orthonormal coordinate system $\{x_i\}_{i=1,\dots,d}$ for X. Assume, without loss of generality, that $\mathcal{L}(C) = \beta(d)$ is the volume of the unit ball B of X. The fundamental idea is to define a map $f : C \to B$ as follows:

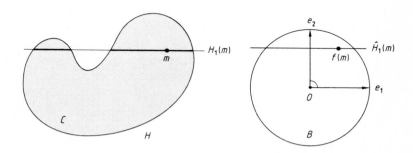

Figure 12.11.4.1

For $m \in C$, we let $H_1(m) = x_1^{-1}(x_1(m))$ be the hyperplane in the direction $x_1 = 0$ containing m, and let $\hat{H}_1(m)$ be the hyperplane in the same direction that partitions B into two subsets with the same volumes as the subsets of C on each side of $H_1(m)$. In other words, $\hat{H}_1(m) = x_1^{-1}(\alpha_1)$, where α_1 is defined by the condition

$$\mathcal{L}\left(C \cap x_1^{-1}([x_1(m), \infty[)\right) = \mathcal{L}\left(B \cap x_1^{-1}([\alpha_1, \infty[)\right).$$

We define two $(d-2)$-dimensional affine subspaces $H_2(m)$ and $\hat{H}_2(m)$ in the direction $x_1 = x_2 = 0$ by an analogous construction, with $C \cap H_1(m)$ instead of C and $B \cap \hat{H}_1(m)$ instead of B. We continue in this fashion until we obtain lines $H_{d-1}(m)$, $\hat{H}_{d-1}(m)$ and finally points $H_d(m) = \{m\}$ and $\hat{H}_d(m) = \{f(m)\}$; this is the definition of $f(m)$. By construction, the Jacobian $Jf(m) = \left(\dfrac{\partial f_i}{\partial x_j}(m)\right)$ is of the form

$$\begin{pmatrix} \lambda_1(m) & ? & \cdots & ? \\ 0 & \lambda_2(m) & \cdots & ? \\ \vdots & \vdots & \ddots & \vdots \\ 0 & 0 & \cdots & \lambda_d(m) \end{pmatrix},$$

where entries below the diagonal are zero and entries above the diagonal don't matter to us. (The differentiability of f is problematic on the "inner folds" of the boundary, cf. proof of unicity below, but we shall ignore these problems.)

By construction and Fubini's theorem we easily verify that f preserves volume, that is,

$$\prod_{i=1}^{d} \lambda_i(m) = 1.$$

Now consider f as a vector field on C; its norm satisfies $\|f(m)\| \leq 1$ since $f(m)$ belongs to the unit ball B. Apply Stokes' theorem to f, C and $H = \partial C$:

$$\int_C \operatorname{div} f(m)\, dm = \int_H \left(f(h) \mid \nu(h)\right) dh,$$

where dm is the Lebesgue measure on X (and C), dh is the measure of H as a differentiable hypersurface, and $\nu(m)$ denotes the unit normal vector to H pointing outwards. We have

$$\operatorname{div} f(m) = \sum_{i=1}^{d} \frac{\partial f_i}{\partial x_i}(m) = \sum_{i=1}^{d} \lambda_i(m),$$

$$\sum_{i=1}^{d} \lambda_i(m) \geq d \left(\prod_{i=1}^{d} \lambda_i(m)\right)^{1/d} = d,$$

by 11.8.11.6. Since $\|f\| \leq 1$, we always have $\left|(f \mid \nu)\right| \leq 1$. We finally get

$$\beta(d) = \mathcal{L}(C) = \int_C dm \leq d \int_C \operatorname{div} f\, dm = d \int_H \left(f(h) \mid \nu(h)\right) dh$$

$$\leq d \int_H dh = d\; \mathcal{A}(H).$$

We conclude that $\mathcal{A}(H) \geq d\beta(d) = \alpha(d)$, by 12.10.8. This is what we wished to show.

Suppose from now on that equality holds. This means, first, that for every point $m \in C$ the entries $\lambda_i(m)$ are equal and have the value 1 (again by 11.8.11.6). Next, $(f \mid \nu) = 1$ at every point of the boundary H of C. This precludes figures like 12.11.3.2, since the points of the thickened part of the frontier (the "inner folds") have their image in the interior of B. In particular, all lines $H_{d-1}(m)$ intersect H in only two points.

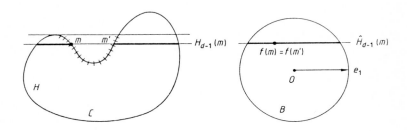

Figure 12.11.4.2

Since $\dfrac{\partial f_i}{\partial x_i} = 1$, the map f, after a translation of C if necessary, takes the form

$$f(x_1, \ldots, x_d) = (x_1, x_2 + a(x_1), x_3 + b(x_1, x_2), \ldots).$$

Using the fact that $(f \mid \nu) = 1$ on H, we get $f = \nu$. Now consider the section $K \cap C$ of C by an affine plane K in the direction $x_3 = x_4 = \cdots = x_d = 0$. The relation above for H and the condition $f = \nu$ show that

$$\frac{2x_1}{x_1} = \frac{2(x_2 + a(x_1))(1 + a'(x_1))}{x_2 + a(x_1)},$$

so that $a'(x_1) = 0$. After translating to eliminate a, we see that the restriction of f to $C \cap K$ is the identity, so that the section $K \cap C$ is identical with the disc $K \cap B$. But the choice of orthonormal coordinates is arbitrary, so every section of C by an affine plane is a disc of radius ≤ 1.

Since f is surjective, there exists at least one such disc D of radius 1. Take two diametrically opposed points m_1 and m_2 on D, and consider an arbitrary affine plane containing m_1 and m_2. Since $P \cap C$ is a disc of radius ≤ 1, it must be a disc of radius 1 on which m_1 and m_2 are diametrically opposed. By varying P we conclude that C must be the ball of radius 1 centered at the midpoint of m_1 and m_2. $\qquad\square$

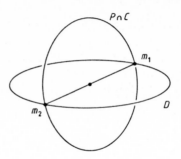

Figure 12.11.4.3

12.11.5. Notes.

12.11.5.1. Even if 11.8.8.6 is known, the technique of the second proof is not sufficient to show that equality only holds for balls, since it includes a passage to the limit.

12.11.5.2. Does the isoperimetric inequality hold if we relax the convexity condition? As we have seen in 12.11.4, the answer is yes for domains whose boundary is a differentiable submanifold of X; also, the second proof (12.11.3) shows that for any compact set the area $\mathcal{M}^-(\cdot)$ of 12.10.9.3 satisfies

$$\mathcal{M}^-(K) \geq d\big(\beta(d)\big)^{1/d}\mathcal{L}(K)^{(d-1)/d}.$$

But without a condition like convexity or regularity of the frontier, equality can be attained for compact sets other than balls: just take a ball in dimension three and add to it lots of hair. It's easy to see that neither the volume nor the area change. In dimension two the situation is different, since hairs have positive length! See 12.11.7.

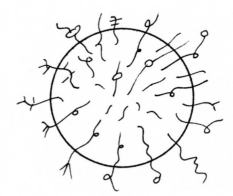

Figure 12.11.5.2

12.11.5.3. There exist inequalities generalizing 12.11.1 to other spaces, like the sphere and hyperbolic space (to be studied in chapters 18 and 19). The point is that these spaces have as many hyperplanes and hyperplane reflections as Euclidean space, so the Kugelungsatz still holds. For proofs of these generalizations, see the references to E. Schmidt and A. D. Alexandrov in [HR, 304] and the nice book [LW].

12.11.5.4. In the case $d = 2$, the proof of 12.11.1 for a convex set whose frontier is a differentiable curve is very elementary, cf. [B–G, 9.3]. For an arbitrary two-dimensional compact set, see below. The situation is more difficult in dimension greater than two for several reasons: first, the notion of area is subtler than that of length (cf. 9.12.7); second, this kind of inequality tends to be easier in the quadratic case, cf. for example the Cauchy–Schwarz and the Hölder inequalities, 11.8.11.9.

12.11.6. THE TWO-DIMENSIONAL CASE. Here convexity is no longer necessary, cf. 12.11.5.2, and we will obtain a more general theorem.

12.11.6.1. Definition. In a topological vector space X, let a *simple closed curve* be any subset $\Gamma \subset X$ such that there exists a map $f : S^1 \rightarrow X$ taking the circle S^1 homeomorphically onto its image $f(S^1) = \Gamma$. If X is a metric space, consider on S^1 points $(t_i)_{i=0,1,\dots,n}$ such that $t_0 = t_n$ and t_i lies between t_{i-1} and t_{i+1} for any i (cf. 9.9.8); we associate to $f : S^1 \rightarrow X$ the sum $\sum_{i=0}^{n-1} d\big(f(t_i), f(t_{i+1})\big)$. The upper bound of such sums does not depend on the simple closed curve Γ, as follows from 9.9.2 once we know that homeomorphisms of the circle preserve "betweenness" (this is because continuous bijections between intervals are strictly monotonic). This upper bound is called the *length* of Γ and *denoted* by $\mathrm{leng}(\Gamma)$, and Γ is called *rectifiable* if its length is finite.

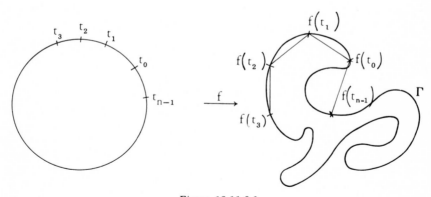

Figure 12.11.6.1

12.11.6.2. Assume that X is a Euclidean plane. If $C \in \mathcal{C}$, its frontier $\Gamma = \mathrm{Fr}\, C$ is a simple closed curve and $\mathrm{leng}(\Gamma) = \mathcal{A}(C)$, cf. 12.10.4.2. If Γ is a simple closed curve, its convex hull $\mathcal{E}(C)$ is in $\mathcal{C}^\bullet$ and figure 12.11.6.1 shows that $\mathrm{leng}(\Gamma) \geq \mathcal{A}\big(\mathcal{E}(\Gamma)\big)$.

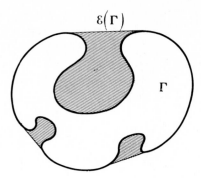

$$\mathcal{E}(\Gamma)$$

$$\Gamma$$

Figure 12.11.6.2

12.11.7. THEOREM. *Every simple closed curve Γ in a Euclidean plane satisfies*

$$\operatorname{leng}(\Gamma) \geq 2\sqrt{\pi \mathcal{L}(\mathcal{E}(\Gamma))},$$

and equality is achieved if and only if Γ is a circle.

Proof. From 12.11.1 and 12.11.6.2 we get

$$\operatorname{leng}(\Gamma) \geq \mathcal{A}(\mathcal{E}(\Gamma)) \geq 2\sqrt{\pi \mathcal{L}(\mathcal{E}(\Gamma))};$$

the only non-trivial part is that Γ is a circle in case of equality. Let this be the case. Then 12.11.6.2 implies that $\Gamma = \operatorname{Fr} C$, where $C = \mathcal{E}(\Gamma)$. For a fixed $x \in \Gamma$, we verify by continuity that there exists $y \in \Gamma$ such that the two curves $\Gamma_1 = \operatorname{Fr} C_1$, $\Gamma_2 = \operatorname{Fr} C_2$ have the same length, where C_1, C_2 are the parts of C lying on each half-plane determined by $\langle x, y \rangle$.

We first show that $\mathcal{L}(C_1) = \mathcal{L}(C_2)$ (figure 12.11.7.1). If $\mathcal{L}(C_1) \leq \mathcal{L}(C_2)$, say, we reflect C_1 through $\langle x, y \rangle$ to obtain an arc C_2' satisfying

$$\mathcal{L}\big(\mathcal{E}(C_1 \cup C_2')\big) \geq \mathcal{L}(C_1 \cup C_2') = \mathcal{L}(C_1) + \mathcal{L}(C_2') > \mathcal{L}(C),$$
$$\mathcal{A}\big(\mathcal{E}(C_1 \cup C_2')\big) \leq \mathcal{A}(C_1) + \mathcal{A}(C_2) = \mathcal{A}(C),$$

contradicting the first part of the theorem.

We finally show that C_1, for example, is a half-circle with diameter $\{x, y\}$, that is, $\langle p, x \rangle \perp \langle p, y \rangle$ for any $p \in \Gamma_1$. For assume there were $p \in \Gamma_1$ with $\langle p, x \rangle$ not perpendicular to $\langle p, y \rangle$; then we can rotate the shaded region in figure 12.11.7.2 around the point p until $\langle p, x_1 \rangle$ becomes orthogonal to $\langle p, y \rangle$. By 10.3.3, the triangle $\{p, x_1, y\}$ has area strictly larger than $\{p, x, y\}$, and we can double the new figure by reflection through $\langle x_1, y \rangle$ to get a compact set $C_2' \cup C_2'$ such that $\mathcal{L}(C_1' \cup C_2') > \mathcal{L}(C)$ and $\operatorname{leng}\big(\operatorname{Fr}(C_1' \cup C_2')\big) \leq \operatorname{leng}\big(\operatorname{Fr}(C)\big)$. This is a contradiction. $\square$

Another consequence of this proof is the result below, due to Dido. Dido was the daughter of the king of Tyre, and the wife of her uncle Acerbas, who was murdered for his riches. Dido fled to Cyprus with the treasure of Acerbas and from there she went on to Africa, landing near Sicily. She told the local

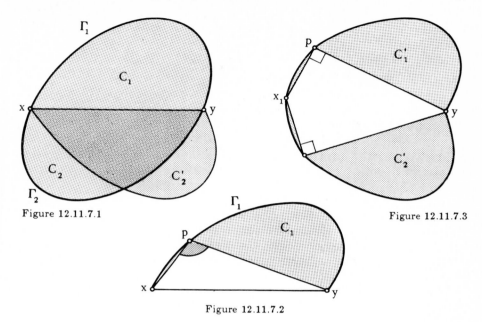

Figure 12.11.7.1

Figure 12.11.7.3

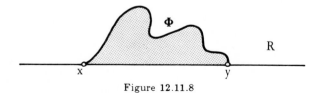

Figure 12.11.7.2

chief that she would like to acquire a tract by the sea, as large as could be contained in an oxhide; whereupon the chief graciously agreed and provided her with a large hide. Dido proceeded to cut the hide into very thin strips, which she tied together to form a long string. She was thus faced with the situation of the following

12.11.8. COROLLARY. *Let R be a Euclidean half-plane and Φ a string of fixed length. If Φ is placed on R so that its enpoints x, y are on $\operatorname{Fr} R$, the area bounded between Φ and $\operatorname{Fr} R$ will be maximal when Φ is a half-circle of diameter $\{x, y\}$.*

Thus was Carthage founded!

Figure 12.11.8

12.11.9. OTHER INEQUALITIES.

12.11.9.1. In 9.13.8 we have already encountered an inequality in the same spirit as 12.11.1. As the reader may suspect, mathematicians have demonstrated a large number of such inequalities, often motivated, incidentally, by conjectures made by physicists. The reference [P–S] is remarkable; others are

[PN], [HR], [BLA1], [B–F], [HS], [GS]. See also [KJ1], [KJ2] and the books
[BU–ZA] and [LW].

Here are some examples:

12.11.9.2. As in 12.10.5, we can start by studying isoperimetric theorems
for a particular kind of convex sets. The first case is that of n-sided polygons,
for n fixed. Among those, the ones with largest area for a fixed perimeter are
the regular ones (and only they). See 12.12.16 or [GR, 191], [FT1, 9].

For polytopes the situation is certainly not as simple, but a theorem due
to Lindelöf asserts that, among three-dimensional polytopes whose volume is
fixed and whose faces have fixed directions the ones of least area are inscribed
in a sphere. See [AW2, 317] or [HR, 276].

12.11.9.3. Another possibility is to study the *deficit*

$$\mathcal{A}(C) - d(\beta(d))^{1/d}\mathcal{L}(C)^{(d-1)/d};$$

when mathematicians cannot prove that two expressions are equal, they settle
for studying their difference. There are few completely general results; one,
the theorem of Dinghas–Hadwiger–Bonnesen, says that

$$\left(\mathcal{A}(C)\right)^{d} - d^{d}\beta(d)\mathcal{L}(C)^{d-1} \geq \left((\mathcal{A}(C))^{1/(d-1)} - (d\beta(d))^{1/(d-1)}r(C)\right)^{d(d-1)},$$

where $r(C)$ is the inner radius of C, that is, the radius of the largest sphere
inscribed in C (cf. 11.9.12). See [HR, 296]. The technique of the proof is
interesting, utilizing the convex sets $C(t)$ "parallel to" and inside C, lying
at a distance t from its boundary. One uses 12.10.7, among other results, to
show that $\mathcal{L}\big(C(t)\big)$ is differentiable and that

$$\frac{d\mathcal{L}\big(C(t)\big)}{dt} = -\mathcal{A}\big(C(t)\big).$$

A particular case of this result will be found in 12.12.16. The notion of a set
parallel to and interior to a convex set was used in an essential way in [C–G].

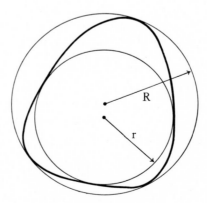

Figure 12.11.9

12.11.9.4. In the two-dimensional case, on the other hand, a lot is known about the deficit. The nicest result is Bonnesen's theorem, that says that, if $R(C)$ is the outer radius of C, that is, the lower bound of radii of balls containing C (cf. 11.5.8), the following relation holds:

$$\left(\mathcal{A}(C)\right)^2 - 4\pi\mathcal{L}(C) \geq \pi\left(R(C) - r(C)\right)^2,$$

where $r(C)$ is still the inner radius of C. See 12.12.15, [EN, 108] or [DE4, volume IV, p. 353, problem 6]. (In the latter reference the deficit is estimated as the sum of a series of terms which have a geometric interpretation; Bonnesen's inequality is obtained by neglecting all but one.) See also [OS1], and the fascinating new result [G–H], which explains how to make a curve rounder and rounder!

Another evaluation of the deficit is obtained through Fourier series, a natural tool in the study of plane curves, which can be seen as pairs of periodic functions (cf. 12.12.12 or [HZ]). So far it has apparently been impossible to use spherical harmonics in lieu of Fourier series in dimension $d \geq 3$ to give an easy proof of the isoperimetric inequality 12.11.1.

12.11.9.5. The proof in [DE4, volume 4, p. 353] is based on "integral geometry"; on this interesting discipline, see also [HR, chapter 6], [SO1], [FR, 173]. See also the important reference [G–G–V] and the recent one [RO], as well as [SO2].

12.11.9.6. For formulas on compact convex sets generalizing Cauchy's and Steiner–Minkowski's formulas, consult [DE4, volume III, p. 182], [HR], [B–F], [BLA1], [LW] or [BU–ZA].

12.12. Exercises

12.12.1. Show that the Stein symmetrization of a polytope is again a polytope.

* **12.12.2.** Show that if E is an ellipsoid in the Euclidean vector space X, containing the origin O in its interior, then its polar body E^* is an ellipsoid with the same property. Their volumes satisfy $\mathcal{L}(E)\mathcal{L}(E^*) \geq \left(\beta(d)\right)^2$, and equality takes place if and only if O is the center of E (for the definition of $\beta(d)$, see 9.12.4).

12.12.3. Prove 12.2.4.

* **12.12.4.** Justify the construction for the regular pentagon given in 12.4.6.

12.12.5. DOWKER'S THEOREMS. Consider a compact convex subset C of a Euclidean plane, such that $\overset{\circ}{C} \neq \emptyset$. For each $n \geq 3$, let T_n (resp. U_n) be the maximum (resp. minimum) area of an n-sided polygon inscribed in C (resp. circumscribed around C). Show that the function $n \mapsto T_n$ is concave and the function $n \mapsto U_n$ is convex, that is,

$$T_n + T_{n+2} \leq T_{n+1} \qquad \text{and} \qquad U_n + U_{n+2} \geq U_{n+1}$$

for all n.

12.12.6. Find the following data for each regular polytope (the last two as functions of the radius R of the circumscribed sphere): the number of k-faces $(k = 0, 1, \ldots, d-1)$, the volume of a k-face $(k = 0, 1, \ldots, d)$, and the dihedral angle between two adjacent faces.

12.12.7. OTHER DEFINITIONS FOR REGULAR POLYTOPES. Show that either of the definitions below is equivalent to the one given for regular polytopes:

  i) A d-dimensional polytope d is called regular is all its faces are isometric regular $(d-1)$-dimensional regular polytopes and if all its dihedral angles are equal.

  ii) A d-dimensional polytope d is called regular is all its faces are regular $(d-1)$-dimensional regular polytopes and if, for any vertex x of P, the other endpoints of edges of P containing x all lie in the same hyperplane and form, in this hyperplane, a regular $(d-1)$-dimensional regular polytope.

12.12.8. Define and classify star regular polytopes, cf. 12.6.10.5.

12.12.9. Find examples of compact sets K for which the expression

$$\frac{\mathcal{L}(B(K, \lambda)) - \mathcal{L}(K)}{\lambda}$$

does not have a limit as λ approaches zero. What happens to this ratio when K is the non-rectifiable curve of 9.9.3.3?

12.12.10. Show that the order of the group of isometries of a regular polyhedron is equal to four times the number of its edges.

12.12.11. Let $n \geq 3$ be an integer and P an n-sided convex polygon, and let $r(P)$ and $R(P)$ be the inner and outer radii of P (cf. 12.11.9). Show that we have

$$n \tan \frac{\pi}{n} \left(r(P) \right)^2 \leq \mathcal{L}(P) \leq \frac{1}{2} n \sin \frac{\pi}{n} \left(R(P) \right)^2,$$

$$2n \tan \frac{\pi}{n} r(P) \leq \mathcal{A}(P) \leq 2n \sin \frac{\pi}{n} R(P),$$

and that equality takes place if and only if P is regular (see [FT1, 6]).

12.12.12. Let C be a compact convex plane set, and P an n-sided convex polygon inscribed in C, maximal under these two conditions. Then

$$\mathcal{L}(P) \geq \frac{n}{2\pi} \sin \frac{2\pi}{n} \mathcal{L}(C),$$

and equality takes place if and only if Fr C is an ellipse (cf. [FT1, 36]). (Hint: write the boundary of C under the from $t \mapsto (\cos t, e(t) \sin t)$, where e is a periodic function of period 2π. In the case of equality, use Fourier series.)

12.12.13. Let $n \geq 3$ be an integer, and $(a_i)_{i=1,\ldots,n}$ be n points on a fixed circle. Show that the function

$$\sum_{i \leq i < j \leq n} \frac{1}{a_i a_j}$$

takes on its minimum if and only if the n points form a regular polygon.

* **12.12.14.** THE BLASCHKE ROLLING THEOREMS. Let C be a compact convex set in the plane whose boundary is a biregular curve (cf. [B–G], 8.2) of class C^2. Let A (resp. a) be a point on the boundary of C where the curvature is maximal (resp. minimal). Show that the osculating circle γ at a can roll all around the boundary, always staying inside C, and the boundary can roll all around the osculating circle Γ at A. Is this still true if we replace γ by the largest circle contained in C or Γ by the smallest circle containing C? See [BLA1, 116] if necessary.

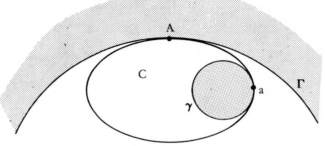

Figure 12.12.14

12.12.15. THE EULER EQUATION AND BONNESEN'S THEOREM. Let C be a compact plane convex set, with boundary Γ of class C^2. Fix an orthonormal frame with origin $O \in \overset{\circ}{C}$, and let $h(t)$ $(t \in \mathbf{R})$ be the distance from O to the tangent to Γ perpendicular to the vector $(\cos t, \sin t)$ and such that $(\cos t, \sin t)$ and C lie in the same side of that tangent (cf. 11.8.12.3). Show that the function $h : \mathbf{R} \to \mathbf{R}$ is of class C^2. Use h to calculate the length and curvature of Γ and the area of C.

Let C' be another convex set, also with C^2 boundary; assume that, for any $m \in \Gamma$ and $m' \in \Gamma'$, there exists a rigid motion f such taking m to m', the tangent to Γ at m to the tangent to Γ' at m', and C inside C'. Show that

$$\mathcal{A}(C)\mathcal{A}(C') \leq 2\pi\big(\mathcal{L}(C) + \mathcal{L}(C')\big).$$

Deduce from this Bonnesen's inequality: if $r(C)$ and $R(C)$ are the inner and outer radii of C we have (cf. 12.11.9.4)

$$\big(\mathcal{A}(C)\big)^2 - 4\pi\mathcal{L}(C) \geq \pi^2\big(R(C) - r(C)\big)^2.$$

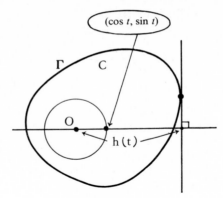

Figure 12.12.15

12.12.16. ISOPERIMETRIC INEQUALITIES FOR POLYGONS. Let P be a convex polygon, and call P' the convex polygon circumscribed around the unit circle and whose sides are parallel to those of P. Prove Lhuillier's inequality:

$$\big(\mathcal{A}(P)\big)^2 \geq 4\mathcal{L}(P)\mathcal{L}(P^*).$$

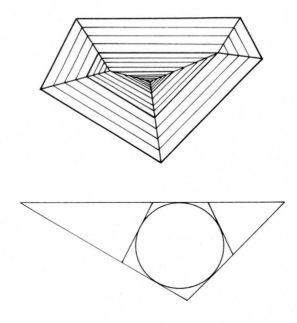

Figure 12.12.16

To do this, refer to figure 12.12.16, and consider the family of polygons $P(t)$ lying inside P, with sides parallel to the sides of P and lying at distance

t from them. Show that, if t is small,

$$\mathcal{L}(P) = \mathcal{L}(P(t)) + t\mathcal{A}(P(t)) + t^2\mathcal{L}(P^*),$$
$$\mathcal{A}(P) = \mathcal{A}(P(t)) + 2t\mathcal{L}(P^*).$$

Deduce from this that $\mathcal{L}(P(t)) = t^2\mathcal{L}(P^*) - t\mathcal{A}(P) + \mathcal{L}(P)$. Show that

$$\left(\mathcal{A}(P)\right)^2 - 4\mathcal{L}(P)\mathcal{L}(P^*) \geq \left(\mathcal{A}(P) - 2\mathcal{L}(P^*)r(P)\right)^2,$$

where $r(P)$ denotes the inner radius of P. Compare with 12.11.9.4.

Show that, among all n-sided polygons with a given radius, regular polygons, and only they, have maximal area (first reduce the problem to the case of polygons circumscribed around a circle).

12.12.17. Comment on [R–C, volume II, p. 234–236].

12.12.18. Let $C \in \mathcal{C}^\bullet$, and show that the following relations hold for the inner and outer radii $r(C)$ and $R(C)$ of C:

$$d\frac{\mathcal{L}(C)}{R(C)} \leq \mathcal{A}(C) \leq d\frac{\mathcal{L}(C)}{r(C)}.$$

Study when equality occurs.

12.12.19. Prove that $\int_{\xi \in \mathcal{S}} |(\xi \mid u)| \sigma = 2\beta(d-1)$ (cf. 12.3.3.2).

12.12.20. Formulas involving areas and volumes.

12.12.20.1. Calculate the length and area of the hypo- and epicycloids studied in 9.14.34.

12.12.20.2. Calculate the volume of a *spherical zone*, that is, the portion of a sphere in $\mathbf{R}^3$ comprised between two parallel planes. In particular, the lateral area of a spherical zone only depends on its thickness.

12.12.20.3. Calculate the volume of the solid bounded by two cylinders of revolution with same radius r and whose axes intersect at an angle α. Calculate the volume of the solid bounded by three cylinders of revolution with mutually orthogonal axes.

12.12.20.4. Calculate the volume of the solid bounded between a paraboloid of revolution and a plane not parallel to its axis.

12.12.20.5. Calculate the volume and lateral area of a *cylindrical wedge* (figure 12.12.20.5).

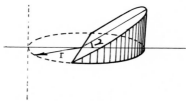

Figure 12.12.20.5

12.12.20.6. Calculate the volume of *Viviani's window*, the set of points of $\mathbf{R}^3$ defined by

$$\{ (x, y, z) \mid x^2 + y^2 + z^2 \leq 1 \text{ and } x^2 + y^2 \leq x \}.$$

12.12.20.7. Formula of the three levels. Let K be a compact set in a three-dimensional Euclidean space, and assume that, for $a \leq z \leq b$, the area $S(z)$ of the section $K \cap H(z)$ of K by a plane with a fixed z-coordinate is a polynomial in z, of degree three or less. Show that the volume of z between $H(a)$ and $H(b)$ is given by the formula of the three levels:

$$\frac{b - a}{6} \left(S(a) + 4S \left(\frac{a + b}{2} \right) + S(b) \right).$$

This gives another proof for the volume of a spherical zone (12.12.20.2). Apply also to truncated cones. Show that the condition of the statement is always satisfied when K is bounded on the sides by a ruled surface (in particular, if K is a polytope).

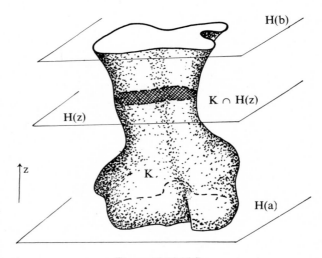

Figure 12.12.20.7

12.12.20.8. Find the area of the ellipsoid of revolution with equation

$$\frac{x^2}{a^2} + \frac{y^2}{a^2} + \frac{z^2}{c^2} = 1.$$

* **12.12.20.9. Theorems of Guldin.** Consider a compact set K of a plane P in the three-dimensional Euclidean space E. Show that the volume of the compact set C of E, generated by rotating K around a line D of P which does not intersect K, is given by the formula

$$\mathcal{L}_E(C) = 2\pi \cdot d(g, D)\mathcal{L}_P(K),$$

where $g = \operatorname{cent}'(K)$ denotes the centroid of K (see 2.7.5.2).

If the boundary of K is considered as a homogeneous wire and h is the center of mass of this wire (in the usual sense), show that the area of C is given by the formula

$$\mathcal{A}_E(C) = 2\pi \cdot d(h, D) \mathcal{A}_P(K).$$

(Both areas are understood in the sense of differentiable manifolds.)

Find applications of this formula, as well as special cases of volumes or areas already known.

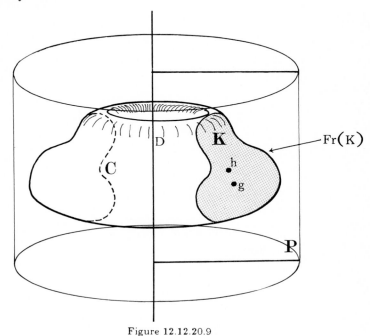

Figure 12.12.20.9

12.12.21. Let A be a compact convex set with non-empty interior in a n-dimensional Euclidean space E, and let x be the center of A (cf. 2.7.5.5). For any unit vector $\xi \in \vec{E}$, denote by $\delta(\xi)$ the distance from x to the supporting hyperplane to A perpendicular to ξ and such that $(\overrightarrow{xy} \mid \xi) > 0$ for any y in this hyperplane (cf. 11.8.12.3). Also, denote by $B(\xi)$ the width of A in the direction of ξ (cf. 11.5.6.3). Show that, for any ξ,

$$\frac{B(\xi)}{n+1} \leq \delta(\xi) \leq \frac{nB(\xi)}{n+1}$$

(cf. [VE, 190, proposition 12.5]).

12.12.22. Show that a quadrilateral with fixed side lengths has maximal area when it can be inscribed in a circle. Generalize for n-sided polygons. See [LEV] for this still partly open problem.

Chapter 13
Quadratic forms

The theory of Euclidean spaces is based on one particular positive definite quadratic form. Other quadratic forms abound in the mathematical zoo, as well as in mechanics and physics; Dieudonné has said that there is hardly a mathematical theory that does not involve a bilinear form. We mention a few:
— in analysis, Hilbert and Sobolev spaces;
— in algebraic topology, the cup product gives a quadratic (or alternating) form on the n-th cohomology group of a $2n$-dimensional compact manifold;
— in arithmetic, the decomposition of integers into sums of squares;

— in differential geometry, Riemannian metrics and Lorentz metrics (the latter occurring in relativity theory);

— in mechanics, the Liouville form and the whole machinery of symplectic geometry, as well as torsors.

In this book we deal with quadratic forms within several frameworks: Euclidean spaces, conics and quadrics, the geometry of spheres (chapter 20) and hyperbolic geometry (chapter 19); they also make a brief appearance in section 14.5.5, about projective correlations. This chapter deals only with certain aspects of the theory of quadratic forms, mostly geared towards the geometric applications mentioned. For more information, see [AN], [S–T], [BI2] and [SE2].

Section 13.1 contains basic definitions and examples, including Artinian spaces, which are repeatedly used in this book. Next we deal with phenomena characteristic of non-Euclidean quadratic forms, to wit, isotropic vectors and singular subspaces (section 13.2); this study is simplified by the introduction of a linear map from the space where the form is defined onto its dual. Section 13.3 extends the concept of orthogonality to non-Euclidean spaces, and includes the proof of a technical result, the non-singular completion of an arbitrary subspace, that will be essential later. In section 13.4 we show that every quadratic form admits an orthogonal basis, and this leads to the classification of quadratic forms on real and complex vector spaces. Section 13.5 is devoted to the important and classical problem of finding the axes of a quadratic form defined on a Euclidean space. In section 13.6 we introduce the group $O(q)$ of a quadratic form q; this is a natural generalization of the orthogonal group of a Euclidean vector space. We also prove some technical results about Artinian spaces, which in a sense are the spaces that contrast most with Euclidean spaces. In section 13.7 we show that the group of a quadratic form is generated by hyperplane reflections and that it acts transitively on isometric subspaces; the proof here is harder than in the Euclidean case. Finally, section 13.8 introduces the counterparts for real Artinian planes of the notions of Euclidean angles and oriented angles—a prelude to the study of hyperbolic geometry, which will be the object of chapter 19.

In this chapter E is a vector space of finite dimension n, over a commutative field K with characteristic different from 2 (cf. 3.3.2). A vector subspace of E is simply called a subspace.

13.1. Definitions and examples

13.1.1. DEFINITION. *A quadratic form over a vector space E is an element q of $P_2^\bullet(E)$ (cf. 3.3.1). The unique symmetric bilinear form giving rise to q is denoted by P and called the polar form of q. We set $Q(E) = P_2^\bullet(E)$.*

13.1.2. NOTES. We have (cf. 3.3.2.1):

$$q(x) = P(x, x), \quad P(x, y) = \frac{1}{2}\big(q(x + y) - q(x) - q(y)\big), \quad q(\lambda x) = \lambda^2 q(x).$$

For $K = \mathbf{R}$ or $\mathbf{C}$, any quadratic form $q : E \to K$ is of class C^∞ and satisfies the Euler identity

$$q(x) = \frac{1}{2}q'(x)(x),$$

since $P(x, y) = \frac{1}{2}q'(x)(y)$. In particular, in a given coordinate system $\{x_i\}$, we obtain

$$P(x, y) = \sum_{i=1}^{n} y_i \frac{\partial q}{\partial x_j}(x_1, \ldots, x_n).$$

Finally,

$$P(x, y) = \frac{1}{2}q''(0)(x, y).$$

These formulas hold whenever $q : E \to K$ is homogeneous of degree two $\big($that is, satisfies $q(\lambda x) = \lambda^2 q(x)\big)$ and has a second derivative at 0; but these two conditions together imply that q is a quadratic form. All these assertions follow from differentiating $q(\lambda x) = \lambda^2 q(x)$ with respect to λ and applying the standard rules (cf. [CH1], for example).

13.1.3. EXAMPLES

13.1.3.1. Euclidean structures (see 8.1.1).

13.1.3.2. The quadratic form $x^2 + y^2 + z^2 - t^2$ is fundamental in relativity.

13.1.3.3. Take $\phi \in E^*$ (the dual of E); then the map $q = \phi^2$ defined by $q(x) = \big(\phi(x)\big)^2$ is a quadratic form. Its polar form is $P(x, y) = \phi(x)\phi(y)$. More generally, for $\phi, \psi \in E^*$, we define $q = \phi\psi$ by $q(x) = \phi(x)\psi(x)$, with polar form $\frac{1}{2}\big(\phi(x)\psi(y) + \phi(y)\psi(x)\big)$. We can also define $q = \sum_i k_i \phi_i^2$, for $k_i \in K$ and $\phi_i \in E^*$.

13.1.3.4. If $\{e_i\}$ is a basis for E and $\{e_i^*\}$ is the dual basis, we can consider the following quadratic forms:

i) $q = \sum_{i=1}^{r}(e_i^*)^2$, for $1 \le r \le n$, generally denoted by $q = \sum_{i=1}^{r} x_i^2$;

ii) $q = \sum_{i=1}^{r} x_i^2 - \sum_{i=r+1}^{r+s} x_i^2$, for $1 \le r \le r+s \le n$.

If $n = 2p$ and $\{e_i\}_{i=1,\dots,p} \cup \{h_i\}_{i=1,\dots,p}$ is a basis for E, with corresponding coordinates x_i and y_i, we can consider the quadratic form

iii) $q = 2\sum_{i=1}^{p} x_i y_i$.

13.1.3.5. If $F \subset E$ is a subspace of E and $q \in Q(E)$ is a quadratic form on E, the *restriction* $q|_E$ is a quadratic form on E.

13.1.3.6. Expression in a given basis and associated matrix. Let $q \in Q(E)$ be a quadratic form with polar form P, and $\{e_i\}$ a basis for E. We know that P (hence Q) is determined by giving $P(e_i, e_j)$. We define the *matrix of q with respect to the basis* $\{e_i\}$ as the $n \times n$ matrix $A = (P(e_i, e_j)) = (a_{ij})$; this correspondence is *denoted by* $q \leftrightarrow A$. Setting

$$x \cong X = \begin{pmatrix} x_1 \\ \vdots \\ x_n \end{pmatrix}, \qquad y \cong Y = \begin{pmatrix} y_1 \\ \vdots \\ y_n \end{pmatrix},$$

we have

13.1.3.7
$$\begin{cases} P(x,y) = {}^t\!XAY = \sum_{i,j} a_{ij} x_i y_j = \sum_i a_{ii} x_i y_i + \sum_{i<j} a_{ij}(x_i y_j + x_j y_i), \\ q(x) = {}^t\!XAX = \sum_{i,j} a_{ij} x_i x_j = \sum_i a_{ii} x_i^2 + 2\sum_{i<j} a_{ij} x_i x_j). \end{cases}$$

From these expressions we see how to obtain P from q, without having to use 3.3.2.1: just replace each x_i^2 by $x_i y_i$ and each $x_i x_j$ by $\frac{1}{2}(x_i y_j + x_j y_i)$. Thus, in examples 13.1.3.4, we have the following matrices:

(i) $A = \begin{pmatrix} I_r & 0 \\ 0 & 0 \end{pmatrix}$,   (ii) $A = \begin{pmatrix} I_r & 0 & 0 \\ 0 & -I_s & 0 \\ 0 & 0 & 0 \end{pmatrix}$,   (iii) $A = \begin{pmatrix} 0 & I_p \\ I_p & 0 \end{pmatrix}$,

where I_t denotes the $t \times t$ identity matrix.

If we change from the basis $\{e_i\}$ to the basis $\{e_i'\}$, and the matrix giving the coordinates of the e_i' in the basis $\{e_i\}$ is S (that is, $S = M(f)$, the matrix in the basis $\{e_i\}$ of the linear map f such that $f(e_i) = e_i'$ for all i), we have

13.1.3.8 $A' = {}^t\!SAS,$

where A' is the matrix of q in the basis $\{e_i'\}$.

13.1.3.9. Pullbacks. Let E and E' be vector spaces (over the same field), let $f \in L(E; E')$ be a morphism and $q' \in Q(E')$ a quadratic form on E'. The *pullback of q' by f*, denoted by f^*q', is the quadratic form $q \in Q(E)$ defined by

$$q(x) = (f^*q')(x) = q'(f(x)).$$

The polar form of q is $P = f^*P'$, defined by

$$P(x,y) = P'(f(x), f(y)),$$

where P' is the polar form of q'. In particular, we have an action $f \mapsto \{q \mapsto f^*q\}$ of the linear group $\mathrm{GL}(E)$ on $Q(E)$ (actually, since $(g \circ f)^*q = f^*(g^*q)$, the group law on $\mathrm{GL}(E)$ must be the opposite of the usual one).

If f has matrix U with respect to the bases $\{e_i\}$ and $\{e'_i\}$ of E and E', respectively, and if $q' \leftrightarrow A'$, we have

13.1.3.10 $f^*q' \leftrightarrow {}^tU A'U,$

as seen from 13.1.3.7:

$$(f^*q')(x) = {}^tX A X = q'(f(x)) = {}^tf(X)A'f(X) = {}^tX {}^tU A'U X. \qquad \square$$

13.1.4. EQUIVALENCE. CLASSIFYING QUADRATIC FORMS.

13.1.4.1. Definition. *Two quadratic forms q on E and q' on E' are called equivalent if there exists an isomorphism $f \in (E; E')$ such that $q = f^*q'$. We also say that the structures (E, q) and (E', q') are isometric (cf. 8.1.5) or isomorphic. The classification of quadratic forms over a field K is the problem of finding the equivalence classes of quadratic forms on finite-dimensional vector spaces over K.*

13.1.4.2. Thus, classifying quadratic forms over K is the same as determining, for each n, the orbits of the action of $\mathrm{GL}(K^n)$ on $Q(K^n)$.

13.1.4.3. Examples. A quadratic form is called *neutral* when it is equivalent to 13.1.3.4 (iii) (in particular, dim E must be even). For example, for $K = \mathbf{C}$ and $n = 2p$, the form $q = \sum_{i=1}^n x_i^2$ is neutral, since one can write

$$\sum_{i=1}^{2p} x_i^2 = \sum_{l=1}^{p}(x_l^2 + x_{l+p}^2) = \sum_{l=1}^{p}(x_l + ix_{l+p})(x_l - ix_{l+p}).$$

Similarly, the form in 13.1.3.4 (ii) will be neutral if $r = s$ and $n = 2r$.

13.1.4.4. Definition. *An Artinian space is a pair (E, q) where q is a neutral quadratic form on E. A $2n$-dimensional Artinian space is generally denoted by Art_{2p}. When $p = 1$ Artinian spaces are called Artinian planes.*

It is clear that Artinian spaces of same dimension are isomorphic. A few books use the expression "hyperbolic plane" instead of Artinian plane; this terminology is undesirable, since by hyperbolic plane we mean something very different, to be studied in chapter 19.

13.1.4.5. The classification of quadratic forms over $\mathbf{C}$ and $\mathbf{R}$ is simple and will be carried out in 13.4.6 and 13.4.7. On the other hand, for K arbitrary the situation is immensely more complicated, and has not been studied enough. Solutions exist for $\mathbf{Q}$ (the whole chapter IV of [SE2]), for finite fields ([AN, 143–148]) and for algebraic number fields (almost the whole of [OM]). The one-dimensional classification over arbitrary fields is also very simple, as we shall see in the sequel.

The reader will find throughout [S–T] a number of elementary results on quadratic forms.

13.1.4.6. The discriminant. It follows from 13.1.3.8 that $\det A$ depends on the choice of a basis, but its image in the quotient $K/(K^*)^2$ does not. This image, the *discriminant* of the quadratic form q, is *denoted* by $\mathrm{disc}(q)$. From 13.1.3.10 we see that equivalent quadratic forms have the same discriminant: $\mathrm{disc}(f^*q) = \mathrm{disc}(q)$. The converse is false; for example, for $K = \mathbf{R}$, $n = 4$, the forms $q = x^2 + y^2 + z^2 + t^2$ and $q' = x^2 + y^2 - z^2 - t^2$ have the same discriminant but are not equivalent, as 13.4.7 shows.

13.1.4.7. If $\dim E = 1$, the forms q and q' are equivalent if and only if $\mathrm{disc}(q) = \mathrm{disc}(q')$. This is immediate from 13.1.3.10.

For instance, for $K = \mathbf{C}$ we have $\mathbf{C}/(\mathbf{C}^*)^2 = \{0,1\}$ and there are two classes of quadratic forms, $q = 0$ and $q = x^2$. For $K = \mathbf{R}$ we have $\mathbf{R}/(\mathbf{R}^*)^2 = \{-1,0,1\}$ and there are three classes: 0, x^2 and $-x^2$. This is a particular case of 13.4.6 and 13.4.7.

From now on (E, q) will *denote* a quadratic form q on the vector space E. The polar form of q will be *denoted* by P.

13.2. Isotropy, singularity, radicals and degeneracy

13.2.0. As in 8.1.8.1, we *associate* to P a morphism $\phi \in L(E; E^*)$ by setting $\phi(x)(y) = P(x,y)$. If $\{e_i\}$ is a basis for E and $\{e_i^*\}$ is the dual basis, the matrix of ϕ with respect to the bases $\{e_i\}$, $\{e_i^*\}$ is the same as the matrix A of q with respect to $\{e_i\}$, for we have

$$a_{ij} = P(e_i, e_j) = \phi(e_i)(e_j) = e_j^*(\phi(e_i)).$$

Recall that $\phi \in \mathrm{Isom}(E; E^*)$ is equivalent to $\mathrm{Ker}\,\phi = 0$.

13.2.1. DEFINITION. *The isotropic cone (or light cone) of q is $q^{-1}(0)$. A vector $x \in E$ is said to be isotropic if $x \in q^{-1}(0)$. The form q is called anisotropic if $q^{-1}(0) = 0$.*

The radical of q is $\mathrm{rad}(q) = \mathrm{Ker}\,\phi = \{ x \in E \mid P(x,y) = 0 \text{ for any } y \in E \}$; the rank of q is defined as the rank of ϕ, that is, $\mathrm{rank}(q) = \dim(\phi(E))$. The form q is said to be non-degenerate if $\mathrm{rad}(q) = 0$; this is equivalent to $\mathrm{rank}(q) = \dim E$, and to $\phi \in \mathrm{Isom}(E; E^*)$. The form q is degenerate if $\mathrm{rad}(q) \neq 0$.

If $F \subset E$ is a subspace, the radical of F is defined as $\mathrm{rad}(F)$

$$\mathrm{rad}(F) = \mathrm{rad}(q|_F) = \{ x \in F \mid P(x,y) = 0 \text{ for all } y \in F \}.$$

The subspace F is said to be non-singular (resp. singular) if $q|_F$ is non-degenerate (resp. degenerate). It is completely singular if $q|_F = 0$, that is, $\mathrm{rad}(F) = F$.

Figure 13.2.1.1 represents the form $q = x^2 - y^2$ on $\mathbf{R}^2$, and figures 13.2.1.2 represent $q = -x^2 - y^2 + z^2$ on $\mathbf{R}^3$. For a quadratic form on $\mathbf{R}^4$

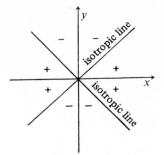

Figure 13.2.1.1

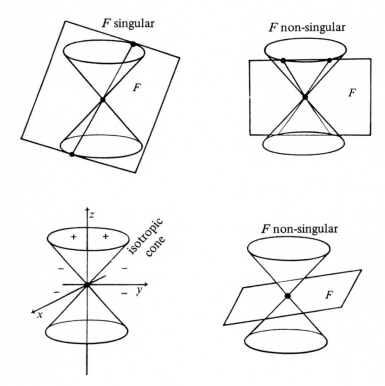

Figure 13.2.1.2

like $x^2 + y^2 - z^2 - t^2$, which cannot be directly visualized, one can draw (a subset of) the projection of the isotropic cone on $P^3(\mathbf{R})$; see figures 13.7.10 and 14.4.6.

13.2.2. REMARKS

13.2.2.1. The word isotropic is consistent with 8.8.6.1. One frequently hears the terms isotropic and completely isotropic applied to singular and completely singular subspaces, but this nomenclature is confusing.

13.2.2.2. From elementary linear algebra we have

$$\text{rank}(q) + \dim(\text{rad}(q)) = \dim E.$$

13.2.2.3. Watch out for the difference between non-degeneracy and anisotropy; the latter implies the former, but not the other way around. For example, (E, q) for $E = \mathbf{R}^2$ and $q = x^2 - y^2$ is non-degenerate but not anisotropic. If $K = \mathbf{C}$ (or any algebraically closed field) no quadratic form is anisotropic in dimension greater than one; this follows from 13.7.6.

13.2.2.4. From 13.2.0 and 13.1.4.6 we see that q is degenerate if and only if $\text{disc}(q) = 0$; in particular, if and only if $\det A = 0$ in some basis.

13.2.2.5. Contrary to the Euclidean case, E can have singular subspaces even when q is non-degenerate; see figure 13.2.1.2, or take $F = Kx$, for x isotropic.

13.2.2.6. If $E = \text{rad}(q) \oplus G$ is a direct sum, the restriction $q|_G$ is non-degenerate (cf. 13.9.1).

13.2.3. EXAMPLES

13.2.3.1. If $(E, q) = \text{Art}_2$, the isotropic cone $q^{-1}(0)$ consists of two distinct lines of E, called *isotropic lines;* one cannot distinguish between the two in the absence of further structure (cf. 13.7.10 and 13.8.2.1). They determine q up to a scalar constant: in a basis consisting of one vector from each line, q has the form kxy.

13.2.3.2. If $q \leftrightarrow A$ (cf. 13.1.3.6), we have $\text{rank}(q) = \text{rank}(A)$. In 13.1.13.4, ranks are as follows: (i) r; (ii) $r + s$; (iii) $2p$. In particular, (iii) is always non-degenerate, (i) is non-degenerate if and only if $r = n$, and (ii) if and only if $r + s = n$.

13.2.3.3. Consider again the examples in 13.1.3.4. In (i), if $K = \mathbf{C}$, the space

$$F = \mathbf{C}(e_1 + ie_2) \oplus \mathbf{C}(e_3 + ie_4) \oplus \cdots \oplus \mathbf{C}(e_{2p-1} + ie_{2p}),$$

where $p = [n/2]$, is completely singular; its dimension is p. In (ii), if $K = \mathbf{R}$ and $r < s$,

$$F = \mathbf{R}(e_1 + e_{r+1}) \oplus \mathbf{R}(e_2 + e_{r+2}) \oplus \cdots \oplus \mathbf{R}(e_r + e_{2r})$$

is completely singular, and has dimension r. Finally, in (iii) and for arbitrary K, the spaces $F = Ke_1 \oplus \cdots \oplus Ke_p$ and $F' = Kh_1 \oplus \cdots \oplus Kh_p$ are p-dimensional and completely singular. In 13.7.6 and 13.3.4.2 we shall encounter a converse for this.

We now state a technical lemma.

13.2.3.4. Lemma. *Let (E, q) be arbitrary and $x \notin \text{rad}(q)$ isotropic and non-zero. There exists a plane $P \subset E$ containing x and such that $(P, q|_P)$ is an Artinian plane. In particular, any pair (E, q), where E is two-dimensional and q is non-degenerate an not anisotropic, is an Artinian plane.*

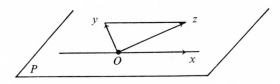

Figure 13.2.3

Proof. Since $x \notin \operatorname{rad}(q)$ the map $\phi(x) \in E^*$ is non-zero (cf. 13.2.0); thus there exists z such that $P(x,z) = 1$. Then we can find $y \in Kz + Kx$ such that $q(y) = 0$, say $y = z - \frac{1}{2}q(z)x$. But this means that the matrix of $q|_{Ky+Kx}$ is $\begin{pmatrix} 0 & 1 \\ 1 & 0 \end{pmatrix}$, as was to be shown.

13.2.4. THE ADJOINT OF AN ENDOMORPHISM. Consider a pair (E, q), with q non-degenerate; since $\phi \in \operatorname{Isom}(E; E^*)$, we can imitate 8.1.8.6. To each $f \in L(E; E) = \operatorname{End}(E)$ we associate a map $^t f \in L(E; E)$, the *adjoint* of f, defined by

13.2.4.1 $P(f(x), y) = P(x, {}^t f(y))$ for any $x, y \in E$,

that is, $^t f = \phi^{-1} \circ f^* \circ \phi$, where f^* denotes the transpose of f. If $\{e_i\}$, $\{e_i^*\}$ are dual bases, $q \leftrightarrow A$ and U is the matrix of f in the basis $\{e_i\}$, we have

13.2.4.2 $U^* = \text{matrix of } {}^t f \text{ in the basis } \{e_i\} = A^{-1}{}^t U A,$

since the matrix of f^* in the basis $\{e_i^*\}$ is $^t U$. In particular, $\det {}^t f = \det f^* = \det f$.

13.3. Orthogonality. The non-singular completion of a subspace

Following the Euclidean case, we have the following definition:

13.3.1. DEFINITION. *Let A be a subset of E. Set*

$$A^{\perp} = \left\{ x \in E \mid P(x, y) = 0 \text{ for any } y \in A \right\} = \bigcap_{y \in A} \operatorname{Ker}(\phi(y)),$$

and call this set (in fact a subspace) the orthogonal *of A. If A and B are subsets, we say that A and B are* orthogonal, *and denote this fact by $A \perp B$, if $B \in A^{\perp}$, that is, if $P(x, y) = 0$ for any $x \in A$ and $y \in B$. The relation $A \perp B$ is symmetric. A direct sum $E = \bigoplus_i E_i$ is called* orthogonal *if $E_i \perp E_j$ for all $i \neq j$; we write this as $E = \bigoplus_i^{\perp} E_i$. A basis $\{e_i\}$ is called* orthogonal *if $e_i \perp e_j$ for any $i \neq j$, and* orthonormal *if it is orthogonal and satisfies $q(e_i) = 1$ for all i. For example, $\operatorname{rad}(q) = E^{\perp}$. In general, if F is a subspace, $\dim F + \dim F^{\perp} \neq \dim E$, because q can be degenerate. On the other hand, the following results (derived, like 8.1.8.3, by using 2.4.8.1) still hold:*

13.3.2. PROPOSITION. *For every subset A we have $A^\perp = \langle A \rangle^\perp$ and $A \subset (A^\perp)^\perp$. Suppose further that q is non-degenerate. For any subspace F of E we have $(F^\perp)^\perp = F$, $\dim F + \dim F^\perp = \dim E$ and $\operatorname{rad} F = F \cap F^\perp$. The subspace F is completely singular if and only if $F \subset F^\perp$. The following equivalences hold: F is non-singular $\iff F \cap F^\perp = 0 \iff E = F \oplus F^\perp \iff F^\perp$ is non-singular. For arbitrary subspaces F, F' of E we have*

$$(F \cap F')^\perp = F^\perp + F'^\perp, \qquad (F + F')^\perp = F^\perp \cap F'^\perp.$$

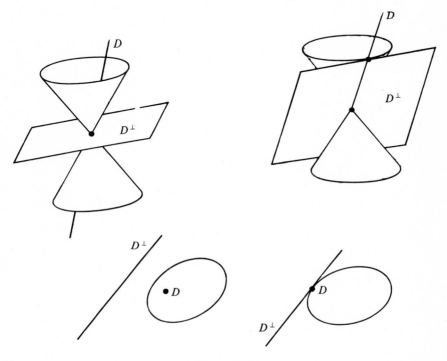

Figure 13.3.2

Figure 13.3.2 illustrates this situation. The top pictures show the isotropic cone, a subspace and its orthogonal; the bottom pictures are sections of the top ones by an affine plane. If they make you think of the duality between a point and its polar line with respect to a conic, that's exactly what they're about: historically, duality (or orthogonality) with respect to a quadratic form arose from duality with respect to a conic. In this book we shall do things the other way around: in 14.5 we define duality with respect to a conic by projectivizing duality with respect to the form that defines the conic.

13.3.3. EXAMPLES

13.3.3.1. If (E, q) is such that $E = \bigoplus_i^\perp E_i$, we can say that (E, q) is the direct sum of the $(E_i, q|_{E_i})$. Conversely, if we are given $E = \sum_i E_i$ and on each E_i a quadratic form q_i, we define $q = \bigoplus_i^\perp q_i$ via its polar form:

$$P\left(\sum_i x_i, \sum_i y_j\right) = \sum_i P_i(x_i, y_i),$$

where P_i denotes the polar form of q_i. We then have $E = \bigoplus_i^\perp E_i$ and $q|_{E_i} = q_i$ for any i.

13.3.3.2. In the sense above, we have, for any Artinian space Art_{2p},

$$\mathrm{Art}_{2p} = \underbrace{\mathrm{Art}_2 \overset{\perp}{\oplus} \cdots \overset{\perp}{\oplus} \mathrm{Art}_2}_{p \text{ terms}};$$

this is seen by taking $E_i = Ke_i \oplus Kh_i$ for all i, in the notation of 13.1.3.4 (iii).

13.3.4. THE NON-SINGULAR COMPLETION. This technical result is simple but will be fundamental in the sequel.

13.3.4.1. Proposition. *Let q be non-degenerate. Let F be a subspace of E, $s = \dim(\mathrm{rad}(F))$, G an arbitrary complement for $\mathrm{rad}(F)$ in E and $\{x_i\}_{i=1,\ldots,s}$ a basis for $\mathrm{rad}(F)$. There exist s planes $P_i \subset E$ such that: (P_i, e_i) is an Artinian plane for all i; $x_i \in P_i$ for all i; $P_i \perp P_j$ for all $i \neq j$; and $G \perp P_i$ for all i. Furthermore, the orthogonal direct sum*

$$\overline{F} = G \overset{\perp}{\oplus} P_i \overset{\perp}{\oplus} \cdots \overset{\perp}{\oplus} P_s$$

is a non-singular subspace of E, and $(\overline{F}, q|_F)$ is isometric to $G \oplus^\perp \mathrm{Art}_{2s}$. Such an $\overline{F}$ is called a non-singular completion of F.

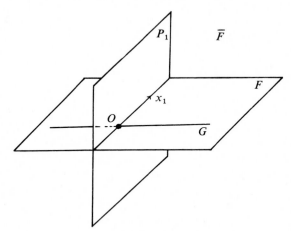

Figure 13.3.4

Proof. An easy induction on $s = \dim(\mathrm{rad}(F))$. For $s = 1$, set $\mathrm{rad}(F) = Kx_1$ and $\overline{F} = G \oplus^\perp Kx_1$. By 13.3.2, the subspace $G^\perp$ is non-singular; since $x_1 \in G^\perp$, we can apply 13.2.3.4 to conclude the existence of a plane P_1 of E containing x_1 and such that $(P_1, q|_{P_1})$ is an Artinian plane. $\square$

For the uniqueness of this construction, see 13.9.3.

13.3.4.2. Corollary. *Let q be non-degenerate. For every subspace $F \subset E$ we have*

$$\dim F + \dim\big(\mathrm{rad}(F)\big) \leq n = \dim E.$$

In particular, if F is completely singular its dimension satisfies $\dim F \leq n/2$, and $\dim F = n/2$ only when $E = \mathrm{Art}_{2p}$. ☐

13.3.5. The rest of this chapter is laid out according to the following plan: an elementary discussion in which we classify quadratic forms over $\mathbf{C}$ and $\mathbf{R}$ and prove the simultaneous diagonalization of two quadratic forms is followed by a "geometric algebra" discussion of the group $O(q)$ (the generalization of the orthogonal group of chapter 8), dealing with simple generators, transitivity of the action of $O(q)$ on subspaces, and the simplicity of $O(q)$ (without proof).

13.4. Orthogonalization. Classification of real and complex quadratic forms

13.4.1. LEMMA. *Every (E, q) admits an orthogonal basis $\{e_i\}$.*

Proof. If $q = 0$ this is trivial; else let x be non-isotropic. Then $\phi(x) \in E^*$ (cf. 13.2) is non-zero, so $x^\perp$ is a hyperplane and $E = Kx \oplus^\perp x^\perp$. This brings us down to a lower-dimensional space $x^\perp$ and we complete the proof by induction. ☐

13.4.2. COROLLARY. *Every (E, q) admits a basis in which $q = \sum_{i=1}^r \lambda_i x_i^2$ for $\lambda_i \in K^*$. Every (E, q) over $K = \mathbf{C}$ admits a basis in which $q = \sum_{i=1}^r x_i^2$. Every (E, q) over $K = \mathbf{R}$ admits a basis in which*

$$q = \sum_{i=1}^r x_i^2 - \sum_{i=r+1}^{r+s} x_i^2.$$

Proof. If $\{e_i\}$ is an arbitrary orthogonal basis for (E, q), we have

$$q(x_1, \ldots, x_n) = P\left(\sum_i x_i e_i, \sum_j x_j e_j\right) = \sum_{i,j} x_i x_j q(e_i, e_j) = \sum_i x_i^2 q(e_i, e_j);$$

just set $\lambda_i = q(e_i, e_i)$ whenever $q(e_i, e_i) \neq 0$. If $K = \mathbf{C}$, replace e_i by $(1/\sqrt{\lambda_i})e_i$; if $K = \mathbf{R}$, replace e_i by $(1/\sqrt{\lambda_i})e_i$ if $\lambda_i > 0$ and by $(1/\sqrt{-\lambda_i})e_i$ if $\lambda_i < 0$. ☐

13.4.3. NOTE. It is clear that one can replace the assumption $K = \mathbf{C}$ by "K is a field where every element has a square root", and in particular by "K is algebraically closed". Similarly, $K = \mathbf{R}$ can be replaced by "K is an ordered field where every positive element has a square root".

13.4.4. EXAMPLE. If $K = \mathbf{C}$ and $n = 2p$, any non-degenerate form q is equivalent to a neutral form, that is, (E, q) is isometric to Art_{2p}.

13.4.5. What we have been saying may give the impression that with 13.4.1 the classification of quadratic forms is almost complete; but in fact, as Artin has it, we have barely scratched the surface! Already when $K = \mathbf{R}$ we will show that $x^2 + y^2 - z^2 - t^2$ and $x^2 + y^2 + z^2 - t^2$ are not equivalent; and for the general case, see 13.1.4.5. But notice the following immediate consequence of 13.4.1:

13.4.6. THEOREM. *Two quadratic forms over* $\mathbf{C}$ *are equivalent if and only if the dimension* n *of the space and the rank* r *of the form are the same. In other words, for a given* E, *the action of* $\mathrm{GL}(E)$ *on* $Q(E)$ *has exactly* $n + 1$ *orbits, each corresponding to a different rank* $r = 0, 1, \ldots, n = \dim E$. *In particular, all non-degenerate quadratic forms are equivalent.* $\square$

13.4.7. THEOREM (SYLVESTER'S LAW OF INERTIA). *Two quadratic forms over* $\mathbf{R}$ *are equivalent if and only if the dimension* n *of the space and the two numbers* r *and* s *in 13.4.2 are the same. In other words, for a given* E, *the action of* $\mathrm{GL}(E)$ *on* $Q(E)$ *has exactly* $(n + 1)(n + 2)/2$ *orbits. The pair* (r, s), *which depends only on* q *and not on the basis used to define it (13.4.2), is called the* signature *of* q, *and can be characterized as follows:* r *(resp.* s) *is the maximal dimension of a subspace* F *of* E *such that* $q|_F$ *is positive (resp. negative) definite (cf. 8.1.1; this means that* $q(x) > 0$ *(resp.* < 0) *for every* $x \in F \setminus 0$).

Proof. The only thing to show is that if q, q' are equivalent the numbers r and s are the same. This follows from the geometric characterization at the end of the statement, which we proceed to prove. Set

$$r' = \sup \{ \dim F \mid q|_F \text{ is positive definite} \},$$
$$s' = \sup \{ \dim F \mid q|_F \text{ is negative definite} \},$$

and let $\{e_i\}$ be a basis where

$$q = \sum_{i=1}^{r} x_i^2 - \sum_{i=r+1}^{r+s} x_i^2.$$

Clearly $r' \geq r$ and $s' \geq s$ because we can take $F = \mathbf{R}e_1 \oplus \cdots \oplus \mathbf{R}e_r$ and $F = \mathbf{R}e_{r+1} \oplus \cdots \oplus \mathbf{R}e_{r+s}$ in the definition of r' and s', respectively. For the converse, let F be such that $q|_F$ is positive definite, and set $G = \mathbf{R}e_{r+1} \oplus \cdots \oplus \mathbf{R}e_n$. By assumption, $F \cap G = 0$, so $\dim F + \dim G \leq n$, and $r' + (n - r) \leq n$, that is, $r' \leq r$. That $s' \leq s$ follows analogously. $\square$

13.4.8. GAUSS'S METHOD. This is a diagonalization procedure affording a constructive proof for 13.4.1; its philosophy is in some sense dual to that of 13.4.1, in that it involves linear forms. This procedure is very often used in practice, for explicit calculations (cf. 13.9.8); if you have never done so, you should take the time to work out a few examples (see 13.9.6 and 13.9.8). For an application to differential geometry, see [B–G, 146]. We work by induction, and explain the first step. Let $q = \sum_{i.j} a_{ij} x_i x_j$ be a quadratic form.

First case. There exists i such that $a_{ii} \neq 0$; we may as well assume it's $i = 1$. Set $\lambda_1 = a_{11} \neq 0$, and write

$$q = \lambda_1 x_1^2 + 2Ax_1 + B = \lambda_1 \left(x_1 + \frac{A}{\lambda_1} \right)^2 + \left(B - \frac{A^2}{\lambda_1} \right),$$

where A (resp. B) is a linear (resp. quadratic) form in $x_2, \ldots, x_n$. Now repeat the process for $B - \dfrac{A^2}{\lambda_1}$, a quadratic form in $n - 1$ variables.

Second case. All the a_{ii} are zero, but there exists a non-zero a_{ij} (otherwise $q = 0$!) Assume it's $\lambda = a_{12}$ that is non-zero. Write

$$q = \lambda x_1 x_2 + Ax_1 + Bx_2 + C = \lambda \left(x_1 + \frac{B}{\lambda} \right) \left(x_2 + \frac{A}{\lambda} \right) + \left(C - \frac{AB}{\lambda} \right),$$

where A and B are linear forms and C is a quadratic form, all involving only $x_3, \ldots, x_n$. Then set $u = x_1 + \dfrac{B}{\lambda}$, $v = x_2 + \dfrac{A}{\lambda}$, replace uv by

$$\frac{(u+v)^2 - (u-v)^2}{4},$$

thus diagonalizing the first two variables, and repeat the process for $C - \dfrac{AB}{\lambda}$, a quadratic form in $n - 2$ variables.

This calculation also gives the rank (cf. 13.9.5).

13.5. Simultaneous orthogonalization of two quadratic forms

13.5.1. Let q and q' be quadratic forms on E. Is there a basis $\{e_i\}$ orthogonal for both q and q'? In general, the answer is no, for example, $q = x^2 - y^2$ and $q' = 2xy$ on $E = \mathbf{R}^2$. Let's see what goes wrong: let $\{e_i\}$ be a basis orthogonal for q and q', and introduce the associated linear maps $\phi, \phi' : E \to E^*$ (cf. 13.2.0). If q is non-degenerate, we can consider the endomorphism $f = \phi^{-1} \circ \phi' : E \to E$. Now e_i is an eigenvector of f for every i, since $\left(\phi(e_i) \right)^{-1}(0)$ is, by assumption, the q-orthogonal hyperplane to e_i, and $\left(\phi'(e_i) \right)^{-1}(0)$ is the q'-orthogonal hyperplane to e_i, so $\phi'(e_i)$ must be a scalar multiple of $\phi(e_i)$ (cf. 2.4.8.2). But in the case of $q = x^2 - y^2$ and $q' = 2xy$, the matrix of $\phi^{-1} \circ \phi'$ (in the canonical basis) is

$$\begin{pmatrix} 0 & 1 \\ 1 & 0 \end{pmatrix} \begin{pmatrix} 1 & 0 \\ 0 & -1 \end{pmatrix} = \begin{pmatrix} 0 & -1 \\ 1 & 0 \end{pmatrix}$$

(cf. 13.2.0). This is the matrix of a rotation of $\mathbf{R}^2$, and rotations have no real eigenvalues.

That is the only obstruction:

13.5.2. PROPOSITION. *If q is non-degenerate and $\phi^{-1} \circ \phi'$ has n distinct eigenvalues, there exists a basis diagonalizing both q and q'.*

Proof. From the assumption, there exists a basis $\{e_i\}$ of eigenvalues of $\phi^{-1}\circ\phi'$. Denoting the eigenvalues by k_i, we have

$$\phi'(e_i)(e_j) = k_i\phi(e_i)(e_j) = P'(e_i, e_j) = k_i P(e_i, e_j);$$

but, since P and P' are symmetric,

$$P'(e_i, e_j) = k_j\phi(e_i)(e_j) = k_i\phi(e_i)(e_j),$$

which forces $P(e_i, e_j) = 0$ for $i \neq j$ since $k_i \neq k_j$. $\square$

13.5.3. EXAMPLE. If K is algebraically closed, two *generic* quadratic forms will be simultaneously diagonalizable; if $K = \mathbf{C}$, this happens with probability one.

13.5.4. NOTES

13.5.4.1. For a complete study of the problem, see [KG2].

13.5.4.2. Geometric interpretation. We shall see in 14.5.4.1 that the existence of an orthogonal basis common to q and q' is equivalent to the existence of a simplex self-polar with respect to the quadrics of $P(E)$ associated with both given quadratic forms. In dimension two, we shall classify all pairs of conics and thus meet again cases where simultaneous orthogonalization is impossible (see 16.4.10).

13.5.5. THEOREM. *Let E be a Euclidean space and q' an arbitrary quadratic form on E. There exists an orthonormal basis for E that diagonalizes q'.*

13.5.6. NOTES. This result is of considerable practical importance, since it is tantamount to the existence of axes for Euclidean conics and quadrics. In mechanics this provides an approach to stability problems, since it means that ellipsoids of inertia have axes (the so-called axes of inertia), and that every small motion admits symmetries (harmonic oscillator).

There exist extensions of this result to infinite-dimensional (Hilbert) spaces, assuring the existence, for instance, of orthogonal bases of eigenvalues of differential operators (see [WR, 254]). Such extensions are fundamental in analysis.

The history of theorem 13.5.5 is relevant, since the need for such a theorem arose from applications: see [BI2, 189–190].

There are practical methods to find orthogonal bases as stated, or, equivalently, to find the eigenvectors of a symmetric matrix; see [KE, 192].

13.5.7. PROOF OF 13.5.5. One classical proof uses the trick of extending E into a Hermitian space (see [DR, 58], for example). Since we haven't studied such spaces, we give two other proofs.

13.5.7.1. First proof. Set $f = \phi^{-1}\circ\phi'$, as in 13.5.1, where q is the Euclidean structure of E. By 7.4.3 there exists a subspace V of E such that $f(V) \subset V$

and $\dim V = 1$ or 2. If $\dim V = 1$, we have found an eigenvector x for f; if $\dim V = 2$, we shall find one in a minute.

Observe that the matrix of f is symmetric in any orthonormal basis, since that matrix is $A^{-1}A'$ (13.2.0), where A' is the matrix of q' and A is the matrix of q, which is the identity in an orthonormal basis. Thus the matrix of $f|_V$ will be of the form $\left(\begin{smallmatrix} a & b \\ b & c \end{smallmatrix}\right)$, and its eigenvalues, the roots of

$$\lambda^2 - (a+c)\lambda + ac - b^2 = 0,$$

must be real. This shows the existence of an eigenvector x.

Now take $e_1 = x/\|x\|$, and set $H = e_1^{\perp}$. We have $f(H) \subset H$ and $\dim H < \dim E$, and by induction there exists an orthonormal basis $\{e_i\}$ consisting of eigenvectors for f. Then

$$P'(e_i, e_j) = \phi'(e_i)(e_j) = k\phi(e_i)(e_j) = kP(e_i, e_j) = k(e_i \mid e_j) = 0$$

for all $i \neq j$, and $\{e_i\}$ is indeed orthonormal for q'. $\qquad\square$

13.5.7.2. A calculus proof. Consider the unit sphere S in E, and the function $f : E \setminus 0 \to \mathbf{R}$ given by

$$f(x) = \|x\|^{-2} q'(x).$$

Let x be such that $f(x) = \sup \{ f(x) \mid x \in S \}$. Such an x exists by continuity; in fact, f is even C^{∞} on $E \setminus 0$. Since $f(\lambda x) = f(x)$ for all $\lambda \neq 0$, the point x is a maximum for f over $E \setminus 0$, and we must have $f'(x) = 0$. But the derivative of f is

$$f'(x)(y) = 2\|x\|^{-4} \big(P'(x, y)\|x\|^2 - q'(x)(x \mid y) \big).$$

In particular, $P'(x, y) = 0$ for any y such that $(x \mid y) = 0$. We continue by induction. $\qquad\square$

13.5.8. The reader will find in 8.2.6 and exercise 8.12.1 a useful application for 13.5.5.

From now till the end of the chapter, (E, q) will be non-degenerate.

13.6. The group of a quadratic form. Generalities

We now extend to the general case a number of notions defined for Euclidean spaces.

13.6.1. DEFINITION. *The group of q, or orthogonal group, is the group $\{ f \in \mathrm{GL}(E) \mid f^*(q) = q \}$ of isometries of (E, q). This group is denoted by $O(E, q)$, $O(E)$ or $O(q)$. As special cases we set $O(n, K) = O\big(K^n, \sum_{i=1}^{n} x_i^2\big)$ and $O(r, s) = O\big(\mathbf{R}^{r+s}, x_1^2 + \cdots + x_r^2 - x_{r+1}^2 - \cdots - x_{r+s}^2\big)$. The group $O(1, 3)$ is called the Lorentz group.*

13.6.2. REMARKS. If (E, q) and (E', q') are isometric, their groups $O(q)$, $O(q')$ are isomorphic. The condition $f \in O(q)$ is equivalent, by 13.2.4, to ${}^t f = f^{-1}$; in particular, $\det f = \pm 1$. We can thus introduce the following

13.6.3. NOTATION. We set $O^+(E) = \{ f \in O(E) \mid \det f = 1 \}$; the elements of $O^+(E)$ are called *rotations*. In particular, $O^+(E)$ is normal in $O(E)$, and $O(E)/O^+(E) \cong \mathbf{Z}_2$. The multiplication rules stated in 8.2.3.2 apply.

13.6.4. MATRIX NOTATION. If $\{e_i\}$ is a basis for E, the map $f : L(E; E)$ has matrix S and $q \leftrightarrow A$ we deduce from 13.1.13.10 that

13.6.5 $$f \in O(q) \iff {}^t S A S = A.$$

For example, $f \in O(n, K)$ is equivalent to ${}^t S S = I$ or $S^{-1} = {}^t S$ (see 13.9.12). For $(E, q) = \mathrm{Art}_2$ we have

$$\begin{pmatrix} a & b \\ c & d \end{pmatrix} \in O(\mathrm{Art}_2) \iff \begin{pmatrix} a & b \\ c & d \end{pmatrix} \begin{pmatrix} 0 & 1 \\ 1 & 0 \end{pmatrix} \begin{pmatrix} a & c \\ b & d \end{pmatrix} = \begin{pmatrix} 0 & 1 \\ 1 & 0 \end{pmatrix}$$

$$\iff ac - bd = 0 \text{ and } ad + bc = 1$$

(see details in 13.8.2).

One always has $\pm \mathrm{Id}_E \in O(E)$, and $-\mathrm{Id}_E \in O^+(E)$ if and only if $\dim E$ is even.

13.6.6. INVOLUTIONS OF $O(q)$. In the non-Euclidean case, one must watch out for changes in 8.2.9, 8.2.10 and 8.2.11. Recall that any involution f of a vector space E is of the form $f|_S = \mathrm{Id}_S$, $f|_T = -\mathrm{Id}_T$, where $E = S \oplus T$ is a direct sum. This immediately yields the following

13.6.6.1. Proposition. *For an involution of $\mathrm{GL}(E)$ to be in $O(q)$ it is necessary and sufficient that $E = S \oplus^{\perp} T$, that is, we must have S non-singular and $T = S^{\perp}$. We say that f is a reflection through S (a hyperplane reflection if $\dim T = 1$). Every hyperplane reflection lies in $O^-(E)$, and every codimension-two reflection lies in $O^+(E)$.* $\square$

The non-singularity condition complicates a lot the generalizations of 8.2.12, 8.2.13, 8.2.11 and 8.2.7. For example, 8.2.11 becomes

13.6.6.2. Lemma. *Let x, y be vectors in E such that $q(x) = q(y)$ and $x - y$ is non-isotropic. There exists a hyperplane reflection h taking x to y.*

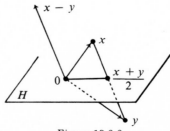

Figure 13.6.6

Proof. By 13.3.2, the subspace $H = (x - y)^\perp$ is a non-singular hyperplane; let h be the reflection through $S = H$. From $P(x - y, x + y) = q(x) - q(y) = 0$ we get $x + y \in H$, so the decomposition

$$x = \frac{x+y}{2} + \frac{x-y}{2}, \qquad y = \frac{x+y}{2} - \frac{x-y}{2}$$

shows that $h(x) = y$. $\qquad\square$

Thus the proof of 8.2.12 cannot be carried over without change. In order to prove an analog, we introduce some technical results.

13.6.7. EXAMPLES OF ROTATIONS. TECHNICAL LEMMAS FOR $O(\mathrm{Art}_{2p})$

13.6.7.1. Lemma. *Let $E = \mathrm{Art}_{2s}$, let $F \subset E$ be an s-dimensional totally singular subspace of E, and let $f \in O(E)$ satisfy $f(F) = F$. Then $f \in O^+(E)$.*

Proof. By 13.3.4 there exists a basis suitable for a (non-orthogonal) direct sum decomposition $E = F \oplus F'$ such that $q \leftrightarrow A$, with $A = \left(\begin{smallmatrix} 0 & I \\ I & 0 \end{smallmatrix}\right)$. The matrix of f in that basis is then of the form $\left(\begin{smallmatrix} U & 0 \\ V & W \end{smallmatrix}\right)$, since $f(F) = F$. Condition 13.6.5 translates as

$$\begin{pmatrix} {}^tU & 0 \\ {}^tV & {}^tW \end{pmatrix} \begin{pmatrix} 0 & I \\ I & 0 \end{pmatrix} \begin{pmatrix} U & V \\ 0 & W \end{pmatrix} = \begin{pmatrix} 0 & {}^tUW \\ {}^tWU & {}^tVW + {}^tWV \end{pmatrix} = \begin{pmatrix} 0 & I \\ I & 0 \end{pmatrix},$$

whence ${}^tUW = I$, and $\det f = (\det U)(\det W) = 1$. $\qquad\square$

13.6.7.2. Lemma. *Let F be a subspace such that $E = \overline{F}$ is a non-singular completion of F. If $f \in O(E)$ and $f|_F = \mathrm{Id}_F$, we have $f \in O^+(E)$.*

Proof. By 13.3.4.1, we have $E = G \oplus \mathrm{Art}_{2s}$, where $\mathrm{Art}_{2s} = M \oplus M'$, $F = G + M$ and $M = \mathrm{rad}(F)$. By assumption $f|_G = \mathrm{Id}_G$, so that $f(\mathrm{Art}_{2s}) = \mathrm{Art}_{2s}$. By 13.6.7.1 we have $f|_{\mathrm{Art}_{2s}} \in O^+(\mathrm{Art}_{2s})$, implying that

$$\det f = \det(f|_G)\det(f|_{\mathrm{Art}_{2s}}) = 1 \cdot 1 = 1. \qquad\square$$

13.6.7.3. Lemma. *Let $f \in O(E)$ be such that $f(x) - x$ is non-zero and isotropic, for every non-isotropic vector x. Then $f \in O^+(E)$ and E is an Artin space.*

The peculiar hypothesis in this lemma is explained as follows: when one tries to extend to the general case the proof of 8.2.12, one verifies that, by 13.6.6.2, induction works if and only if f does not satisfy the hypothesis of the lemma. This fact will be used in 13.7.12.

Proof. Assume first that $n = 2$. The space E contains non-zero isotropic vectors, so $E = \mathrm{Art}_2$ by 13.2.3.4. In 13.8.2 we shall see that the matrix of f can be cast into one of the forms $\left(\begin{smallmatrix} k & 0 \\ 0 & k^{-1} \end{smallmatrix}\right)$ or $\left(\begin{smallmatrix} 0 & k \\ k^{-1} & 0 \end{smallmatrix}\right)$, for $k \in K^*$; in the first case, $\det f = 1$, and in the second, $f(x + ky) = x + ky$, contradicting the assumption.

Now assume that $n \geq 3$. Observe that the subspaces $\mathrm{Ker}(f - \mathrm{Id}_E)$ and $\mathrm{Im}(f - \mathrm{Id}_E)$ are completely singular: for the kernel, this follows from the assumption; for the image, we know that $f(x) - x$ is isotropic for x non-isotropic, and we must show the same thing for x isotropic. Let x be isotropic, and apply 13.3.4.2; the subspace $x^\perp$ cannot be completely singular because $n - 1 \geq n/2$ (since $n \geq 3$), so there exists $y \in x^\perp$ such that $q(y) \neq 0$. Then

$$q(x \pm y) = q(y) \neq 0.$$

Applying the assumption of the lemma to the three vectors y, $x+y$ and $x-y$ gives

$$q\big(f(x) - x + f(y) - y\big) + q\big(f(x) - x - (f(y) - y)\big)$$
$$= 2q\big(f(x) - x\big) + 2q\big(f(y) - y\big) = 2q\big(f(x) - x\big) = 0.$$

Set $U = \mathrm{Im}(f - \mathrm{Id}_E)$; we have just seen that U is completely singular. A direct calculation (as in 9.3.1) shows that

$$\mathrm{Ker}(f - \mathrm{Id}_E) \perp \mathrm{Im}(f - \mathrm{Id}_E),$$

and since

$$\dim\big(\mathrm{Ker}(f - \mathrm{Id}_E)\big) + \dim\big(\mathrm{Im}(f - \mathrm{Id}_E)\big) = \dim E,$$

we get $U^\perp = \mathrm{Ker}(f - \mathrm{Id}_E)$. By 13.3.2 we must have $U = U^\perp$ and $\dim U = \dim E/2$, and 13.3.4.2 shows that $E = \mathrm{Art}_{2s}$. We conclude by noticing that $f(U) = U$, so 13.6.7.1 implies $f \in O^+(E)$.   □

13.6.7.4. This proof does not exhibit maps satisfying the condition of the lemma. We shall not need the existence of such maps, but the reader can consult [S–T, 238 ff.] or solve 13.9.15.

13.6.8. PLAN OF STUDY. In chapter 8 we have seen that the orthogonal group acts transitively on every grassmannian $G_{E.k}$ ($k = 0, 1, \ldots, \dim E$) (cf. 8.2.7), that $O^+(E)$ is abelian and isomorphic to S^1 when $\dim E = 2$ (cf. 8.3.3 and 8.3.6), that $O(E)$ is generated by hyperplane reflections and $O^+(E)$ by codimension-two reflections (cf. 8.2.12 and 8.4.6), that $O^+(E)$ (modded out by its quotient) is simple when $\dim E = 3$ or $\dim E \geq 5$. We also studied in 8.4.1 the structure of a given map $f \in O(E)$.

We shall broach only a few of these questions in the general case. For more complete surveys, the reader can consult [S–T], an elementary but very pleasant book, [BI2] for a very general discussion, [PO] and [AN] for Clifford algebras, and [DE1] for the question of simplicity of orthogonal groups.

The first difference is that $O(E)$ cannot act transitively on $G_{E.k}$, since two subspaces of same dimension are not always isometric. For example, $O(1, 1)$ (cf. 13.6.1) cannot take a line D where $q|_D$ is positive definite into a line D' where $q|_{D'}$ is negative definite. The amazing thing is that the condition that F and F' be isometric is sufficient for the existence of a map $f \in O(E)$ taking F into F'. This is the so-called Witt's theorem (13.7.1). An interesting case, involving orientation, is that of maximal completely singular subspaces.

The case $\dim E = 2$ is quite similar to the Euclidean case: $O^+(E)$ is always abelian (13.8.1), but for $E = \mathrm{Art}_2$ and K the reals, $O^+(E)$ is homeomorphic to the disjoint sum $\mathbf{R} \cup \mathbf{R}$ (cf. 13.8.3).

The group $O(E)$ is still generated by reflections through non-singular hyperplanes, but the proof is now much longer: see 13.7.12, the theorem of Cartan–Dieudonné.

The question of the simplicity of orthogonal groups $O(E)$ has been completely solved, but its solution is long: see [DE1].

13.6.9. NOTE. A more detailed study of these questions requires the introduction of Clifford algebras, objects that can be associated to any quadratic form, and that play a role similar to that of quaternions in chapter 8. See [PO], [AN] and [BI2].

13.7. Witt's theorem and the theorem of Cartan–Dieudonné

13.7.1. THEOREM (WITT, 1936). *Let F, F' be subspaces of E and $f : F \to F'$ an isometry between $(F, q|_F)$ and $(F', q|_{F'})$. There exists $\hat{f} \in O(q)$ such that $\hat{f}|_F = f$. In other words, the orbits of $G_{E.k}$ under the action of $O(q)$ are exactly the sets of isometric subspaces, where each subspace is considered with the form induced by q.*

Proof. We start by reducing the problem to the case where F and F' are non-singular. Let $\overline{F}, \overline{F}'$ be the non-singular completions of F, F'; by 13.3.4.1, we have

$$\overline{F} = G \overset{\perp}{\oplus} P_1 \overset{\perp}{\oplus} \cdots \overset{\perp}{\oplus} P_s \qquad \text{and} \qquad \overline{F}' = G' \overset{\perp}{\oplus} P'_1 \overset{\perp}{\oplus} \cdots \overset{\perp}{\oplus} P'_s,$$

where (e_i, h_i) (resp. (e'_i, h'_i)) is a basis for P_i (resp. P'_i) and $e'_i = f(e_i)$ $(1 \le i \le s)$. We can extend f into a map $\overline{f} : \overline{F} \to \overline{F}'$ by setting $\overline{f}(h_i) = h'_i$, and $\overline{f}$ is still an isometry. This completes the reduction, and we assume from now on that F and F' are non-singular.

Consider first the case $\dim F = \dim F' = 1$. Take $x \in F$, $y = f(x) \in F'$ and $x \ne 0$. By assumption, $q(x) = q(y)$. Either $x + y$ or $x - y$ is non-isotropic, otherwise we'd have $q(x+y) + q(x-y) = 2q(x) + 2q(y) = 4q(y) = 0$, contradiction. Let g be the reflection through $(x+y)^\perp$ or $(x-y)^\perp$ (whichever is a hyperplane); g exists by 13.6.2. But then $g(x) = \pm x$, so $\pm g \in O(q)$ and we have found an isometry taking F onto F'.

When $\dim F$ is arbitrary, we work by induction on $\dim F$. Since F is non-singular, we can write $F = F_1 \oplus^\perp F_2$, where $\dim F_1, \dim F_2 < \dim F$ (for example, take $F_1 = K \cdot x$, where x is non-isotropic). Set $F'_1 = f(F_1)$; by the induction hypothesis, there exists $g \in O(q)$ such that $g|_{F_1} = f$. Now consider $F_1'^\perp$; this space contains $f(F_2)$ and $g(F_2)$, and $f \circ g^{-1} : g(F_2) \to f(F_2)$ is an isometry. Again by the induction hypothesis, there exists $h \in O(F_1'^\perp, q|_{F_1'^\perp})$ such that $h|_{g(F_2)} = f \circ g^{-1}$. We now put everything together by setting $\hat{f} = (\mathrm{Id}_{F'_1} \oplus h) \circ g$. $\qquad\square$

13.7.2. NOTE. Artin once said that Witt's theorem was a scandal, referring to the fact that one had to wait until 1936 to formulate and prove a theorem at once so simple in its statement and underlying concepts, and so useful in divers domains, of which we give two examples:
— geometry (see, for example, corollary 19.4.6.2), and
— arithmetic, since Witt's theorem occurs naturally in the classification of quadratic forms over $\mathbf{Q}$ (cf. [SE2, chapter IV]).

13.7.3. COROLLARY. *If two subspaces F, F' of E are isometric, so are their complements $F^\perp$ and $F'^\perp$.* □

This corollary may seem surprising, and the reader may find it interesting to try to prove it directly, as a means of familiarizing himself with quadratic forms.

13.7.4. EXAMPLES. If $K = \mathbf{C}$, the orbits of $G_{E,k}$ under the action of $O(q)$ are the sets of k-dimensional subspaces restricted to which g has a fixed rank (cf. 13.4.6). In particular, $O(q)$ acts transitively on the set of non-singular k-dimensional subspaces, for any fixed k.

For $K = \mathbf{R}$ the first remark applies, substituting "signature" for "rank" (cf. 13.4.7).

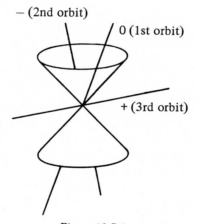

Figure 13.7.4

More specifically (figure 13.7.4), if $E = \mathbf{R}^3$ and $q = x^2 + y^2 - z^2$, we have three orbits in $G_{E,1} = P(E)$, corresponding to the signatures $(0,0)$, $(1,0)$ and $(0,1)$; the first orbit consists of the lines on the isotropic cone, the second of lines lying strictly outside the cone and the third of those strictly inside.

For $O(1,1)$, see 13.8; for $O(n-1,1)$, see 18.10, 19.2 and 20.6.

13.7.5. COROLLARY AND DEFINITION. *All maximal completely singular subspaces have the same dimension, and are taken into one another by the action of $O(q)$. The dimension of a maximal completely singular subspace is called the index of q.*

Proof. Let F, F' be maximal completely singular subspaces. If $\dim F > \dim F'$, write $F = F_1 \oplus^{\perp} F_2$ with $\dim F_1 = \dim F'$, and take $f \in \mathrm{Isom}(F_1, F')$ (any linear map will do!) Extend f into $\hat{f} \in O(q)$; then $\hat{f}(F)$ is a completely singular subspace and $\hat{f}(F)$ properly contains F', contradicting our assumption.   □

13.7.6. EXAMPLES. If $K = \mathbf{C}$ and q is non-degenerate, we have $\mathrm{index}(q) = [n/2]$. In fact, q is equivalent to $\sum_{i=1}^{n} x_i^2$ (cf. 13.4.6), and the subspace F constructed in 13.2.3.3 is maximal.

If $K = \mathbf{R}$ and q is non-degenerate and has signature $\{r, s\}$, we have

$$\mathrm{index}(q) = \inf\{r, s\}.$$

To see this, apply 13.4.7 and remark that the subpace F constructed in 13.2.3.3 is maximal.

13.7.7. REMARKS. Witt's theorem can lead to unexpected results. For example, if F is completely singular, any $f \in \mathrm{GL}(F)$ is an isometry of $(F, q|_F)$, and thus can be extended to $\hat{f} \in O(q)$. In particular, $O(q)$ acts transitively on the *points* of the isotropic cone, and not only on the lines lying in it (see also figure 13.8.4).

It is reasonable to ask whether Witt's theorem subsists when $O(q)$ is replaced by $O^+(q)$. The answer is almost always yes; but an interesting exception, that will lead to new geometric phenomena (14.4), occurs when $E = \mathrm{Art}_{2s}$ and F, F' are completely singular subspaces.

13.7.8. THEOREM (WITT—SHARPENED VERSION) *Theorem 13.7.1 is still valid with $O^+(q)$ instead of $O(q)$, as long as we have*

$$\dim F + \dim(\mathrm{rad}(F)) < \dim E.$$

Otherwise it is false: if F satisfies

$$\dim F + \dim(\mathrm{rad}(F)) = \dim E$$

and $f \in O^-(q)$, there exists no $g \in O^+(q)$ such that $g|_F = f|_F$.

Proof. First consider F such that $\dim F + \dim(\mathrm{rad}(F)) < \dim E$, and form the non-singular completion $\overline{F}$ of F (cf. 13.3.4.1). Given $f \in \mathrm{Isom}(F; F')$, take $\hat{f} \in O(q)$ such that $\hat{f}|_F = f$; this exists from 13.7.1. If $\hat{f} \in O^+(q)$, we're done; otherwise, let g be the reflection through the hyperplane $x^{\perp}$, where $x \in F^{\perp}$ is not isotropic. Then $g|_F = \mathrm{Id}_F$, and, since $g \in O^-(q)$, the composition $f \circ g$ lies in $O^+(q)$, as desired.

Now let $\dim F + \dim(\mathrm{rad}(F)) = \dim E$ and $f \in O^-(q)$. If there were $g \in O^+(q)$ such that $g|_F = f|_F$, the composition $g^{-1} \circ f$ would contradict 13.6.7.2.   □

The result above doesn't yield the orbits of $G_{E.k}$ under $O^+(q)$ yet; for that we need a bit more work:

13.7.9. THEOREM. *The orbits of $G_{E.k}$ under the action of $O^+(q)$ are the same as those under $O(q)$, with the following exception: if $E = \mathrm{Art}_{2_s}$ and $k = s$, the subset of $G_{E.s}$ consisting of completely singular subspaces is formed of two orbits under $O^+(q)$ (instead of one).*

Proof. Assume first that we are not in the exceptional case; then, by 13.3.4.2, there exists a non-isotropic vector $x \in F^\perp$. Take $f \in \mathrm{Isom}(F; F')$, and locate $\hat{f} \in O(q)$ such that $\hat{f}|_F = f$, using 13.7.1. If $\hat{f} \in O^+(q)$, we're done; otherwise compose $\hat{f}$ with the reflection h through the hyperplane $x^\perp$, and observe that $\hat{f} \circ h \in O^+(q)$ and $(\hat{f} \circ h)(F) = \hat{f}(h(F)) = \hat{f}(F) = F'$.

On the other hand, if $E = \mathrm{Art}_{2_s}$ and F is completely singular, the existence of two orbits is derived from 13.6.7.1.   □

13.7.10. EXAMPLES. The first case to consider is when $E = \mathrm{Art}_2$; by 13.2.3.1, the maximal completely singular subspaces are the isotropic lines, denoted by I and J. It is obvious that $I \cup J = q^{-1}(0)$ is invariant under $O(q)$; but by 13.7.9 we have the following characterization for $O^+(q)$ (resp. $O^-(q)$):

$$f \in O^+(q) \quad (\text{resp. } O^-(q)) \quad \text{if and only if} \quad f(I) = I \quad (\text{resp. } f(I) = J),$$

that is, orientation-preserving isometries leave each isotropic line invariant, and orientation-reversing ones switch the two. Compare with 8.8.6.1, and see also section 13.8.

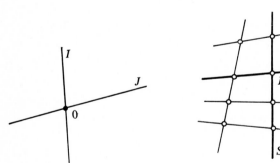
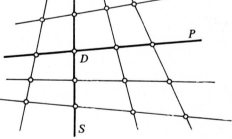

Figure 13.7.10   Figure 13.7.11.0

Our second example is $E = \mathrm{Art}_4$, where an interesting phenomenon occurs, expressed in the corollary below. Interpreted projectively, this is a result about the two systems of so-called *generating lines* of the quadric $p(q^{-1}(0)) \subset P(E)$ (cf. 14.4.1). Figure 13.7.11.0 shows the situation in $P(E)$.

13.7.11. COROLLARY. *Let Γ be the set of completely singular planes of $E = \mathrm{Art}_4$, and let Π and Σ be the two orbits of Γ under the action of $O^+(q)$. Then:*

i) *any isotropic (i.e., completely singular) line D of E is contained in a unique plane $P \in \Pi$ and in a unique $S \in \Sigma$;*

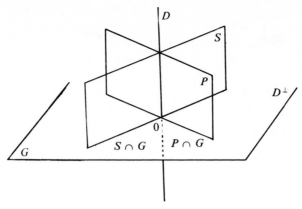

Figure 13.7.11.1

ii) *for any $P \in \Pi$ and any $S \in \Sigma$, the intersection $P \cap S$ has dimension one (and so must be an isotropic line);*

iii) *for any planes P and $P' \in \Pi$, either $P \cap P' = 0$ or $P = P'$; the same for planes in Σ.*

13.7.11.1. Proof using 13.7.9. Every isotropic line D is contained in some $P \in \Pi$ and some $S \in \Sigma$, by an application of 13.7.8 to $F = D$ and an arbitrary isotropic line F' contained in a given plane $P_0 \in \Pi$ or $S_0 \in \Sigma$. To prove uniqueness, write $D^\perp$ as a direct sum $D^\perp = D \oplus^\perp G$. The plane G contains isotropic vectors, for example, $G \cap P$, for P as above; but then 13.2.3.4 implies that G is an Artinian plane, and so has only two isotropic lines, which must be $P \cap G$ and $S \cap G$.

Consider two lines $P \in \Pi$ and $S \in \Sigma$, and assume that $P \cap S = 0$. Take $x \in P \setminus 0$; since $\dim x^\perp = 3$, there exists a non-zero $y \in S \cap x^\perp$. The plane $T = K \cdot x + K \cdot y$ is completely singular; since $P \cap T \neq \emptyset$, part (i) implies that $T \in \Sigma$, and, similarly $T \in \Pi$, which is absurd.

Statement (iii) follows from (i). $\qquad\square$

13.7.11.2. Elementary proof. By definition, $q = 2xy + 2zt$, where the coordinates in E are denoted by x, y, z and t. There are two sets of completely singular planes: one of the form $\lambda x = -\mu z$, $\mu y = \lambda t$, and another of the form $\lambda x = -\mu t$, $\mu y = \lambda z$, where $(\lambda, \mu) \neq (0,0)$. Properties (i)–(iii) are easily checked. $\qquad\square$

13.7.12. THEOREM (CARTAN–DIEUDONNÉ). *Every isometry $f \in O(q)$ is the product of at most $n = \dim E$ hyperplane reflections.*

Proof. We adapt the proof of 8.12.12 and use induction on n. If there exists a non-isotropic vector x such that $f(x) = x$, or if there exists x such that $f(x) - x$ is non-isotropic, we pass from $n - 1$ to n by using reflection through the hyperplane $x^\perp$ (cf. 13.6.6.1) or mimicking 13.6.6.2.

The remaining case is when $f(x) - x$ is isotropic for every non-isotropic x. By 13.6.7.3, this implies that $f \in O^+(q)$ and $E = \mathrm{Art}_{2s}$; in particular,

$n = \dim E$ is even. Let $t \in O^-(q)$ be an arbitrary hyperplane reflection; since $t \circ f \in O^+(q)$, we apply the induction hypothesis to write $t \circ f = t_1 \cdots t_k$ for some $k \leq n$. But, in fact, k must be strictly less than n, for the determinant of $f \in O^+(q)$ is $1 = (-1)^{k+1}$, and n is even. Thus the composition $f = tt_1 \cdots t_k$ has at most n terms. $\square$

13.7.13. COROLLARY. *If* $\dim E = 2$ *the set* $O^-(q)$ *consists of all reflections through non-isotropic lines.* $\square$

13.7.14. NOTES.

13.7.14.1. It is possible to extend 8.4.6: see 13.9.14 or [S–T, 332].

13.7.14.2. It is natural to ask whether $O^+(q)$ (modulo its center) is still simple; see [DE1]. The commutator subgroup and the center of $O^+(q)$ are discussed in [DE1] and [S–T], as are generalizations of the notion of similarity.

13.7.14.3. The appropriate generalizations of the results in 8.10 are valid for the groups $O(n, \mathbf{C})$ and $O(r, s)$ (cf. 13.6.1). This is no big deal once we show that that $O(n, \mathbf{C})$ is homeomorphic to the product

$$O(n) \times \mathbf{R}^{n(n-1)/2}$$

and $O(r, s)$ to the product

$$O(r) \times O(s) \times \mathbf{R}^{rs}.$$

These homeomorphisms can be obtained in an elementary way (see [CY, 16] for the first); but they are particular cases of a general theorem that says that every Lie group is homeomorphic to the product of a compact Lie group and a vector space: see [HN, 270 and notes on page 279] for this general result. The elementary case of $O(1, 1)$ will be studied in 13.8.

13.7.14.4. By analogy with 8.4.5 and 8.4.6, one can also try to pin down the smallest number of reflections (or codimension-two reflections) needed to make up an arbitrary $f \in O(q)$; see [S–T, 255 and 260] and [AN, 186]. As mentioned in 13.6.9, a detailed study of the groups $O(q)$ requires the use of Clifford algebras, which are, in any case, fundamental objects in contemporary mathematics, occurring in algebraic topology, K-theory and differential operators. For a general discussion of Clifford algebras, see [PO], [AN] and [BI2]; for recent applications, see [A–B–S] and [HU, 144 ff.]

13.8. Isometries of Artinian planes

13.8.1. PROPOSITION. *Let* q *be a non-degenerate quadratic form on a two-dimensional space* E. *Then:*

  i) $O^+(E)$ *is abelian;*

  ii) $O^-(E)$ *consists of reflections through non-isotropic lines; and*

  iii) *for every* $f \in O^+(E)$ *and* $g \in O^-(E)$ *we have* $gfg^{-1} = f^{-1}$.

Proof. Statement (ii) is just 13.7.13. Take $f \in O^+(E)$ and $g \in O^-(E)$; since $gf \in O^-(E)$, we can write $f = gg'$, with $g, g' \in O^-(E)$. By (ii) we have $g^2 = g'^2 = \mathrm{Id}_E$; then

$$gfg^{-1} = ggg'g^{-1} = g'g = g'^{-1}g^{-1} = (gg')^{-1} = f^{-1}.$$

Finally, take f and $f' \in O^+(E)$, and an arbitrary $g \in O^-(E)$. By (iii) we have

$$ff' = (gf^{-1}g^{-1})(gf'^{-1}g^{-1}) = g(f^{-1}f'^{-1})g^{-1} = g(f'f)^{-1}g^{-1} = f'f. \qquad \square$$

13.8.2. PROPOSITION. *Let $E = \mathrm{Art}_2$ and take a basis of E such that the matrix of q is $\begin{pmatrix} 0 & 1 \\ 1 & 0 \end{pmatrix}$ (cf. 13.1.3.8). If $f \in O(q)$, the matrix A of f in that basis is of the form $\begin{pmatrix} k & 0 \\ 0 & k^{-1} \end{pmatrix}$, for some $k \in K^*$, if and only if $f \in O^+(E)$, and of the form $\begin{pmatrix} 0 & k^{-1} \\ k & 0 \end{pmatrix}$ if and only if $f \in O^-(E)$. In particular, $O^+(q)$ is isomorphic to K^*.*

13.8.2.1. Notice that this isomorphism depends on the basis; to eliminate the ambiguity it is enough to fix one of the isotropic lines, or, equivalently in the case $K = \mathbf{R}$, to orient E. The reason for the equivalence is that a pair $\{x, y\}$, with $x \in I$ and $y \in J$, is a positively oriented basis if and only if the sector between $\mathbf{R} \cdot x$ and $\mathbf{R} \cdot y$ contains half-lines on which $q > 0$ (cf. 8.7.5.4).

Proof. Use 13.7.10: if $f \in O^+(q)$, we have $f(I) = I$ and $f(J) = J$, whence $A = \begin{pmatrix} k & 0 \\ 0 & h \end{pmatrix}$; but then $h = k^{-1}$, since $\det f = 1$. Similarly, if $f \in O^-(q)$, we have $A = \begin{pmatrix} 0 & h \\ k & 0 \end{pmatrix}$ and $h = k^{-1}$. It is obvious that matrices of the given forms satisfy 13.6.5. $\qquad \square$

We shall now find out how 8.6 and 8.7 work out for Art_2, under the additional assumption $K = \mathbf{R}$; in other words, we'll be studying the group $O(1, 1)$ (up to isomorphism—cf. 13.6.1). For a study of $O(q)$ in dimension two, without the assumption $K = \mathbf{R}$, see [S–T, 271–310] or [BI2, 160–175].

$$\boxed{\text{From now on we set } (E, q) = \mathrm{Art}_2 \text{ and } K = \mathbf{R}.}$$

13.8.3. THE TOPOLOGY OF $O(E)$. ORTHOCHRONOUS ROTATIONS. It is obvious that the maps

$$\mathbf{R}^* \ni k \mapsto \begin{pmatrix} k & 0 \\ 0 & k^{-1} \end{pmatrix} \in \mathbf{R}^4 \quad \text{and} \quad \mathbf{R}^* \ni k \mapsto \begin{pmatrix} 0 & k \\ k^{-1} & 0 \end{pmatrix} \in \mathbf{R}^4,$$

introduced in 13.8.2, are homeomorphisms onto their images. This implies that $O(E)$ has four connected components, each homeomorphic to $\mathbf{R}^*_+$. We denote by $O^{++}(E)$ the subgroup of $O(E)$ formed by the connected component of the identity. Its elements, called *orthochronous rotations*, are the transformations whose matrix, expressed in a basis as in 13.8.2, is of the form $\begin{pmatrix} k & 0 \\ 0 & k^{-1} \end{pmatrix}$ for $k > 0$.

13.8.4. ORBITS. Since q can take both strictly positive and strictly negative values, it is clear that $O(E)$ does not act transitively on $G_{E.1}$ (cf. 13.6.8); on the other hand, we know that $O^+(E)$ acts transitively on the set of lines where $q > 0$ and on the set of those where $q < 0$ (this comes from 13.7.9, or can be seen directly in this simple particular case). Figure 13.8.4 represents the orbits of the action of $O^{++}(E)$ on E; the arrows indicate the direction of flow when the number k above runs from 0 to ∞ (after we have fixed a basis; see 13.2.3.1 or 13.8.5). The fact that almost all those orbits are hyperbolas gave rise to the infelicitous name "hyperbolic plane" for the Artinian plane (cf. 13.1.4.4); but it also justifies the nomenclature "hyperbolic geometry" for the situation we shall study in chapter 19 (cf. also 13.8.9), which derives from a generalization of this same structure.

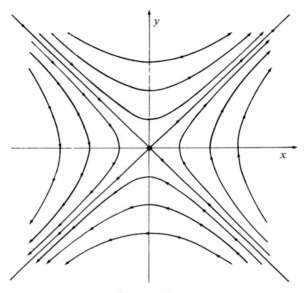

Figure 13.8.4

It is clear from the figure, and also from an elementary calculation, that the action of $O^{++}(E)$ on $G_{E.1}$ has four orbits: the two isotropic lines, the set of lines where $q > 0$ and the set of lines where $q < 0$. The action of $O^{++}(E)$ on half-lines has eight orbits, the four isotropic half-lines, and the four open quadrants determined by them. Further, $O^+(E)$ acts simply transitively on the set of half-lines where $q > 0$ (say), and $O^{++}(E)$ acts simply transitively on the set of half-lines of a fixed quadrant, and on the set of lines where $q > 0$.

13.8.5. ANGLES. We can thus recast 8.7 and establish a theory of oriented angles (in E) between half-lines or lines, for the orthochronous group $O^{++}(E)$; we leave this task to the reader. But the analogue of 8.3 deserves detailed attention, since it will be used in chapter 19. We want to measure angles in E between lines where $q > 0$; in other words, we want to find a group homomorphism $\mathbf{R} \to O^{++}(E)$ (where $\mathbf{R}$ is considered with its additive group structure). Things are more straightforward here than in 8.3: once we fix a basis for E in which the matrix of q is $\left(\begin{smallmatrix} 0 & 1 \\ 1 & 0 \end{smallmatrix}\right)$, or, equivalently, once we choose an ordering for the isotropic lines, a map $f \in O^{++}(E)$ is well-determined by the number $k \in \mathbf{R}_+^*$ appearing in the matrix of f. Thus the desired homomorphism goes from $\mathbf{R}$ to $\mathbf{R}_+^*$, and, naturally enough, should be the exponential map $t \mapsto e^t$. Here, contrary to 8.3.7, the angle map is bijective.

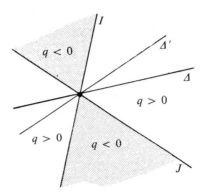

Figure 13.8.5

If E is oriented, we define the *measure* of the angle $\widehat{\Delta\Delta'}$ between two lines Δ and Δ' where $q > 0$ to be the unique real number t such that $\Delta' = f(\Delta)$, where f is the map with matrix $\left(\begin{smallmatrix} e^t & 0 \\ 0 & e^{-t} \end{smallmatrix}\right)$. The interesting thing is that this measure can be calculated in a purely projective setting, using the cross-ratio:

13.8.6. PROPOSITION. *Let I and J be the isotropic lines, in an order compatible with the orientation of E. For any pair of lines Δ, Δ' on which $q > 0$, the measure of $\widehat{\Delta\Delta'}$ is equal to $\frac{1}{2}\log\left([\Delta, \Delta', J, I]\right)$, where the cross-ratio is taken on the projective line $P(E)$.*

Proof. Like Laguerre's formula (8.8.7.2), this is immediate, following directly from the notation $\left(\begin{smallmatrix} k & 0 \\ 0 & k^{-1} \end{smallmatrix}\right)$ and 6.6.3. $\qquad\square$

13.8.7. HYPERBOLIC DISTANCE AND NON-ORIENTED ANGLES. What is the Artinian analogue of the non-oriented angles introduced in 8.6? Following 8.8.7.4, we set, for two lines Δ and Δ' where $q > 0$,

13.8.8 $$\overline{\Delta\Delta'} = \frac{1}{2}\left|\log\left([\Delta, \Delta', J, I]\right)\right|.$$

By 6.3.1, this definition does not depend on the choice of the order of the isotropic lines. We always have $\overline{\Delta\Delta'} = |\widehat{\Delta\Delta'}|$, and the reader can check that $\overline{}$ defines a metric on the set of lines where $q > 0$; furthermore,

$$\overline{\Delta\Delta'} + \overline{\Delta'\Delta''} = \overline{\Delta\Delta''}$$

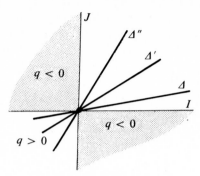

Figure 13.8.8

if and only if Δ' lies between Δ and Δ'' in the sense of 8.7.5.2 (see 6.8.1, and compare with 8.7.5.3).

Here's a simpler way to define $\overline{\Delta\Delta'}$, by analogy with 8.6.3, and using hyperbolic trigonometry (for the definition of Arccosh, see 0.5):

13.8.9. PROPOSITION. *Let Δ and Δ' be lines where $q > 0$. For $\xi \in \Delta \setminus 0$ and $\xi' \in \Delta' \setminus 0$, the number*

$$d = \frac{P(\xi, \xi')}{\sqrt{q(\xi)}\sqrt{q(\xi')}}$$

is ≥ 1 and satisfies $\overline{\Delta\Delta'} = $ Arccosh d.

Proof. Choose a basis in which $q = 2xy$, and let $\xi = (a, b)$ and $\xi' = (a', b')$ in this basis. Then

$$\frac{P(\xi, \xi')}{\sqrt{q(\xi)}\sqrt{q(\xi')}} = \frac{ab' + a'b}{2\sqrt{aa'bb'}}$$

is clearly ≥ 1. If $\Delta' = f(\Delta)$, the matrix of f being $\left(\begin{smallmatrix} e^t & 0 \\ 0 & e^{-t} \end{smallmatrix}\right)$, we have $a' = e^t a$, $b' = e^t b$, whence

$$\frac{P(\xi, \xi')}{\sqrt{q(\xi)}\sqrt{q(\xi')}} = \frac{e^t + e^{-t}}{2} = \cosh t. \qquad \square$$

13.9. Exercises

13.9.1. Let q be a quadratic form on E, and $\mathrm{rad}(q)$ its radical. Show that q defines a non-degenerate quadratic form on the vector space $E/\mathrm{rad}(q)$.

13.9.2. If $\{\phi_i\}$ is a finite family of vectors in E^*, and $k_i \in K^*$ are corresponding scalars, study the rank and the radical of the quadratic form $\sum_i k_i \phi_i^2$.

13.9.3. Let F be a subspace of E. Show that every non-singular space V such that $V \supset F$ and $\dim V = \dim F + \dim\left(\mathrm{rad}(F)\right)$ is of the form stated in 13.3.4.1.

* **13.9.4.** Reduce the classification of quadratic forms to that of anisotropic forms.

13.9.5. Show that the algorithm in 13.4.8 always yields linearly independent linear forms, and that the number of such forms is the rank of q.

13.9.6. Diagonalize $q = x_1 x_2 + x_2 x_3 + \cdots + x_{n-1} x_n \pm x_n x_1$ and discuss its rank.

13.9.7. Decide whether 13.4.8 can be reduced to the proof of 13.4.1, and show how to do it.

13.9.8. Diagonalize the following quadratic forms on $\mathbf{R}$ (where λ and μ intervene, study how the diagonalization depends on their values):

$$x^2 + y^2 + 2(z^2 + t^2) + xz + xt + zt;$$
$$\lambda t^2 + yx + zx + xy + \mu(x + y + z)t;$$
$$4x^2 + 3y^2 + 9z^2 + 8xz + 4xy + 4yt + 8zt + \lambda t^2.$$

Write down their polar forms.

$*$ **13.9.9.** Show that if $K = \mathbf{Q}$ and $n = \dim E = 1$ there are infinitely many non-isometric forms (E, q).

$*$ **13.9.10.** Show that if $n = 1$ and K is finite, there are exactly three classes of quadratic forms.

13.9.11. Show that if $K = \mathbf{R}$, $n \geq 3$ and the two quadratic forms q and q' satisfy $q(x) + q'(x) > 0$ for every $x \neq 0$, there is a basis that diagonalizes q and q' simultaneously.

13.9.12. Write down the conditions that characterize the 16 elements of a matrix in $O(3, 1)$.

13.9.13. Study the center of $O(q)$ and the center of $O^+(q)$.

13.9.14. Show that $O^+(q)$ is generated by codimension-two reflections.

$*$ **13.9.15.** Show that if $f \in O(E)$ satisfies the assumptions of 13.7.6.3 we necessarily have $E = \mathrm{Art}_{4p}$. Show, conversely, that for $E = \mathrm{Art}_{4p}$ there are isometries satisfying those assumptions.

$*$ **13.9.16.** Show that $O(E)$ is never commutative unless $E = \mathrm{Art}_2$ over the field with three elements.

Chapter 14
Projective quadrics

Projective quadrics arose, historically speaking, as the projective completion of affine quadrics, the first examples having been the Euclidean conics. This order is reversed here: we define projective quadrics first, as subsets of a projective space associated with quadratic forms on the parent vector space.

In a certain sense, the theory of projective quadrics is derived from the theory of quadratic forms by means of a dictionary, whose entries will be discussed as the chapter unfolds. Although this chapter and the next present little difficulty, they are indispensable: a dictionary is not enough for someone to speak a language fluently. Already a result like 14.5.4.3 requires some familiarity. In chapter 16, we shall use the tools developed here to prove even more intricate results.

The idea behind the definitions in section 14.1 is that any notion involving a vector space and a quadratic form can be carried over to the corresponding projective space if it is invariant under multiplication by non-zero scalars. Section 14.2 uses the fact that the set of quadrics in a projective space is itself a projective space; in particular, the notion of a pencil of quadrics will be fundamental in chapter 16. Section 14.3 studies the physical appearance of quadrics in real and complex projective spaces, along the lines of section 4.3, where such spaces themselves are studied.

In section 14.4 we discuss the three-dimensional case and the so-called ruled quadrics, which can be geometrically defined in a very simple way. Sections 14.5 and 14.6 are devoted to the duality relation given by polarity with respect to a quadric. This notion, fundamental in applications, has already been encountered in 11.1.5.

Conics are studied in greater detail in chapter 16.

In this whole chapter E denotes a vector space of finite dimension $n + 1$ over a commutative field K of characteristic different from 2. The projective space associated with E is denoted by $P(E)$, and the canonical projection by $p : E \setminus 0 \to P(E)$. The vector space of quadratic forms over E is denoted by $Q(E)$; if q is a quadratic form on E, its polar form will be written P. We assume that $\dim P(E) = n \geq 1$.

14.1. Definitions and examples

Take $q \in Q(E)$. The projection $p\big(q^{-1}(0) \setminus 0\big)$ of the isotropic cone of q (minus the origin) into $P(E)$ does not change when we replace q by kq, where $k \in K^*$; this means that such a projection is actually associated to a point of the the the projective space $PQ(E) = P\big(Q(E)\big)$. The rank, index and radical of a quadratic form also do not change when we multiply it by k (cf. 13.2.1 and 13.7.5). This leads to the following

14.1.1. DEFINITION. *A (projective) quadric in $P(E)$ is an element $\alpha \in PQ(E)$. The image of α, denoted by $\mathrm{im}(\alpha)$, is the set*

$$\mathrm{im}(\alpha) = p\big(q^{-1}(0) \setminus 0\big),$$

where $p(q) = \alpha$. The projective space of quadrics in $P(E)$ is $PQ(E)$. A quadric is called a conic *when $n = 2$. If $p(q) = \alpha$, we say that q is an* equation for α, *or for $\mathrm{im}(\alpha)$. The quadric α is called* proper *if it has a non-degenerate equation; otherwise it is called* degenerate. *The rank and index of α are the rank and index of any of its equations.*

14.1.2. NOTATION. We will not spell out each time that q is an equation for α; we assume that q, q', and so on stand for equations for α, α', and so on.

The rank of a quadric is always ≥ 1 because the case $q = 0$ is excluded.

14.1.3. EXAMPLES

14.1.3.1. If q is anisotropic, $\mathrm{im}(\alpha) = \emptyset$, as in the case of a Euclidean space with the form $q = \|\cdot\|$. This, however, does not mean we shouldn't discuss α, and especially its polar form P (cf. 14.5.2.0). On the other hand, if K is algebraically closed, we have $\mathrm{im}(\alpha) \neq \emptyset$ for any $\alpha \in PQ(E)$, as can be seen from 13.7.6, for example.

14.1.3.2. Proposition. *Let α be a quadric with $n = 1$. If $\mathrm{rank}(\alpha) = 1$ the image of α has exactly one point. If $\mathrm{rank}(\alpha) = 2$ the image of α is empty or contains exactly two distinct points. If $\mathrm{im}(\alpha) \neq \emptyset$ and $\alpha' \in P(E)$ is another quadric satisfying $\mathrm{im}(\alpha) = \mathrm{im}(\alpha')$, we have $\alpha = \alpha'$.*

Proof. It suffices to apply 13.2.3.4 and 13.2.3.1 (figure 14.1.3). $\qquad\square$

Figure 14.1.3

14.1.3.3. Heredity. Let $\alpha \in P(E)$ be a quadric and $S = p(F)$ a projective subspace of $P(E)$. If F is not completely singular under q, we can define the *intersection* $\alpha \cap S$ as the quadric having the restriction $q|_F$ as equation. And of course

$$\mathrm{im}(\alpha \cap S) = \mathrm{im}(\alpha) \cap S.$$

On the other hand, if F is completely singular, we have $S \subset \mathrm{im}(\alpha)$, and conversely.

14.1.3.4. Intersection with a line. We now combine 14.1.3.2 and 14.1.3.3. Take $\alpha \in PQ(E)$ and let $D = p(F)$ be a line in $P(E)$. Then $D \subset \mathrm{im}(\alpha)$ if and only if F is completely singular; $D \cap \mathrm{im}(\alpha)$ contains one point if and only if F is singular but not completely singular; otherwise $D \cap \mathrm{im}(\alpha)$ is empty or contains two points. In particular, if $D \cap \mathrm{im}(\alpha)$ has more than two points we must have $D \subset \mathrm{im}(\alpha)$.

14.1.3.5. Definition. *A subspace $S = p(F)$ of $P(E)$ is said to be tangent to a quadric α if F is singular under q; we also say that S is tangent to α at m, for every $m \in \mathrm{im}(\alpha) \cap S$. If α is a proper quadric, the tangent hyperplane to α at $m \in \mathrm{im}(\alpha)$ is the projective hyperplane $p(x^{\perp})$, for some $x \in p^{-1}(m)$* (cf. 13.3.1).

14.1.3.6. By 14.1.3.4, a necessary and sufficient condition for the projective line D to be tangent to α is that $D \cap \mathrm{im}(\alpha)$ contain exactly one or more than two points. If α is proper, the tangent hyperplane to α at x is the union of all lines tangent to α and containing x: if fact, by 14.1.3.4, we have either $D \subset \mathrm{im}(\alpha)$ or $D \cap D^{\perp} = \{x\}$, and $D \cap D^{\perp} \ni x$ in either case. By 13.3.2 this implies $(D \cap D^{\perp})^{\perp} \subset x^{\perp}$, hence $\langle D^{\perp} \cup D \rangle \subset x^{\perp}$, and in particular $D \subset x^{\perp}$.

14.1.3.7. We now recouch 9.5.5 in the language of quadrics. If X is a Euclidean affine space, the complexification N^C of the norm of $\vec{X}$ is a quadratic form in $Q(\vec{X}^C)$, and thus defines a quadric in $P(\vec{X}^C)$. The image of this quadric is the umbilical locus. This quadric is always proper; when $\dim X = 2$, in particular, it consists of two points (cf. 14.1.3.1 and 14.1.3.2), called the cyclic points of X.

14.1.3.8. Image under an isomorphism. Let E, E' be vector spaces, $\bar{f} \in \mathrm{Isom}(E; E')$ a vector isomorphism, $f \in \mathrm{Isom}(P(E); P(E'))$ the associated projective object (cf. 4.5), and $\alpha \in PQ(E)$ a projective quadric with equation q. The *image* of α under f is the quadric in $P(F')$ having $((\bar{f})^{-1})^{*}(q)$ for equation (cf. 13.1.3.9); this quadric is *denoted* by $f(\alpha)$. By definition,

$$\mathrm{im}(f(\alpha)) = f(\mathrm{im}(\alpha)).$$

In particular, the projective group $GP(E)$ of $P(E)$ acts on $PQ(E)$ (here we don't have to take the opposite law of composition as in 13.1.3.9).

14.1.4. EXPLICIT CALCULATIONS.

14.1.4.1. We shall use homogeneous coordinates (cf. 4.2.3) associated with a basis $\{e_i\}$ of E. If A is the matrix of q in that basis (cf. 13.1.3.6), its image $\operatorname{im}(\alpha)$ is the set of points $X = {}^t(x_0, x_1, \ldots, x_n)$ such that

$$ {}^tXAX = \sum_{i,j} a_{ij}x_i x_j = 0. $$

14.1.4.2. Saying that α is proper is equivalent to saying that $\det A \neq 0$, by 13.2.2.4. It follows that, in $PQ(E)$, degenerate quadrics form an $(n+1)$-dimensional algebraic hypersurface.

14.1.4.3. The action of a map $\overline{f} \in GL(E)$ on quadrics is described as follows: If α has an equation q with matrix A and f has matrix S, the image $f(\alpha)$ has an equation with matrix ${}^tS^{-1}AS^{-1}$ (all matrices being taken with respect to an arbitrary, fixed basis).

14.1.5. CLASSIFICATION OF QUADRICS IN REAL AND COMPLEX PROJECTIVE SPACES. The *classification* consists in finding the orbits of $PQ(E)$ under the action of $GP(E)$ defined in 14.1.3.8. Since 13.4.6 and 13.4.7 give the orbits of the action of $GL(E)$ on $Q(E)$, and since $GP(E) = GL(E)/K^* \operatorname{Id}_E$ (cf. 4.5.9), there remains to check the effect of $K^* \operatorname{Id}_E$ on the orbits of $Q(E)$. In the complex case, since the rank is invariant under the map $q \mapsto kq$ $(k \in \mathbf{C})$, there are exactly as many orbits in $PQ(E)$ as in $Q(E)$. In contrast, for $K = \mathbf{R}$, a form with signature (r, s) acquires the opposite signature (s, r) when multiplied by a negative scalar. We thus have the following

14.1.5.1. Theorem. *If $K = \mathbf{C}$, the action of $GP(E)$ on $P(E)$ has exactly $n+1$ orbits, classified by the rank k, where $1 \leq k \leq n+1$. If $K = \mathbf{R}$, the orbits are classified by pairs (r, s) such that*

$$ 1 \leq s \leq r \leq n+1; $$

in particular, there exist $\left[\frac{1}{2}n(n+1)\right] + 1$ types of proper quadrics, where $[x]$ denotes the largest integer $\leq x$.

14.1.5.2. Notes. The remark made in 13.4.3 clearly holds true here.

In the real case, quadrics can also be classified by rank and index, because, by 13.7.6, $\operatorname{rank}(q) = r + s$ and $\operatorname{index}(q) = s$ if $s \leq r$.

For a recent work of classification, see [NW].

14.1.5.3. Examples. If K is algebraically closed, proper quadrics lie all in the same orbit.

Let $K = \mathbf{R}$. If $n = 1$, there are three types of quadrics: $(0, 1)$, a point; $(0, 2)$, the empty set; and $(1, 1)$, two points. If $n = 2$, there are five types: $(0, 1)$, a line (said to be *double*); $(0, 2)$, a point; $(1, 1)$, two lines; $(0, 3)$, the empty set; and $(1, 2)$, a proper conic. If $n = 3$, there are three types of proper quadrics: $(0, 4)$, $(1, 3)$ and $(2, 2)$, the first being empty and the other

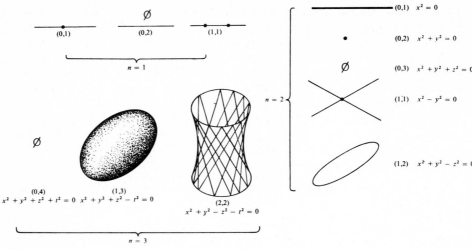

Figure 14.1.5

two non-empty and distinct. For the topology of quadrics of type $(1, 2)$ for $n = 2$ and $(1, 3)$ and $(2, 2)$ for $n = 3$, see 14.3.3.

14.1.6. The Nullstellensatz

14.1.6.1. Consider the map $\text{im}(\cdot) : \text{PQ}(E) \to \mathcal{P}(P(E))$ taking projective quadrics to subsets of $P(E)$. It would be nice if that map were injective, since that would mean that the algebraic and geometric definitions of a quadric were equivalent; unfortunately, this is not the case. For example, if $E = \mathbf{R}^3$ the quadrics with equations $q = x^2 + y^2$ and $q = x^2 + 2y^2$ have the same image, the point with projective coordinates $(0, 0, 1)$; but the two quadrics are distinct. But we have the following result:

14.1.6.2. Theorem. *If K is algebraically closed, the map*

$$\text{im}(\cdot) : \text{PQ}(E) \to \mathcal{P}(P(E))$$

is injective.

Proof. Let q and q' be such that $p(q^{-1}(0) \setminus 0)$ and $p(q'^{-1}(0) \setminus 0)$ are equal. By 14.1.3.1 there exists $x \in E \setminus 0$ such that $q(x) \neq 0$. We can assume that $q'(x) = q(x)$, by multiplying q' by a constant. Let $y \in E \setminus 0$ be such that $p(y) \neq p(x)$, and let D be the projective line $\langle p(x), p(y) \rangle$. By 14.1.3.2, since $D \cap \text{im}(p(q))$ and $D \cap \text{im}(p(q'))$ coincide, we must have $\alpha \cap D = \alpha' \cap D$, and, in particular,

$$\frac{q'(y)}{q(y)} = \frac{q'(x)}{q(x)} = 1. \qquad \square$$

The name "Nullstellensatz" refers to a general theorem about algebraic varieties, of which 14.1.6.2 is a very special case; see [FN, 21], for example. For the case $K = \mathbf{R}$, see 14.8.9.

14.1.7. REDUCTION TO THE NON-DEGENERATE CASE. Here we translate 13.2.2.6 into geometric terms. Call *singular* any point of a quadric α that belongs to $p(\text{rad}(q))$; the set of singular points of α is a projective subspace, called its *radical*. More concretely, if $q \leftrightarrow A$ in some basis $\{e_i\}$, singular points can be described as column vectors X such that $AX = 0$, or as eigenvectors of A with eigenvalue 0, or as solutions of $\sum_j a_{ij}x_j = 0$ for all i, or again as solutions of $\partial q / \partial x_i = 0$ (cf. 13.1.2).

Let $E = \text{rad}(q) \oplus G$ be a direct sum, $A = p(\text{rad}(q))$ the radical of α and B the image of the quadric with equation $q|_G$ into the subspace $p(G)$ of $p(E)$. The sets A and B both sit in $P(E)$. If $B = \emptyset$, we have $\text{im}(\alpha) = A$. If $B \neq \emptyset$, we have the following result:

14.1.7.1. Proposition. *The image of α is the conoid with vertex A and base B, that is, the union in $P(E)$ of all projective lines $\langle m, n \rangle$, for $m \in A$ and $n \in B$.*

Proof. Take $x \in E \setminus 0$ such that $p(x) \in \text{im}(\alpha)$, that is, $q(x) = 0$, and write $x = y + z$, for $y \in \text{rad}(q)$ and $z \in G$. Then

$$0 = q(x) = q(y + z) = q(y) + 2P(y, z) + q(z) = q(z),$$

so y can be any point in $\text{rad}(q)$ and z any point in $q^{-1}(0)$. $\square$

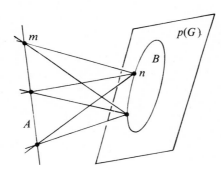

Figure 14.1.7.1

14.1.7.2. For example, if $\dim(r(q)) = 1$, the set $\text{im}(\alpha)$ is a true cone in $P(E)$, the vertex being $p(\text{rad}(q))$ and the base the image of the proper quadric B in $P(E)$. Here is a particular case of this example: let α be a proper quadric in $P(E)$ and m a point in $\text{im}(\alpha)$; we want to know what is $\alpha \cap H$, the intersection of α with the hyperplane H, tangent to α at m (cf. 14.1.3.3 and 14.1.3.5). By 13.3.2 the radical of $q|_{p^{-1}(H)}$ is one-dimensional, so $\text{im}(\alpha \cap H)$ is a cone in H, with vertex m and base some proper quadric.

Thus, if $n = 2$, the image of $(\alpha \cap H)$ is just the point m; if $m = 3$, it is either m or two distinct lines of H intersecting at m. This is a consequence of 14.1.3.2. Those two cases occur for the quadrics of type $(1, 3)$ and $(2, 2)$ in figure 14.1.5.

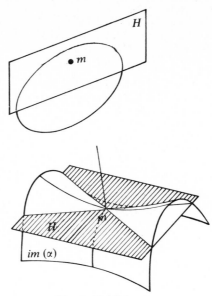

Figure 14.1.7.2

14.2. Subspaces of $PQ(E)$. Pencils of quadrics

It follows from 13.1.3.6 that $\dim Q(E) = (n+1)(n+2)/2$ (see also 3.7.10). Thus

14.2.1 $$\dim PQ(E) = \frac{n(n+3)}{2};$$

in particular, for $n = 1$, 2 and 3 the dimension is 2, 5 and 9, respectively.

14.2.2. PROPOSITION. *Take $m \in P(E)$; then*

$$H(m) = \big\{\, \alpha \in PQ(E) \mid m \in \mathrm{im}(\alpha) \,\big\}$$

is a hyperplane.

Proof. Write $m = p(x)$, $\alpha = p(q)$. The condition $m \in \mathrm{im}(\alpha)$ is equivalent to $q(x) = 0$; this condition is linear in q, and non-trivial since $x \neq 0$.   □

14.2.3. COROLLARY. *Let $\{m_i\}_{i=1\ldots d}$ be points in $P(E)$. Then the set*

$$\big\{\, \alpha \in PQ(E) \mid m_i \in \mathrm{im}(\alpha) \text{ for all } i \,\big\}$$

is a subspace of $P(E)$, with dimension at least $\dfrac{n(n+3)}{2} - d$. In particular, given any $n(n+3)/2$ points of $P(E)$ there exists at least one quadric passing through these points.

14.2.4. EXAMPLES. Five distinct points in a projective plane are contained in at least one conic; uniqueness will be discussed in 16.1.4.

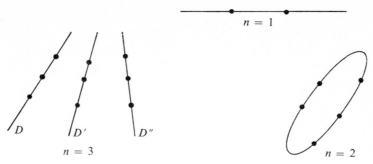

Figure 14.2.4

The next example is even neater: Assume $n = 3$ and let D, D' and D'' be lines in $P(E)$. There exists at least one quadric α such that im(α) $\supset$ $D \cup D' \cup D''$. For we can consider three distinct points on each line; the result follows from 14.2.3 and 14.1.3.4.

14.2.5. WARNING. Not every hyperplane in PQ(E) is of the form 14.2.2. We will return to this question in 14.5.4.6.

In spite of its simplicity, the definition below gives rise to a number of geometric applications, for example, 16.5.3 and 17.5.

14.2.6. DEFINITION. *A pencil (of quadrics) is a line in* PQ(E).

14.2.7. EXAMPLES.

14.2.7.1. In practice (cf. 4.6.7) a pencil of quadrics $\mathcal{F}$ is defined by two distinct quadrics α and α', that is, $\mathcal{F} = \langle \alpha, \alpha' \rangle$. If $\alpha = p(q)$, $\alpha' = p(q')$, the set of equations of quadrics in $\mathcal{F}$ is given by

$$\{\, \lambda q + \lambda' q' \mid (\lambda, \lambda') \neq (0,0) \,\}.$$

14.2.7.2. By 14.2.3, if $\{m_i\}_{i=1,\dots,n(n+3)/2-1}$ are points in $P(E)$, the set

$$\{\, \alpha \in PQ(E) \mid m_i \in \mathrm{im}(\alpha) \text{ for all } i \,\}$$

will be a pencil, at least in general. For example, if $n = 2$ and ϕ, ψ, ξ and η are equations (linear forms in E) for the lines $\langle m_1, m_2 \rangle$, $\langle m_3, m_4 \rangle$, $\langle m_1, m_3 \rangle$ and $\langle m_2, m_4 \rangle$, respectively, the pencil $\mathcal{F}$ of all conics going through m_1, m_2, m_3 and m_4 will be defined by the equations $\lambda \phi \psi + \lambda' \xi \eta$ (cf. 13.1.3.3).

Figure 14.2.7.2

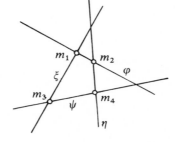

14.2.7.3. Definition *The base of the pencil $\mathcal{F}$, denoted by* base$(\mathcal{F})$, *is the subset of $P(E)$ given by $\bigcap_{\alpha \in \mathcal{F}} \mathrm{im}(\alpha)$.*

14.2.7.4. By 14.2.2 and 4.6.12, given a pencil $\mathcal{F}$ and a point m of $P(E)$, there is exactly one quadric of $\mathcal{F}$ going through m, at least in general.

14.2.7.5. We now use 14.1.4.2 to study the degenerate quadrics in a pencil $\mathcal{F}$. Let q, q' be quadrics defining $\mathcal{F}$, and assume $q \leftrightarrow A$ and $q' \leftrightarrow A'$. The quadric with equation $\lambda A + \lambda' A'$ will be degenerate if and only if $\det(\lambda A + \lambda' A') = 0$; since this determinant is a homogeneous polynomial of degree $n + 1$ in the variables λ, λ', the equation either has at most $n + 1$ distinct roots in the projective line $P(K^2)$, or is identically satisfied. Thus, either all the quadrics in $\mathcal{F}$ are degenerate, or $\mathcal{F}$ contains at most $n + 1$ degenerate quadrics. For instance, if K is algebraically closed, $\mathcal{F}$ will typically contain $n + 1$ degenerate quadrics. For a complete discussion in the case $n = 2$, see 16.5.

A pencil is called *degenerate* if all its quadrics are degenerate.

14.2.8. THE ONE-DIMENSIONAL CASE

14.2.8.1. Proposition. *Let $\mathcal{F}$ be a pencil of quadrics in $D = P(E)$, where $n = \dim P(E) = 1$ (that is, D is a projective line). Then $\mathcal{F}$ is not degenerate, and falls into one of the following two categories:*

i) *There exists $m_0 \in D$ such that*

$$\mathcal{F} = \big\{\, \alpha \in \mathrm{PQ}(E) \mid \mathrm{im}(\alpha) = \{m_0, m\} \text{ and } m \in D \,\big\}$$

   (recall from 14.1.3.2 that for $n = 1$ a quadric is determined by its image if it is non-empty).

ii) *There exists an involution f of D (cf. 6.7) such that, for any $\alpha \in \mathcal{F}$ and any $m \in P(E) \cap \mathrm{im}(\alpha)$, we have*

$$D \cap \mathrm{im}(\alpha) = \big\{m, f(m)\big\}.$$

Conversely, for any involution f of D, there exists a pencil $\mathcal{F}$ of quadrics of $P(E)$ such that, for any $\alpha \in \mathcal{F}$ and any $m \in P(E) \cap \mathrm{im}(\alpha)$, we have $D \cap \mathrm{im}(\alpha) = \{m, f(m)\}$.

Proof. Let $\alpha, \alpha' \in \mathcal{F}$ be degenerate and distinct; by 14.1.3.2, we can write $\alpha = p(\phi^2)$, $\alpha' = p(\phi'^2)$ for some $\phi, \phi' \in E^*$. Then the quadric with equation $\phi^2 + \phi'^2$ is proper and belongs to $\mathcal{F}$, showing that $\mathcal{F}$ cannot be degenerate.

Let $\alpha \in \mathcal{F}$ be proper and assume that $\mathrm{im}(\alpha) \neq \emptyset$ (such a quadric exists by 14.2.7.4). Take homogeneous coordinates (x, y) on D such that

$$\mathrm{im}(\alpha) = \big\{p(1,0),\, p(0,1)\big\},$$

so that $2xy$ is an equation for α. Let $ax^2 + 2bxy + cy^2$ be an equation for a quadric $\alpha' \in \mathcal{F}$ distinct from α (take $c \neq 0$, for example). We just need to study the set $D \cap \mathrm{im}(\beta)$, for $\beta \in \mathcal{F} \setminus \alpha$, which is the same as studying the equation

$$ax^2 + 2bxy + cy^2 + \lambda(2xy) = ax^2 + 2(b + \lambda)xy + cy^2 = 0,$$

for $\lambda \in K$. Since $c \neq 0$, the pair $(0, 1)$ is never a solution, and we can just look for solutions of the form $(1, t)$, namely, the roots of

$$a + 2(b + \lambda)t + ct^2.$$

The two roots t, t' satisfy $tt' = a/c$, and thus belong to the involution f of D defined by $f(t) = a/ct$ (plus $f(0) = \infty$ and $f(\infty) = 0$, cf. 5.2.4).

Conversely, the proof of 6.7.3 shows that, in the appropriate homogeneous coordinates, an arbitrary involution f is of the form $f(t) = k/t$ for some $k \in K^*$. The calculation above shows that the desired pencil $\mathcal{F}$ can be defined by the two equations $q = 2xy$ and $q' = kx^2 + y^2$. $\qquad\square$

14.2.8.2. Notes. In case (ii) the fixed points of the involution are exactly the degenerate quadrics in the pencil; there are either 0 or 2 of them. In case (i) there is always exactly one degenerate quadric, with image $\{m_0\}$.

14.2.8.3. Corollary (Desargues's theorem). *For* dim E *arbitrary, $\mathcal{F}$ a pencil of quadrics in $P(E)$ and D a line in $P(E)$ not intersecting the base of $\mathcal{F}$, assume that $\{\alpha \cap D \mid \alpha \in \mathcal{F}\}$ is a pencil in D (by 14.1.3.3 this is the same as saying that there is no $\alpha \in \mathcal{F}$ such that $D \subset \mathrm{im}(\alpha)$). Then, for any $m \in D$, there exists a unique quadric $\alpha_m \in \mathcal{F}$ such that $m \in \mathrm{im}(\alpha_m)$, and the map $f : D \to D$ defined by $\mathrm{im}(\alpha_m) = \{m, f(m)\}$ for all $m \in D$ is an involution of D. The fixed points of f are the intersections with D of the $\alpha \in \mathcal{F}$ to which D is tangent (cf. 14.1.3.5); in particular, there are either zero or two quadrics of $\mathcal{F}$ tangent to D.* $\qquad\square$

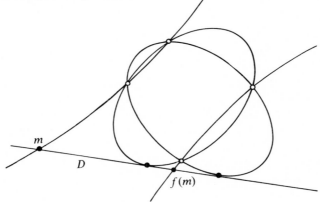

Figure 14.2.8.3

14.3. Topology of real and complex quadrics

In this section K is equal to $\mathbf{R}$ or $\mathbf{C}$.

By section 4.3, the projective space $P(E)$ has a natural topological structure. We study here the induced topological structure on the sets $\mathrm{im}(\alpha) \subset P(E)$, for $\alpha \in PQ(E)$.

14.3.1. PROPOSITION. *The image of any quadric is compact and, for $n \geq 2$, path-connected.*

Proof. Compactness is proved in the same way as for projective spaces (see 4.3.3.2), and we use the same notation as in that proof. If q is an equation for the quadric $\alpha \in PQ(E)$, we have

$$\text{im}(\alpha) = p(q^{-1}(0) \setminus 0) = p(q^{-1}(0) \cap S(E)).$$

Now $q^{-1}(0) \cap S(E)$ is compact, since $S(E)$ is compact and q is continuous; this implies that $p(q^{-1}(0) \cap S(E))$ is compact, since $P(E)$ is Hausdorff (cf. 4.3.3).

As for connectedness, we see from 14.1.3.2 that $\text{im}(\alpha)$ cannot be connected when $n = 1$, so assume $n \geq 2$. We can assume that α is proper since, by 14.1.7.1, $\text{im}(\alpha)$ will be path-connected if A and B are (recall, cf. 4.3.3, that a projective line is path-connected). For $K = \mathbf{R}$ we conclude that $\text{im}(\alpha)$ is path-connected by using 14.3.3. For $K = \mathbf{C}$, take $m, n \in \text{im}(\alpha)$, and consider an arbitrary projective plane S containing m and n. If the quadric $\alpha \cap S$ is proper, it is path-connected by 14.3.6; otherwise it is the union of two lines (cf. 14.1.7.2 and 14.1.3.2), and again path-connected.   □

14.3.2. REAL PROPER QUADRICS. Thanks to 14.1.7.1, the topology of quadrics can be reduced that of proper quadrics; here, surprisingly enough, the real case is simpler than the complex case. By 13.4.2 we can assume that q has an equation of the form

$$\sum_{i=1}^{r} x_i^2 - \sum_{i=r+1}^{r+s=n+1} x_i^2;$$

we *denote* by $C(r, s)$ the image in $P^n(\mathbf{R})$ of the quadric defined by that equation. We have the following

14.3.3. PROPOSITION. *The topological space $C(r, s)$ is homeomorphic to the quotient $(S^{r-1} \times S^{s-1})/\mathbf{Z}_2$ of the product $S^{r-1} \times S^{s-1}$ of spheres by the subgroup of $O(r + s)$ consisting of $\pm \text{Id}_{\mathbf{R}^{r+s}}$. In particular, $C(n, 1)$ is homeomorphic to S^{n-1} and $C(2, 2)$ to $S^1 \times S^1$.*

Proof. Just think of $\mathbf{R}^{r+s} = \mathbf{R}^{n+1}$ as the product of the Euclidean spaces $\mathbf{R}^r$ and $\mathbf{R}^s$; the equation of α becomes simply

$$q(x, y) = \|x\|^2 - \|y\|^2,$$

for $x \in \mathbf{R}^r$ and $y \in \mathbf{R}^s$. As in the proof of 14.3.1, we have $\text{im}(\alpha) = p(q^{-1} \cap S(E))$, and this intersection, characterized by the conditions $\|x\|^2 - \|y\|^2 = 0$ and $\|x\|^2 + \|y\|^2 = 1$, or, alternatively, by $\|x\| = \|y\| = 1/\sqrt{2}$, is simply the product of the spheres of radius $1/\sqrt{2}$ in $\mathbf{R}^r$ and $\mathbf{R}^s$. We next have to study the effect of the projection $p : \mathbf{R}^{r+s} \to P^{r+s-1}(\mathbf{R}) = P(\mathbf{R}^{r+s})$. If $p(x, y) = p(x', y')$, we must have $x' = kx$ and $y' = ky$ for some $k \in \mathbf{R}$, and, since

$$\|x'\| = \|x\| = \|y'\| = \|y\| = \frac{1}{\sqrt{2}},$$

we conclude that $k = \pm 1$.

If $s = 1$, the sphere S^{s-1} consists of two distinct points, say a and b, and the restriction $p : S^{r-1} \times \{a\} \in p(S^{r-1} \times S^{s-1}) = \mathrm{im}(\alpha)$ is bijective, hence a homeomorphism.

If $r = s = 2$, we see from 14.4.2 that $\mathrm{im}(\alpha)$ is homeomorphic to the product of two projective lines, that is, $S^1 \times S^1$, by 4.3.6. $\qquad\square$

14.3.4. EXAMPLES: THE REAL CASE. For $n = 2$, all non-empty images are homeomorphic to the circle S^1 (figure 14.1.5); this homeomorphism would also follow from 16.2.4.

For $n = 3$, there are two homeomorphism classes of images; one is the sphere S^2, clearly seen in figure 14.1.5, and the other is the torus $S^1 \times S^1$, which can be seen with the help of 14.4.2 (this set, $C(2, 2)$, cannot be drawn in an affine space, since it intersects every projective plane). Observe the linking of lines in the same family, cf. 18.8.6 and figure 14.3.4.

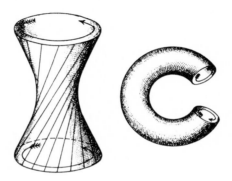

Figure 14.3.4

The case $C(4, 4) \subset P^7(\mathbf{R})$ is particularly neat; by 8.9.8, this set is homeomorphic to the group $O^+(4)$ of rotations of $\mathbf{R}^4$.

14.3.5. THE COMPLEX CASE. When $K = \mathbf{C}$, we see from 13.4.2 that there exists only one homeomorphism class of proper quadrics, represented by the quadric with equation $\sum_{i=1}^{n+1} x_i^2$ in $P^n(\mathbf{C})$; we *denote* this image by $C(n)$. There is no analogue to 14.3.3 that would permit us to express $C(n)$ in terms of known spaces, but there are two particular cases:

14.3.6. PROPOSITION. *The set $C(2)$ is homeomorphic to S^2, and $C(3)$ is homeomorphic to $S^2 \times S^2$.*

Proof. The first statement follows from 16.2.4, and the second from 14.4.2.
$\qquad\square$

14.3.7. NOTES. As the reader may have suspected, the topology of $C(n)$ has been studied, the first historical reference being [CE1]. Although a detailed analysis would be beyond the scope of this book, it is easy to see (cf. 14.8.4) that $C(n)$ is homeomorphic to the real grassmannian of *oriented* lines of $P^n(\mathbf{R})$, which can be identified with the homogeneous space

$$O^+(n+1)/O^+(n-1) \times O^+(2)$$

(cf. 8.2.8). The topology of grassmannians has been very well studied; in particular, it gave rise to "characteristic classes" (cf., for example, [HU, chapter 18]).

The cases $n = 4$ and 5 give rise to special interpretations, due to isomorphisms of the classical groups $O(5)$ and $O(6)$. See [PO, 266].

We now study quadrics from the differentiable viewpoint:

14.3.8. PROPOSITION. *Let α be a proper quadric in $P(E)$. The image $C = \mathrm{im}(\alpha)$ of α is a C^∞ submanifold of $P(E)$, of codimension one if $K = \mathbf{R}$ and two if $K = \mathbf{C}$. Furthermore, for every $m \in C$, the tangent space $T_m C$ to C at m is naturally identified with the quotient vector space $x^\perp/Kx$, where $p(x) = m$ (cf. also 14.1.3.5).*

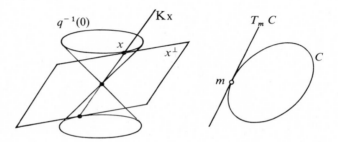

Figure 14.3.8

Proof. Use 4.2.6 and the fact that the charts introduced in 4.2.4.2 are C^∞. More specifically, to show that C is a C^∞ submanifold in a neighborhood of $m = p(x) \in C$, choose a chart

$$(v_1, \ldots, v_n) \mapsto p(v_1, \ldots, v_n, 1)$$

such that $m = p(0, \ldots, 0, 1)$. The equation of C, read in K^n, is of the form

$$f = \sum_{i,j=1}^n a_{ij} v_i v_j + 2 \sum_{i=1}^n a_{i,n+1} v_i + a_{n+1,n+1} = 0$$

(cf. 14.1.4.1), and the derivative $f'(0, \ldots, 0)$ is given by the partials

$$\frac{\partial f}{\partial v_i}(0, \ldots, 0) = 2a_{i,n+1}.$$

These are not all zero because $\det A \neq 0$ (cf. 14.1.4.2), so by the implicit function theorem C is indeed a C^∞ submanifold. (It is even real or complex analytic, cf. 4.2.6.)

The tangent space $T_m C$ is read in K^n as the set $(f'(0, \ldots, 0))^{-1}(0)$, characterized by the equation $\sum_{i=1}^{n} a_{i.n+1} y_i = 0$; on the other hand, $x^\perp$ is given by

$$P(x, y) = \sum_{i=1}^{n} a_{i.n+1} y_i + a_{n+1.n+1} y_{n+1} = 0.$$

This equation reduces to the previous one when $y_{n+1} = 0$, which corresponds to taking the quotient $x^\perp / Kx$ (see also 13.1.2).

If α is not proper, we use 14.1.7.1; see also 14.8.5.   □

For the differential geometry involved here, see [B–G, especially 2.5.7 and 2.6.15], for example.

14.4. Quadrics in Art$_4$

We now study the special case that E is four-dimensional over an arbitrary field and q is neutral. This is done by rephrasing 13.7.11 in terms of $P(E)$. Recall that 13.7.11 has a completely elementary proof (13.7.11.2).

14.4.1. PROPOSITION. *Let (E, q) be a neutral form over a four-dimensional space (cf. 13.1.4.3), α the corresponding quadric in $P(E)$ and $C = \mathrm{im}(\alpha) \subset P(E)$ its image. Then C contains two families of lines, Ξ and Θ, satisfying the following conditions:*
  i) *any $m \in C$ is contained in a unique $X \in \Xi$ and a unique $T \in \Theta$;*
  ii) *any $X \in \Xi$ and $T \in \Theta$ intersect in a unique point of C;*
  iii) *any two distinct lines $X, X' \in \Xi$ have empty intersection, and similarly for Θ.*

These families are called systems of generating lines of α.   □

14.4.2. COROLLARY. *There exist a natural bijection*

$$\phi : C \to P^1(K) \times P^1(K).$$

If $K = \mathbf{R}$ or $\mathbf{C}$, this bijection is a homeomorphism.

Proof. Fix $X_0 \in \Xi$ and $T_0 \in \Theta$, and set $\phi(m) = (T(m) \cap X_0, X(m) \cap T_0)$, where $X(m)$ (resp. $T(m)$) denotes the unique generating line of Ξ (resp. Θ) containing $m \in C$. The map ϕ is well-defined by condition (i) of the proposition, and from condition (ii) we see that ϕ is a bijection from C onto $X_0 \times T_0$, which is isomorphic to $P^1(K) \times P^1(K)$. There remains to check that ϕ is a homeomorphism; but this follows from continuity (formulas in 13.7.11.2), since all spaces involved are compact.   □

Conversely, given a three-dimensional projective space, there are at least three geometric constructions for images of quadrics with neutral equation. Here's the first:

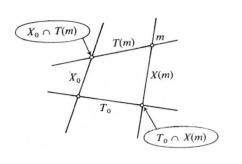

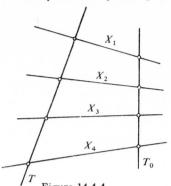

Figure 14.4.2   Figure 14.4.4

14.4.3. PROPOSITION. *Let $P(E)$ have dimension three, and let D, D' and D'' be pairwise disjoint lines of $P(E)$. There exist lines intersecting D, D' and D'', and the union of all such lines is the image of a quadric with neutral equation.*

Proof. By 14.2.4 there exists a quadric α whose image $C = \mathrm{im}(\alpha)$ contains all three lines; we start by verifying that α is proper. Assume there exists $m \in p(\mathrm{rad}(q))$; we have $m \notin D$, for example, and by 14.1.7.1 the image C contains the whole plane $\langle m, D \rangle$. But an arbitrary line intersecting D' and D'' also intersects $\langle m, D \rangle$, so it intersects C in three points, and must be contained in C (cf. 14.1.3.4); this is absurd, since lines intersecting D' and D'' fill the whole of $P(E)$.

We finish off by using 13.3.4.2: since q is non-degenerate and C contains a line, q must be neutral.   □

For the other two constructions, notice first the following fact:

14.4.4. LEMMA. *Let C be as in 14.4.1 and $(X_i)_{i=1,2,3,4} \subset \Xi$. The cross-ratio $[T \cap X_i]$ of the four points $T \cap X_i$ does not depend on $T \in \Theta$. This number is called the cross-ratio of the (X_i), and denoted by $[X_i]$.*

Proof. Fix $T_0 \in \Theta$, and let $P_i = \langle X_i, T_0 \rangle$ be the plane spanned by X_i and T_0. We have $P_i \cap T = X_i \cap T$ for all $T \in \Theta$, so $[\Xi \cap T] = [P_i]$ by 6.5.2. This number clearly does not depend on T.   □

14.4.5. COROLLARY. *Let D and D' be disjoint lines in $P(E)$ (where $P(E)$ has dimension three) and let $f \in \mathrm{Isom}(D; D')$ be a homography between D and D'. Then $\{\langle m, f(m) \rangle \mid m \in D\}$ is the image of a quadric with neutral equation. The dual statement holds: if f is a homography between the pencil $\mathcal{F}$ of planes containing D and the pencil $\mathcal{F}'$ of planes containing D', the set $\{\langle P, f(P) \rangle \mid P \in \mathcal{F}'\}$ is the image of a quadric with neutral equation.*

Proof. The dual assertion follows from the first and from 6.5.2. To prove the first assertion, fix three points $a, b, c \in D$, with images $a' = f(a)$, $b' = f(b)$ and $c' = f(c) \in D'$. We just have to apply 14.4.4 to the quadric containing the three lines $\Delta = \langle a, a' \rangle$, $\Delta' = \langle b, b' \rangle$ and $\Delta'' = \langle c, c' \rangle$ (cf. 14.4.3 and 6.1.5).   □

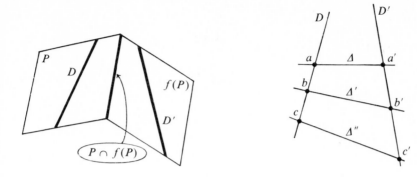

Figure 14.4.5

14.4.6. COROLLARY. *Let D and D' be disjoint lines in a three-dimensional Euclidean affine space X, and let m(t) and m'(t) be points moving along D and D' with constant speed. The affine line ⟨m(t), m'(t)⟩ describes an affine quadric of X (a hyperbolic paraboloid) as t ranges from −∞ to +∞ (cf. 15.3.3.3).* □

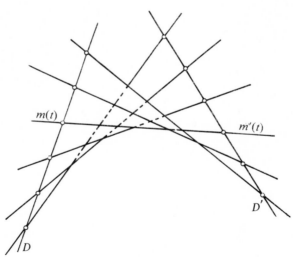

Figure 14.4.6.1

This result has recently found an important practical application, enabling architects to easily design structures in prestressed concrete whose surface is generated by straight lines (corresponding to the tendons). Indeed, the simplest case of this design employs tendons stretched between points uniformly distributed along two straight lines. Figure 14.4.6.3 shows a case where the supporting curves are circles.

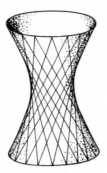
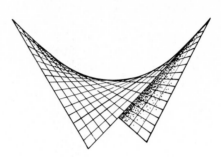

Figure 14.4.6.2

Figure 14.4.6.3 (Source: [JE])

Water tower—Fedala, Spain

For the water tower at Fedala a form was chosen in which straight boards could be used as shuttering and the prestressing steel laid straight. The [one-sheet] hyperboloid, [having] two groups of straight generators, fulfilled these conditions perfectly. It could be prestressed along the two [sets of] generators in two directions. This double prestressing ensured crack-free concrete, the first essential in providing an absolute watertight sealing for the water holder. [The shell is actually made up of two hyperboloids, one above and one below the "neck".]   (Reprinted with permission from [JE].)

14.4.7. NOTES. For $n = 3$ and $K = \mathbf{C}$, we recover the result that the image C of any proper quadric α contains lots of lines, namely, two through each point m, because the tangent plane to C at m intersects C along two distinct lines, by 14.1.7.2. This is in agreement with 13.1.4.3. Conversely, the

Figure 14.4.6.4 (Source: [JE])

Cowl of air-extraction shaft at a foundry in Lohr, Germany —see also figure 15.3.3.2.2

arguments above show that, in the framework of 14.4.1, the tangent plane to C at m is exactly the plane containing $X(m)$ and $T(m)$.

Notice that a Euclidean sphere contains straight lines! In the following sense: let $S \subset \mathbf{R}^3$ be the unit sphere. Complexifying $\mathbf{R}^3$ into $\mathbf{C}^3$ and $\| \cdot \|^2$ into N^{c} (cf. 9.5.5 and 14.1.3.7), we obtain a complexification S^{c} of S defined by $S^{c} = \{ z \in \mathbf{C}^3 \mid N^{c}(z) = 1 \}$. That's an affine quadric in $\mathbf{C}^3$ (cf. 15.1.3.2) that contains two families of generating lines (cf. 14.4.1). Such lines (in S^{c}) are called the *isotropic lines* of S. For applications of this remark, see [DX, 187–234] and [DQ, 147].

14.5. Duality with respect to a proper quadric. Polarity

In this section, unless we say otherwise, α is a *proper* quadric; q is an equation for α, and P its polar form; A is the matrix of q in a chosen basis; and C is the image of α.

14.5.1. DEFINITION. By 13.2.1, α determines, via q, an *isomorphism* $\phi : E \to E^*$, whence an isomorphism $\psi : P(E) \to P(E^*)$ between projective spaces. Following 4.1.3.5, we will identify $P(E^*)$ with $\mathcal{H}(E)$, the set of hyperplanes of $P(E)$. By 4.5.4 and 14.1.1 the map ψ depends only on α, and not on q. For $m \in P(E)$, we call $\psi(m) \in \mathcal{H}(E)$ the *polar (hyperplane)* of m, and we *denote it by* $m^{\perp}$.

More generally, the condition $P(x, y) = 0$ depends only on $m = p(x)$ and $n = p(y)$ (not on x and y) and on α (not on P). We say that m and n are *conjugate with respect to x and y*, and we *denote this fact by* $m \perp n$. Conjugation is a symmetric relation. For every subset $S \subset P(E)$, we *set*

$$S^{\perp} = \{ n \in P(E) \mid m \perp n \text{ for all } m \in S \}.$$

If S and T are subspaces, we have $(S^{\perp})^{\perp} = S$,

$$\dim S + \dim S^{\perp} = \dim P(E) - 1,$$
$$(S \cap T)^{\perp} = \langle S^{\perp} \cup T^{\perp} \rangle,$$
$$\langle S \cup T \rangle^{\perp} = S^{\perp} \cap T^{\perp}.$$

We say that $S^{\perp}$ is the *polar* subspace to S, and that S and $S^{\perp}$ are *conjugate* subspaces (with respect to α); when S is a hyperplane, the point $S^{\perp}$ is called the *pole* of S. All of the facts above follow from translating 13.3.2 into the language of projective spaces.

14.5.2. EXAMPLES.

14.5.2.0. If E is a Euclidean vector space and $\| \cdot \|^2$ is an equation for α, polarity with respect to α is the same as orthogonality in E. This is a typical case where the concept of polarity arises naturally, even though $\text{im}(\alpha) = \emptyset$!

14.5.2.1. The tangent hyperplane to $m \in \text{im}(\alpha)$ is $m^{\perp}$ (cf. 14.1.3.5).

14.5.2.2. If $m \in n^{\perp}$, we have $n \in m^{\perp}$, since conjugation is a symmetric relation.

14.5.2.3. For $D = \langle m, n \rangle$ $(m \neq n)$ a line, the polar $D^{\perp}$ is equal to $m^{\perp} \cap n^{\perp}$.

14.5.2.4. **Heredity.** Let T be a subspace such that $\alpha \cap T$ is still proper (cf. 14.1.3.3). If $m, n \in T$ satisfy $m \perp n$ with respect to α, they also satisfy $m \perp n$ with respect to $\alpha \cap T$.

14.5.2.5. The one-dimensional case. Let D be a projective line, α a proper quadric in D. If the image of α is non-empty, it consists, by 14.1.3.2, of two points a and b. Then $m \perp n$ if and only if $[a,b,m,n] = -1$ (cf. 6.4.1). One way to see this is to take on D homogeneous coordinates (x,y) such that α has $q = 2xy$ for equation (cf. the proof of 14.2.8.1). Then, by 13.1.3.6,

$$(x,y) \perp (x',y') \iff xy' + x'y = 0;$$

since $a = p\big((1,0)\big)$ and $b = p\big((0,1)\big)$, the statement follows from 6.2.3.

14.5.2.6. By combining 14.5.2.4 and 14.5.2.5 we see that, if the line $\langle m,n \rangle$ intersects C in $\{a,b\}$, we have $m \perp n$ if and only if $[a,b,m,n] = -1$. Applying 6.4.4 we deduce the geometric construction given in figure 14.5.2 for the polar hyperplane $m^{\perp}$ of m, assuming that C contains enough points.

14.5.2.7. The points where tangents to α through a point m intersect $\operatorname{im}(\alpha)$ are given by $\operatorname{im}(\alpha) \cap m^{\perp}$.

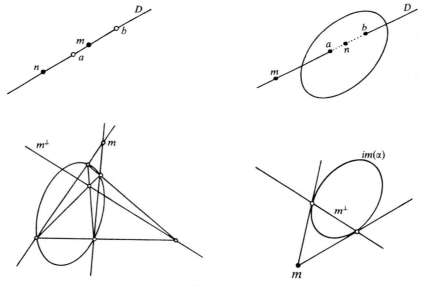

Figure 14.5.2

14.5.3. Explicit calculations. By 13.1.2 and 13.1.3.6, two points $p(x) = m$ and $p(y) = n$ are conjugate if and only if ${}^{t}XAY = 0$, or, equivalently, $\sum_{i.j} a_{ij}x_i y_j = 0$, or again

$$\sum_{i} \frac{\partial q}{\partial x_i}(x_1, \ldots, x_n) \cdot y_i = 0.$$

For instance, the polar hyperplane of x will be $\sum_{i.j} a_{ij}x_i y_j = 0$. In order to find the pole of the hyperplane $\sum_{i.j} a_{ij}\xi_i y_j = 0$ one must calculate the inverse of A; by 13.2.0, the pole x is such that $\phi(x) = \xi = (\xi_1, \ldots, \xi_{n+1})$, so that $x = \phi^{-1}(\xi)$. Thus

$$x \leftrightarrow X = \begin{pmatrix} x_1 \\ \vdots \\ x_{n+1} \end{pmatrix} = A^{-1} \begin{pmatrix} \xi_1 \\ \vdots \\ \xi_{n+1} \end{pmatrix}.$$

One can also solve a homogeneous system of $n + 1$ linear equations (see 15.5.5.2, for example).

Another type of calculation involves the *cone with vertex m and circumscribed around α*. By definition, this is the union of the tangents to α containing m. Set $m = p(x_0)$; then x belongs to this cone if the restriction of α to the line $D = \langle x_0, x \rangle$ is degenerate (cf. 14.1.3.5). In the basis made up of x_0 and x, the matrix of $\alpha \cap D$ is

$$\begin{pmatrix} q(x_0) & P(x_0, x) \\ P(x_0, x) & q(x) \end{pmatrix},$$

and 14.1.4.2 shows that the equation of the cone is $q(x_0)q(x) - \left(P(x_0, x)\right)^2 = 0$. For $n = 2$ we obtain a set of lines. See figure 14.7.4 or 14.5.2.7.

14.5.4. SELF-POLAR SIMPLICES AND HARMONICALLY CIRCUMSCRIBED QUADRICS

14.5.4.1. A *simplex* $\{m_i\}_{i=1,\ldots,n+1}$ of $P(E)$ is a set of $n + 1$ independent points (cf. 4.6.6); a set of $n+1$ dependent points $\{m_i\}_{i=1,\ldots,n+1}$ is sometimes called a *degenerate simplex*. A (possibly degenerate) simplex $\{m_i\}$ is *self-polar* with respect to the proper quadric α if $m_i \perp m_j$ for $i \neq j$; if $\{m_i\}$ is non-degenerate, the polar hyperplane of m_i will be

$$\langle m_1, \ldots, \hat{m}_i, \ldots, m_{n+1} \rangle,$$

for every i. Saying that the simplex $\{m_i\} = \{p(x_i)\}$ is self-polar is the same as saying that $\{x_i\}$ is an orthonormal basis for (E, q); thus self-polar simplices with respect to α always exist (cf. 13.4.1). On the other hand, if α and α' are distinct quadrics, there isn't always a simplex self-polar with respect to both (cf. 13.5). For a complete discussion in the case $n = 2$, see 16.4.10.

14.5.4.2. We now play with two proper quadrics α and α'. Let q, q' be equations for α, α', and let $\phi, \phi' : E \to E'$ be the associated isomorphisms (cf. 14.5.1). It is natural to associate to the two quadrics the isomorphism $f = \phi^{-1} \circ \phi \in \mathrm{Isom}(E)$ (cf. 13.5.1), and to consider the invariants associated with this isomorphism: its trace, determinant, and so on. The condition that the trace (or any invariant other than the determinant) be zero is not affected by modding out by K^*, and it has a simple geometric interpretation:

14.5.4.3. Proposition. *If there exists a simplex $\{m_i\}$ self-dual with respect to α and inscribed in α', that is, satisfying $m_i \in \mathrm{im}(\alpha')$ for all i, the trace of f is zero. Conversely, if K is algebraically closed and $\mathrm{Tr}(f) = 0$, there exists, for any $m \in \mathrm{im}(\alpha') \setminus \mathrm{im}(\alpha)$, a simplex $\{m_i\}$ self-dual with respect to α, inscribed in α' and such that $m_1 = m$.*

Proof. One direction is easy: if $m_i = p(e_i)$, we have $q = \sum_i a_{ii} x_i^2$ (see 14.5.4.1), and, if we set $q' = \sum_{i,j} a'_{i,j} x_i y_j$, the condition $q'(e_i) = 0$ implies $a'_{ii} = 0$ for all i, whence

$$\mathrm{Tr}(f) = \sum_i a'_{ii} a_{ii}^{-1} = 0$$

(cf. 13.2.0). The converse is proved by induction on the dimension. Assume $n = 1$, and choose a basis $\{e_1, e_2\}$ such that $m = p(e_1)$ is the desired point in $\mathrm{im}(\alpha')$ and $n = p(e_2)$ its conjugate with respect to α. The form q will be written as $ax^2 + cy^2$, and the form q' as $2b'xy + c'y^2$; since the trace condition is $c^{-1}c' = 0$, we have $c' = 0$ and $q' = 2b'xy$, whence

$$n = p(e_2) \in \mathrm{im}(\alpha').$$

Now assume the result is true for $n - 1$; take $m \in \mathrm{im}(\alpha')$ and let $H = m^{\perp}$ be its conjugate hyperplane with respect to α. Write $q = \sum_i a_{ii} x_i^2$, with $m = p(e_1)$, and $q' = \sum_{i,j} a_{ij} x_i y_j$; thus $a'_{11} = 0$ and H has equation $x_1 = 0$. The trace of the isomorphism associated with the quadrics $\alpha \cap H$ and $\alpha' \cap H$ (which are proper because $m \notin \mathrm{im}(\alpha)$) is

$$\sum_{i=2}^{n+1} a_{ii}^{-1} a'_{11} = \sum_{i=1}^{n+1} a_{ii}^{-1} a'_{11} = 0;$$

by the induction assumption, there exist points $\{m_i\}_{i=2\ldots n+1}$ in $H \cap \mathrm{im}(\alpha')$ such that $m_i \perp m_j$ with respect to α for all $i \neq j$. Now the family

$$\{m\} \cup \{m_i\}_{i=2\ldots n+1}$$

satisfies all the desired conditions. □

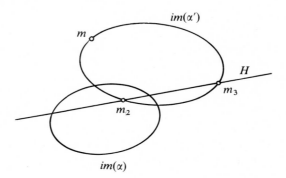

Figure 14.5.4.3

14.5.4.4. Notes. We have just seen an either-or statement in the spirit of 10.10.3 and 16.6: given two quadrics α and α', either there exists no simplex inscribed in α' and self-polar with respect to α, or there exist infinitely many of them (at least if K is closed).

When $\mathrm{Tr}(f) = 0$ we say that α' is *harmonically circumscribed* with respect to α.

For another, metrical, interpretation of the invariants associated with two quadratic forms, see 15.6.4.

14.5.4.5. Examples. Let E be a three-dimensional Euclidean vector space and $q' \in Q(E)$ a non-degenerate form with image $(q')^{-1}(0) \neq 0$ (this image is then a cone with vertex 0—see figure 14.5.4). Set $q = \|\cdot\|^2$. A simplex of $P(E)$ inscribed in $\alpha' = p(q')$ and self-polar with respect to $\alpha = p(q)$ corresponds to a set of three pairwise orthogonal lines on the cone (cf. 14.5.2). The proof of 14.5.4.3 shows that for the existence of such lines it is necessary that, if q' has the form

$$q' = ax^2 + a'y^2 + a''z^2 + 2byz + 2b'zx + 2b''xy,$$

we have $a + a' + a'' = 0$. Conversely, if $a + a' + a'' = 0$, there exists, for any line D on the cone, two other lines D' and D'', also on the cone, orthogonal to one another and to D. See also 14.8.9.

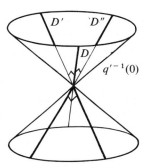

Figure 14.5.4.5

14.5.4.6. Hyperplanes in $PQ(E)$. We can now answer, at least geometrically, the question raised in 14.2.5: what are the hyperplanes of $PQ(E)$? Consider an arbitrary basis for D; by 13.1.3.6 a hyperplane of $PQ(E)$ will have the expression $\sum_{i \neq j} a_{ij}\lambda_{ij} = 0$, where the λ_{ij} are fixed elements of K. Call Λ the matrix with entries $\lambda_{ij} = \lambda_{ji}$; the condition $\sum_{i \neq j} a_{ij}\lambda_{ij} = 0$ is equivalent to $\mathrm{Tr}(A\Lambda) = 0$. If the λ_{ij} are such that $\det \Lambda \neq 0$, we can write $\mathrm{Tr}(A(\Lambda^{-1})^{-1}) = 0$, so the hyperplane will be geometrically defined as the set of all quadrics harmonically circumscribed to the fixed quadric having Λ^{-1} as an equation. See 14.8.8 and 14.8.10.

14.5.5. INVOLUTIVE CORRELATIONS. The polarity correspondence $\psi : P(E) \to \mathcal{H}(E)$ associated with a proper quadric is pregnant with geometric applications, as we shall see in the sequel. One of the most useful properties of this correspondence is 14.5.2.2: if $n \in \psi(m)$, then $m \in \psi(n)$. It is natural to look for other maps $f : P(E) \to \mathcal{H}(E)$ having some of the properties listed in 14.5.1, and 14.5.2.2 in particular. The complete solution is the object of

exercise 14.8.12, or can be looked up in [FL, p.270]. Here we limit ourselves
to a simple case, namely, $f = g$, where $g \in \mathrm{Isom}(E; E^*)$ (cf. 4.5). For $x \in E$,
the orthogonal hyperplane to x is $\{\, y \in E \mid g(x)(y) = 0 \,\}$, and we want to
find the maps $g \in \mathrm{Isom}(E; E^*)$ such that $g(x)(y) = 0$ implies $g(y)(x) = 0$ for
all x and y.

It is convenient to introduce here the (not necessarily symmetric) bilinear
form $P(x, y) = g(x)(y)$. Our condition is equivalent to the two linear forms
$y \mapsto P(x, y)$ and $y \mapsto P(y, x)$ having the same kernel; this implies that,
for every $x \in E$, there exists $k(x) \in K^*$ such that $P(x, y) = k(x)P(y, x)$.
Reasoning as in the proof of 8.8.5.1, we conclude that $k(x)$ does not depend
on x, that is, $P(x, y) = kP(y, x)$ for all $x, y \in E$, where $k \in K^*$. Then
$P(x, y) = k^2 P(x, y)$; since g is an isomorphism, there exist $x, y \in E$ with
$P(x, y) \neq 0$, and we have $k = \pm 1$. If $k = 1$ we have polarity with respect to
the proper quadric with polar form P; the case $k = -1$ is new. It can only
happen when the dimension of E is even, by the properties of alternating
forms. This case gives rise to interesting geometric phenomena, and explains
the miracle of the Möbius tetrahedra observed in 4.9.12. For more details,
see 14.8.12 and [FL, 270].

14.5.6. THE DEGENERATE CASE. If α is no longer proper, we can still
use $\phi : E \to E^*$ to define orthogonality. For example, for $m \in p(x)$, we
have $m^\perp = \{\, p(y) \mid P(x, y) = 0 \,\}$. Here, however, $m^\perp$ will be the whole of
$P(E)$ when $x \in \mathrm{rad}(q)$, and we can have $m^\perp = n^\perp$ even though $m \neq n$. For
the sake of simplicity, we will not discuss this bad duality relation, but the
reader should study it as an exercise, since it occurs naturally: a pencil of
quadrics, for example, generally contains degenerate quadrics (see an example
in 16.4.10).

14.6. Tangential quadrics and tangential equation

14.6.1. Let α be a proper quadric in $P(E)$. Consider, in $P(E^*) = \mathcal{H}(E)$,
the set $\{\, m^\perp \mid m \in \mathrm{im}(\alpha) \,\}$ of tangent hyperplanes to α (cf. 14.5.2.1). What
kind of a set is it? According to 14.5.1, it is exactly $\phi(\mathrm{im}(\alpha))$, so it is the
image of a quadric, the quadric $\alpha^* = \psi(\alpha)$, since the properness of α implies
$\phi \in \mathrm{Isom}(P(E); P(E^*))$.

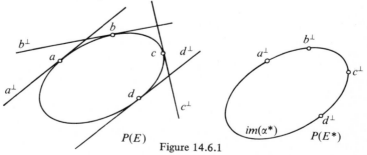

Figure 14.6.1

If A is the matrix of the equation q of α in a basis of E, the matrix of the equation $(\phi^{-1})^*(q)$ of α^* in the dual basis of E^* will be ${}^tA^{-1}AA^{-1} = A^{-1}$, according to 14.1.4.3 and 13.2.0. We say that $(\phi^{-1})^*(q) = q^*$ is a *tangential equation* or *envelope equation* for α^*. Practically speaking, $q^*(\xi) = 0$ is the necessary and sufficient condition for the hyperplane $\xi^{-1}(0)$ to be tangent to α; in the coordinates $\xi = (\xi_1, \ldots, \xi_{n+1})$, this equation reads $\sum_{i,j} a^*_{ij} \xi_i \xi_j$, where the a^*_{ij} are the entries of A^{-1} (cf. 14.8.13).

14.6.2. DEFINITION. *A quadric in $P(E^*)$ is called a tangential quadric. A tangential pencil of quadrics is a line in* $\mathrm{PQ}(E^*)$.

14.6.3. EXAMPLES.

14.6.3.1. For a vector space F, denote by $PPQ(F)$ the set of proper quadrics of F. By 14.6.1, the operation $*$ defines a bijection

$$* : PPQ(E) \to PPQ(E^*)$$

which satisfies $** = \mathrm{Id}_{PPQ(E)}$. There is no natural extension of this map to the whole of $\mathrm{PQ}(E)$, however:

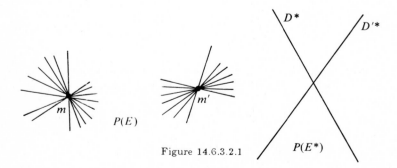

Figure 14.6.3.1

14.6.3.2. Take $E = \mathbf{R}^3$ and let α^* be the conic in $P(E^*)$ having equation $2\xi\eta$, where (ξ, η, ς) are the canonical homogeneous coordinates of $P(E^*)$. The quadric α^* is degenerate and consists of the two lines D^* and D'^* with equations $\xi = 0$ and $\eta = 0$, respectively. In $P(E)$ the set of lines of D^* and

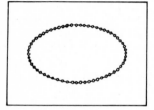
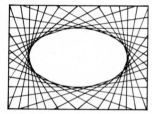

Figure 14.6.3.2.1

D'^* consists of all lines going through either $(1,0,0)$ or $(0,1,0)$; this set can never be the set of tangents to a conic of $P(E)$ (cf. 14.1.5.3).

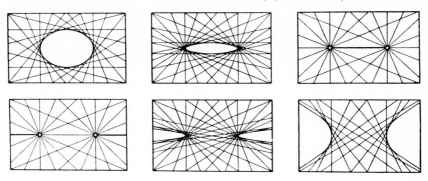

Figure 14.6.3.2.2

14.6.4. One can generalize the situation of 14.6.1 and 14.6.3.1 as follows: take α and β in $PPQ(E)$, and consider the subset $S = \{\, m^\perp \mid m \in \mathrm{im}(\beta)\,\}$, where $m^\perp$ denotes the polar hyperplane to m with respect to α. Is this the image of a tangential quadric? Yes, namely, $S = \mathrm{im}\big(\psi_\alpha(\beta)\big)$, where $\psi_\alpha : P(E) \to P(E^*)$ is the isomorphism associated to α in 14.5.1. We can apply the bijection $*^{-1} = *$ of 14.6.3.1 to $\psi_\alpha(\beta) \in PPQ(E^*)$ to obtain a quadric in $PPQ(E)$, called the *polar quadric* of β with respect to α, and denoted by β_α^*.

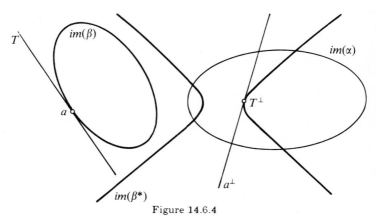

Figure 14.6.4

If A and B denote the matrices of α and β in a fixed basis of E, the matrix of $\phi_\alpha(\beta)$ in the dual basis of E^* will be given by ${}^t A^{-1} B A^{-1}$, by 14.1.4.3. The matrix of β_α^* will be

$$\big({}^t A^{-1} B A^{-1}\big)^{-1} = A B^{-1} A,$$

by 14.6.1.

We have $(\beta_\alpha^*)_\alpha^* = \beta$. If $a \in \mathrm{im}(\beta)$, the hyperplane $a^\perp$ is tangent to $\mathrm{im}(\beta_\alpha^*)$ at the point $T^\perp$, where T is the tangent hyperplane to β at a.

14.6.5. EXAMPLES.

14.6.5.1. In the affine case, we leave to the reader the task of detailing the relation between the above and 11.1.5.

14.6.5.2. If α and β are represented by $q = x^2 + y^2 - z^2$ and $q' = x^2 - y^2 - z^2$, we have $\beta_\alpha^* = \beta$ (figure 14.6.5; see 14.8.14).

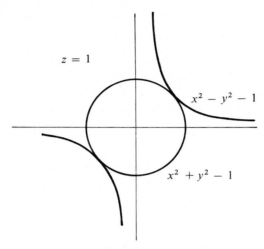

Figure 14.6.5

14.7. The group of a proper quadric

14.7.1. Let α be a proper quadric in $P(E)$. Any map $f \in O(q)$, where q is an equation for α, yields a map $\underline{f} \in \mathrm{GP}(E)$ such that $\underline{f}(\mathrm{im}(\alpha)) = \mathrm{im}(\alpha)$. The converse is false: if $\mathrm{im}(\alpha) = \emptyset$, every map $g \in \mathrm{GP}(E)$ leaves $\mathrm{im}(\alpha)$ invariant! But if K is algebraically closed, we see from 14.1.6.2 that $\underline{f}(\mathrm{im}(\alpha)) = \mathrm{im}(\alpha)$ is indeed equivalent to $f \in O(q)$. This justifies the following

14.7.2. DEFINITION. *The group of the quadric α, denoted by* $\mathrm{PO}(\alpha)$, *is defined as*

$$\mathrm{PO}(\alpha) = \{ \, \underline{f} \in \mathrm{GP}(E) \mid f \in O(q) \, \},$$

where q is an equation for α. We also call $\mathrm{PO}(\alpha)$ *the orthogonal projective group of α.*

The fact that $\mathrm{PO}(\alpha)$ depends only on α and not on q follows from 4.5. From 13.9.13 we deduce that $\mathrm{PO}(\alpha) = O(q)/\pm\mathrm{Id}_E$. When $n+1$ is even we can define $\mathrm{PO}^+(\alpha)$, since then $-\mathrm{Id}_E \in O^+(q)$; this fact is related to the orientability of $P(E)$ in the real case.

14.7.3. From 13.7.5 and 13.7.9 we immediately deduce that all the maximal subspaces contained in $\mathrm{im}(\alpha)$ have same dimension, equal to the index of q minus one, and that they are all conjugate under the action of $\mathrm{PO}(\alpha)$. The same holds for the action $PQ^+(\alpha)$ when the dimension of $P(E)$ is odd, except when q is neutral, in which case there are two conjugacy classes. In particular $PO(\alpha)$ (and also $PO^+(\alpha)$ if it exists) acts transitively on the points of $im(\alpha)$.

This applies to the quadrics of 14.4, for example.

14.7.4. GEOMETRIC REALIZATION OF HYPERPLANE REFLECTIONS. Take a proper quadric $\alpha \in PPQ(E)$ with equation q, and let $f \in O(q)$ be the reflection through the hyperplane $U \subset E$ (cf. 13.6.6), a component of the orthogonal direct sum decomposition $E = U \oplus^\perp D$, where D is necessarily a line. By 6.4.6 we can obtain $\underline{f} \in PO(\alpha)$ geometrically in the following way: if $m = p(D)$ and $H = p(U)$, the image $\underline{f}(t)$ of a point $t \in P(E) \setminus m$ is the point on the line $\langle m, t \rangle$ satisfying $[m, \langle m, t \rangle \cap H, t, \underline{f}(t)] = -1$. Of course, H is exactly the polar hyperplane $m^\perp$ of m with respect to α.

If necessary, we could use 14.5.2.6 to show that $im(\alpha)$ is indeed left globally invariant by $\underline{f}$. Notice that hyperplane reflections generate $PO(\alpha)$, by 13.7.12.

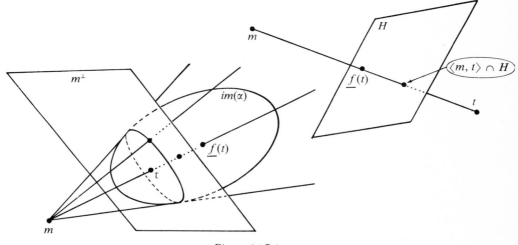

Figure 14.7.4

14.8. Exercises

14.8.1. Show that 14.1.6.2 still holds if $K = \mathbf{R}$ and q satisfies $q(E) = \mathbf{R}$.

14.8.2. Fix a point $m \in P(E)$ and a line D containing m; show that the set

$$\{ \alpha \in PQ(E) \mid m \in im(\alpha) \text{ and } D \text{ is tangent to } \alpha \}$$

is a subspace. Is this still true if $m \notin D$?

14.8.3. Study the orientability of images of quadrics for $K = \mathbf{R}$ and $\mathbf{C}$.

* **14.8.4.** Show that $C(n)$ (cf. 14.3.7) is homeomorphic to the grassmannian of oriented lines of $P^n(\mathbf{R})$.

14.8.5. Study whether the image of a degenerate quadric can be a submanifold (for $K = \mathbf{R}$ or $\mathbf{C}$).

14.8.6. Let $\mathcal{F}$ be a pencil of quadrics in $P(E)$ and α a proper quadric. Show that $\mathcal{F}$ contains at least one quadric harmonically circumscribed around α, and that if it contains two, then all its quadrics have that property.

14.8.7. Consider $n(n + 3)/2$ pairs of points $\{m_i, n_i\}$ in $P(E)$; show that there exists at least one quadric such that $m_i \perp n_i$ for every i (work in the context of 14.5.6).

14.8.8. Study 14.5.4.6 when $\det \Lambda = 0$.

14.8.9. Show that 14.5.4.3 remains valid if $K = \mathbf{R}$, α' is arbitrary and $\mathrm{im}(\alpha) = \emptyset$.

* **14.8.10.** We say that the proper quadric α' is *harmonically inscribed* in α if $\mathrm{Tr}(\phi'^{-1} \circ \phi) = 0$. Interpret this condition geometrically.

* **14.8.11.** Let α be a proper conic, $\{a, b, c\}$ and $\{a', b', c'\}$ two self-polar triangles with respect to α; show that the six points a, b, c, a', b' and c' belong to one single conic.

14.8.12. CORRELATIONS (consult [FL, 260 ff.] if necessary).

14.8.12.1. A *correlation* in $P(E)$ is a bijection $f : \Gamma \to \Gamma$ from the set Γ of subspaces of $P(E)$ into itself satisfying the following condition: for any U and $V \in \Gamma$ such that $U \subset V$ we have $f(U) \supset f(V)$. Show that if m is a point in $P(E)$ then $f(m)$ is a hyperplane of $P(E)$. Show that there exists a canonical correlation in $P(E)$.

14.8.12.2. The correlation f is called *involutive* if $f^2 = \mathrm{Id}_\Gamma$. Apply the fundamental theorem of projective geometry (5.4.8) to show that, if f is an involutive correlation, f is obtained from a map $g : P(E) \to P(E^*)$ satisfying either of the following conditions:

i) there exists an alternating bilinear form P such that, for any $m = p(x) \in P(E)$, we have $g(m) = p(\{ y \in E \mid P(x, y) = 0 \})$;

ii) there exists $P : E \times E \to K$ giving rise to g as in condition (i), but P is sesquilinear for some involutive automorphism $\sigma : K \to K$, that is, P is linear in the first variable and $P(y, x) = \sigma(P(x, y))$ for all $x, y \in E$.

14.8.12.3. Study non-involutive correlations.

14.8.12.4. Show (cf. 4.9.12) that, if $\dim P(E)$ is odd, there exist pairs of simplices of $P(E)$ such that each vertex of the first belongs to a face of the second and vice versa.

14.8.12.5. Let $\mathcal{T}$ be a torsor in a three-dimensional Euclidean affine space X (cf. [LF–AR, chapter 4] or [BE, chapter 3]). At each point $m \in X$, consider the set of lines with moment zero with respect to $\mathcal{T}$ and containing m. Show that these lines form a plane $H(m)$; show that the map $m \mapsto H(m)$ is the restriction to X of an involutive correlation of the projectivization $\hat{X}$ of X (cf. 5.1), satisfying the first condition in 14.8.12.2.

14.8.13. Calculate the envelope equation of the quadric $\sum_{i=1}^{n+1} k_i x_i^2 = 0$, where the k_i are non-zero for all i, and of the quadrics given in 13.9.6 and 13.9.8 when they are proper.

14.8.14. Given two proper quadrics β and γ having a common self-polar simplex, find all proper quadrics α such that $\gamma = \beta_\alpha$. Use 16.4.10 to solve the same problem when β and γ are arbitrary proper conics.

14.8.15. Study whether it is possible to find a proper quadric α tangent to $n(n+3)/2$ given hyperplanes in $P(E)$. Write a complete discussion in the case $n = 2$ (cf. 16.1.4).

14.8.16. GÉRARDIN'S CONSTRUCTION. Let E be a two-dimensional vector space over a field K with characteristic different from two. Denote by $D = P(E)$ the associated projective line, by $\operatorname{End} E$ the vector space of all endomorphisms of E and by $P(\operatorname{End} E)$ the projective space associated with $\operatorname{End} E$. Show that the determinant map $\det : \operatorname{End} E \to K$ is a quadratic form. Denote by $Q(D)$ the associated quadric in $P(\operatorname{End} E)$. Show that $Q(D)$ is a neutral quadric; study the correspondence between its generating lines and the endomorphisms of E.

Study the conic $C(D)$ obtained by cutting $Q(D)$ by the hyperplane associated with the trace linear form $\operatorname{Tr} : \operatorname{End} E \to K$. Is every conic of the form $C(D)$? Same question for $Q(D)$.

For more details, see [GN, chapter IV].

Chapter 15
Affine quadrics

This chapter presents no essential difficulty, the only subtler points being the characterization of quadrics by the existence of diameters in all directions (cf. 15.5.9) and the passage from affine spaces to their projectivizations. Otherwise all results and objects arise from combining chapters 3 and 14.

Affine quadrics, however, have a considerable practical importance: they are the simplest curves and surfaces after lines and planes, and they are found everywhere in two and three dimensions, not only in mathematics, but also in physics and astronomy. For example, the orbits of planets and comets are, to a first approximation, conical sections; the tip of the Eiffel Tower, like that of any tall structure, moves in an elliptic orbit under the action of the wind, and one does well to make sure that the amplitude of that ellipse is small; and the simplest spline that can be used to join two lines is a parabola (cf. 15.7.6).

The particular case of Euclidean conics occupies the whole chapter 17. For more information on affine and projective quadrics, see [DQ], [ML], [PE] and [EE].

We shall employ in this chapter the following notation (cf. chapter 5):
X is an affine space of finite dimension $n \geq 1$ over a commutative field K
   of characteristic different from two.
$\hat{X}$ is the universal vector space associated with X, of dimension $n + 1$.
$\tilde{X} = P(\hat{X})$ is the projectivization of X; we have $X \cong X \cup \infty_X$, where
   $\infty_X = P(\vec{X})$ is the hyperplane at infinity of X.
$Q(X) = P^2(X)$ is the vector space of affine quadratic forms on X (cf. 3.3).
q is the symbol of a quadratic form $q \in Q(X)$; if we vectorialize X at a
   and write $q = q_2 + q_1 + q_0$, with $q_0 \in K$ and $q_1 \in X_a^*$, we can identify q_2
   with $\vec{q}$.
$\hat{q}$ is the element of $Q(\hat{X})$ such that $q = \hat{q}|_X$ (cf. 3.3.14).

15.1. Definitions and notation

The simplest definition for an affine quadric is obtained by translating
14.1.1 into the affine framework; the condition $\vec{q} \neq 0$ guarantees that q is
really of degree two.

15.1.1. DEFINITION. *An (affine) quadric in X is an element α of $P\big(Q(X)\big)$*
such that $\vec{q} \neq 0$, where $q \in Q(X)$ is such that $\alpha = p(q)$. The set of quadrics
in X is denoted by $\mathrm{QA}(X)$. For $n = 2$ quadrics are called conics. The image
of α is $\mathrm{im}(\alpha) = q^{-1}(0)$. We say that q is an equation for α.

15.1.1.1. Note. It is understood, and we will not always mention explicitly,
that q is an equation for α. Also, it is clear that $\mathrm{im}(\alpha)$ only depends on α,
not on q; any notion associated to α from now on should depend only on
$\alpha = p(q)$, not on q.

15.1.2. EXAMPLES

15.1.2.1. If E is a Euclidean affine space, the spheres $S(a, r)$, for $a \in E$ and
$r \geq 0$, are images of quadrics. Just take $q = d^2(a, \cdot) - r^2$. Notice that, for
every $k \in \mathbf{R}$, the equation $q = d^2(a, \cdot) - k$ defines a quadric, but with empty
image if $k > 0$.

15.1.2.2. Let X and X' be affine spaces, $f \in \mathrm{Isom}(X; X')$ and $\alpha \in \mathrm{QA}(X)$.
The *image* of α under f, denoted by $f(\alpha)$, is the quadric having $(f^{-1})^*(q)$
for equation (we recall that $(f^{-1})^*(q)$ is defined by

$$(f^{-1})^*(q)(x') = q\big(f^{-1}(x')\big)$$

for every $x' \in X'$). Image quadrics satisfy $\mathrm{im}\big(f(\alpha)\big) = f\big(\mathrm{im}(\alpha)\big)$. In particular, $\mathrm{GA}(X)$ acts on $\mathrm{QA}(X)$, and the classification of quadrics in X is the
search for the orbits of this action (cf. 15.2 and 15.3).

15.1.3. EQUIVALENT DEFINITION.

15.1.3.1. Take $\alpha \in QA(X)$, and let q be an equation for α; the form $\hat{q} \in Q(\hat{X})$ gives rise to a projective quadric $\tilde{\alpha} = p(\hat{q}) \in PQ(\hat{X})$, since $\widehat{\lambda q} = \lambda \hat{q}$ (by 3.3.13 or 3.3.15, for example). Since $q = \hat{q}|_X$, we have $\mathrm{im}(\alpha) = \mathrm{im}(\tilde{\alpha}) \cap X$. What is the set $\mathrm{im}(\tilde{\alpha}) \cap \infty_X$? Since $\vec{q} = \hat{q}|_{\vec{X}}$ and $\infty_X = P(\vec{X})$, we have $\mathrm{im}(\tilde{\alpha}) \cap \infty_X = \mathrm{im}(\vec{\alpha})$, where $\vec{\alpha} = p(\vec{q}) \in PQ(\vec{X})$. Or again, by 14.1.3.3, $\vec{\alpha} = \tilde{\alpha} \cap \infty_X$.

Conversely, let $P(E)$ be a finite-dimensional projective space, $P(H)$ a hyperplane in $P(E)$ and β a quadric in $P(E)$. If we want a quadric in $X = P(E) \backslash P(H)$ (cf. 5.1), we identify ∞_H with $P(H)$ and, letting r be an equation for β, we conclude from 3.3.14 that $r|_X$ will define a quadric $\beta|_X$ in X if and only if $r|_X \neq 0$. Equivalent conditions are that $\beta \cap P(H)$ be a quadric (cf. 14.1.3.3), or that $\infty_X \not\subset \mathrm{im}(\beta)$. Thus we have the following result:

15.1.3.2. Proposition. *There exists a natural bijection ⁀ between affine quadrics α in X and projective quadrics β in $\tilde{X}$ whose image does not contain ∞_X. We have $\tilde{\alpha} = p(\hat{q})$ for any equation q of α, and $\tilde{\alpha} \cap \infty_X = \vec{\alpha}$ (where $\vec{\alpha} = p(\vec{q})$) is called the quadric at infinity of α (for $n = 2$ we talk about points at infinity, for $n = 3$ about the conic at infinity). We have $\mathrm{im}(\alpha) = \mathrm{im}(\tilde{\alpha}) \cap X$. By definition, $\alpha \in QA(X)$ is proper if $\tilde{\alpha}$ is; similarly, the rank and image of α are those of $\tilde{\alpha}$.*

15.1.4. NOTE. The reader should check, as an exercise, that the inclusion $\mathrm{im}(\beta) \supset \infty|_X$ can only take place in two cases: either $\mathrm{rank}(r) = 1$, hence $\mathrm{im}(\beta) = \infty_X$ and $\mathrm{im}(\beta) \cap X = \emptyset$, or $\mathrm{rank}(r) = 2$ and $\mathrm{im}(\beta) = \infty_X \cup H$, where H is a hyperplane distinct from ∞_X, hence $\mathrm{im}(\beta) \cap X$ is an (affine) hyperplane of X.

15.1.5. With an eye toward practical applications, we will introduce the pair $(\mathrm{rank}(\alpha), \mathrm{rank}(\tilde{\alpha}))$, for $\alpha \in QA(X)$. Observe that this pair is invariant under the action of $GA(X)$ on $QA(X)$ (cf. 5.2.2).

15.1.6. EXPRESSIONS. For a fixed affine frame of X and using a notation similar to that of 3.3.15, we have

15.1.6.1 $$q = \sum_{i,j} a_{ij} x_i x_j + 2 \sum_i b_i x_i + c, \qquad \vec{q} = \sum_{i,j} a_{ij} x_i x_j;$$

in the associated homogeneous coordinates,

15.1.6.2 $$\tilde{q} = \sum_{i,j} a_{ij} x_i x_j + 2 \sum_i b_i x_i t + ct^2.$$

We associate to q two matrices $\hat{A}$ and $\vec{A}$, as follows:

15.1.6.3 $$\vec{A} = (a_{ij}), \qquad \hat{A} = \begin{pmatrix} & & & b_1 \\ & \vec{A} & & \vdots \\ & & & b_n \\ b_1 & \cdots & b_n & c \end{pmatrix}.$$

Then $\operatorname{rank}(\alpha) = \operatorname{rank}(\hat{A})$ and $\operatorname{rank}(\vec{\alpha}) = \operatorname{rank}(\vec{A})$.

15.2. Reduction of affine quadratic forms

15.2.1. From 13.1.4.5 we know that it is hopeless to try to classify affine quadrics (cf. 15.1.2.2). But we can reduce them to three types, as follows. Let $\alpha = p(q)$, with $q \in Q(X)$; by 13.4 there exists a basis of $\vec{X}$ in which $\vec{q}$ has the form $\vec{q} = \sum_{i=1}^{r} a_i x_i^2$, where $r = \operatorname{rank}(\vec{\alpha})$ and $a_i \neq 0$ for $i \leq r$.

In an affine frame associated with this basis, we have, from 15.1.6.1,

$$q = \sum_{i=1}^{r} a_i x_i^2 + 2 \sum_{i=1}^{n} b_i x_i + c, \quad \prod_{i=1}^{r} a_i \neq 0;$$

after performing a change of variables, $x_i \to x_i + b_i/a_i$, we can write q as

$$q = \sum_{i=1}^{r} a_i x_i^2 + 2 \sum_{i=r+1}^{n} b_i x_i + c, \quad \prod_{i=1}^{r} a_i \neq 0.$$

If $r < n$, we can perform another change of variables, $\sum_{i=r+1}^{n} b_i x_i \to b_n x_n$, to get

$$q = \sum_{i=1}^{r} a_i x_i^2 + 2 b_n x_n + c, \quad \prod_{i=1}^{r} a_i \neq 0,$$

and if, in addition, $b_n \neq 0$, we can change x_n into $x_n + c/(2b_n)$ to obtain

$$q = \sum_{i=1}^{r} a_i x_i^2 + 2 x_n, \quad \prod_{i=1}^{r} a_i \neq 0.$$

Thus q is necessarily of one of the types below:

15.2.2
$$\begin{cases} \text{type I: } q = \sum_{i=1}^{r} a_i x_i^2, \quad \prod_{i=1}^{r} a_i \neq 0, \quad 1 \leq r \leq n; \\[2mm] \text{type II: } q = \sum_{i=1}^{r} a_i x_i^2 + 1, \quad \prod_{i=1}^{r} a_i \neq 0, \quad 1 \leq r \leq n; \\[2mm] \text{type III: } q = \sum_{i=1}^{r} a_i x_i^2 + 2 x_n, \quad \prod_{i=1}^{r} a_i \neq 0, \quad 1 \leq r \leq n-1. \end{cases}$$

Table 15.2.3 shows, because of 15.1.5, that quadrics of different types cannot be images of one another under an element of $GA(X)$:

15.2.3.

α arbitrary   α proper

type	rank($\tilde{\alpha}$)	rank($\vec{\alpha}$)
I	r	r
II	$r+1$	r
III	$r+2$	r

type	rank($\tilde{\alpha}$)	rank($\vec{\alpha}$)
II	$n+1$	n
III	$n+1$	$n-1$

15.2.4. NOTES. In certain cases explicit calculations are facilitated by the use of the "equations of the center" (cf. 15.5.4).

For another characterization of the three types of quadrics and for geometric consequences of this reduction, see 15.7.4.

15.3. Classification of real and complex affine quadrics

15.3.1. PROPOSITION. *If $K = \mathbf{C}$, the orbits of $\mathrm{QA}(X)$ under the action of $\mathrm{GA}(X)$ are represented by the forms below (expressed in a fixed basis):*

$$\mathrm{I}(r): \quad \sum_{i=1}^{r} x_i^2, \quad 1 \le r \le n;$$

$$\mathrm{II}(r): \quad \sum_{i=1}^{r} x_i^2 + 1, \quad 1 \le r \le n;$$

$$\mathrm{III}(r): \quad \sum_{i=1}^{r} x_i^2 + 2x_n, \quad 1 \le r \le n-1.$$

In particular, there exist only two orbits of proper quadrics: $\sum_{i=1}^{n} x_i^2 + 1$ and $\sum_{i=1}^{n-1} x_i^2 + 2x_n$.

Proof. This follows immediately from the proof of 13.4.2.   □

For the real case, observe that we can switch r and s in types I and III, since we're working with $\mathrm{QA}(X)$ and not $Q(X)$. All remaining orbits, as stated in the proposition below, are distinct: just apply 13.4.7 to $\vec{q}$ and $\hat{q}$.

15.3.2. PROPOSITION. *If $K = \mathbf{R}$, the orbits of $\mathrm{QA}(X)$ under the action of $\mathrm{GA}(X)$ are represented by the forms below (expressed in a fixed basis):*

$$\mathrm{I}(r,s): \quad \sum_{i=1}^{r} x_i^2 - \sum_{i=r+1}^{r+s} x_i^2, \quad r \ge s, \quad 1 \le r+s \le n;$$

$$\mathrm{II}(r,s): \quad \sum_{i=1}^{r} x_i^2 - \sum_{i=r+1}^{r+s} x_i^2 + 1, \quad 1 \le r+s \le n;$$

$$\mathrm{III}(r,s): \quad \sum_{i=1}^{r} x_i^2 - \sum_{i=r+1}^{r+s} x_i^2 + 2x_n, \quad r \ge s, \quad 1 \le r \le n-1. \qquad □$$

15.3.3. LOW-DIMENSIONAL EXAMPLES.

15.3.3.1. For $K = \mathbf{C}$ and $n = 2$, the image of the conic α with equation $x_1^2 + x_2^2 + 1$ is homeomorphic to the cylinder $\mathbf{R} \times S^1$. In fact, the image of $\tilde{\alpha}$ is homeomorphic to S^2 (cf. 14.3.6) and $\mathrm{im}(\alpha) = \mathrm{im}(\tilde{\alpha}) \setminus \infty_X$ is obtained by removing from $\mathrm{im}(\alpha)$ the set $\mathrm{im}(\alpha) \cap \infty_X$, which consists of two distinct points; it is well-known that a sphere minus two points is homeomorphic to $\mathbf{R} \times S^1$. If α has equation $x_1^2 + 2x_2$, the image is the graph of the equation $x_2 = -x_1^2/2$, which is homeomorphic to $\mathbf{C} = \mathbf{R}^2$.

15.3.3.2. For $K = \mathbf{R}$ and $n = 2$, the proper conics are II(2,0), II(1,1), II(0,2) and III(1,0). The first has empty image; the other three are called *hyperbolas*, *ellipses* and *parabolas*, respectively.

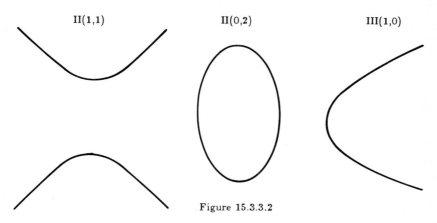

II(1,1) II(0,2) III(1,0)

Figure 15.3.3.2

15.3.3.3. For $K = \mathbf{R}$ and $n = 3$, the proper quadrics with non-empty image are II(2,1), II(1,2), II(0,3), III(2,0) and III(1,1), and are called, respectively, *two-sheet hyperboloids*, *one-sheet hyperboloids*, *ellipsoids*, *elliptic paraboloids* and *hyperbolic paraboloids* (figure 15.3.3.3.5). Quadrics of type II(1,2) and III(1,1) contain affine lines, since for them the form $\tilde{q}$ is neutral (cf. 13.1.4.3, 14.4 and 14.4.6).

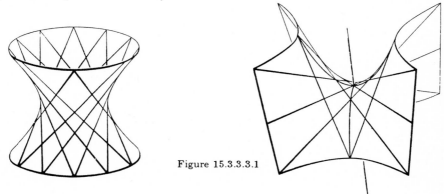

Figure 15.3.3.3.1

Figure 15.3.3.3.2 (Source: [H–C])

Opposite: Figure 15.3.3.3.3 (Source: [JE])

Foundry for the Lohr Ironworks—Lohr, Germany

The unusual shell forms of the foundry at Lohr are the result of efforts to find a characteristic form for this building type. To get rid quickly of the smoke formed during work and
to ensure thorough ventilation, two courses are possible: one implies—as is usual in the
USA—complete air-conditioning, while the other is concerned to achieve a natural system
of ventilation by the form of the building and without elaborate technical equipment. The
clients, in this instance, expressed the wish that the remedy should not be sought in any
over-elaborate technical system. The solution was found in a covering for the hall which
gradually tapers upwards, a form acting like a chimney to achieve complete elimination of
polluted air. The design was studied experimentally, and the result justified expectations
[...] The natural lighting of the hall offered a further problem. As the Northlight type of
factory has proved extraordinarily successful in this respect, the funnel-shaped shell was
designed to include a large North-East roof light. The single "hood" is 44 ft 3 in × 49 ft
1 in on plan, and rest on four supports. In shape it consists of two mirrorlike hyperbolic
paraboloid surfaces with three straight edges, the fourth, along which they are joined, being
arched. The glazed surface is an inclined plane. An air extraction cowl in the form of a
[one-sheet] hyperboloid [of revolution] terminates the design. The low points of the shell
roof are joined together by a bent tie, secured in the middle—just in front of the bending
point—by two diagonals lying in the glazed surface. The forces acting in the shell where
calculated by experiments with models. For this purpose a 1 in 10 scale plaster prototype
was constructed by the engineer, in which the prevailing stresses were ascertained by means
of an electrical tensioning method. Concreting of the shell was carried out on scaffolding
supported on the traversing crane rails. It could therefore be lowered, moved forward, and
used as shuttering for the next section. As the shell's angle of slope varies between 20° and
90°, none of the conventional types of roof covering could be used, and it was consequently
decided to do without. Experiments with coating of special kinds of resinous mastic are
not yet concluded, but the shells were still watertight after two winters, despite the absence
of any roof covering.   (Reprinted with permission from [JE].)

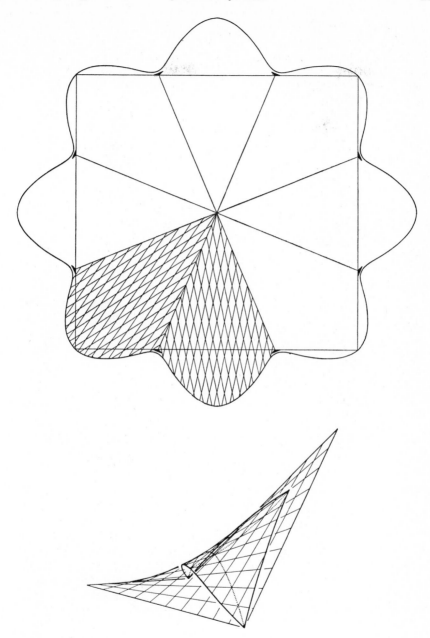

Opposite and above: Figure 15.3.3.3.4 (Source: [JE])

Los Manantiales Restaurant—Xochimilco, D.F., Mexico

The shell is composed of eight equal parts of a hyperbolic paraboloid. The diagram shows the geometry of any one part.

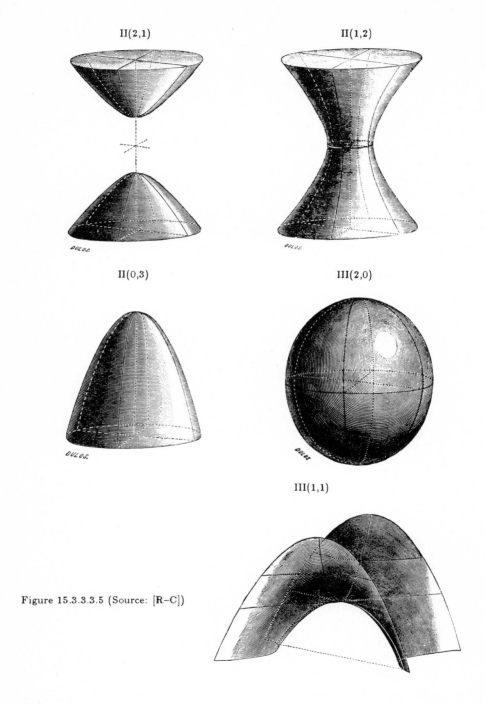

Figure 15.3.3.3.5 (Source: [R–C])

15.4. The topology of real and complex affine quadrics

15.4.1. We shall assume here that $K = \mathbf{R}$ or $\mathbf{C}$ and that $\alpha \in \mathrm{QA}(X)$ is proper, leaving to the reader the task of studying the degenerate case. The complex case is easy only when α is of type III, for then $\mathrm{im}(\alpha)$ is the graph of the equation $x_n = \frac{1}{2} \sum_{i=1}^{n-1} x_i^2$ (cf. 15.3.1), so that the image is homeomorphic to $\mathbf{C}^{n-1} \cong \mathbf{R}^{2(n-1)}$. If α is of type II, $\mathrm{im}(\alpha)$ is equal to $\mathrm{im}(\tilde\alpha) \setminus \infty_X$, where $\mathrm{im}(\tilde\alpha)$ is the set $C(n+1)$ of 14.3.7 (whose topology is not simple), and $\mathrm{im}(\tilde\alpha) \cap \infty_X$ is homeomorphic to $C(n)$. We will not discuss this case further.

15.4.2. On the other hand, for $K = \mathbf{R}$, we shall see that the homeomorphism types of images of proper quadrics are all very simple. On figure 15.4.3 we see the image of $\mathrm{II}(2,1)$, which is homeomorphic to $\mathbf{R}^2 \times \{\text{two points}\}$, that is, to $\mathbf{R}^2 \times S^0$, where S^0 is the zero-dimensional sphere in $\mathbf{R}$. Similarly, $\mathrm{II}(1,2)$ is homeomorphic to $\mathbf{R} \times S^1$ and $\mathrm{II}(0,3)$ to S^2. This is a general phenomenon:

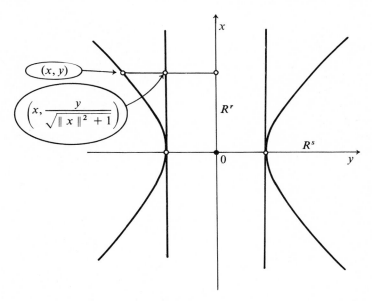

Figure 15.4.3

15.4.3. PROPOSITION. *The image of a proper affine quadric of type* $\mathrm{II}(r,s)$ *is homeomorphic to* $\mathbf{R}^r \times S^{s-1}$.

Proof. Choose a basis in which α has the expression

$$\sum_{i=1}^{r} x_i^2 - \sum_{i=r+1}^{n} x_i^2 + 1,$$

identify X with $\mathbf{R}^{r+s}$ via this basis, and consider $\mathbf{R}^{r+s}$ as a product $\mathbf{R}^r \times \mathbf{R}^s$ of Euclidean spaces, so a point in X can be written (x, y), with $x \in \mathbf{R}^r$ and $y \in \mathbf{R}^s$. Then $\mathrm{im}(\alpha)$ is the set of solutions of the equation $\|x\|^2 - \|y\|^2 + 1 = 0$. Now the map

$$(x, y) \mapsto \left(x, \frac{y}{\sqrt{\|x\|^2 + 1}} \right)$$

is a C^∞ diffeomorphism f of $\mathbf{R}^{r+s}$, and it transforms points (x, y) such that $\|x\|^2 - \|y\|^2 + 1 = 0$ into points such that $\|y\| = 1$; this shows that $f(\mathrm{im}(\alpha)) = \mathbf{R}^r \times S^{s-1}$. □

15.4.4. As to quadrics of type III, we have already remarked in 15.4.1 that their images, being graphs of continuous functions, are homeomorphic to $\mathbf{R}^{n-1}$.

15.4.5. IMAGES AS SUBMANIFOLDS. It is clear (cf. [B–G, 56], for example) that the equations in 15.3.1 and 15.3.2, in the case of proper quadrics, define submanifolds, of real codimension one for $K = \mathbf{R}$ and two for $K = \mathbf{C}$. The tangent space to α at $m \in \mathrm{im}(\alpha)$ can be identified with the polar hyperplane $m^\perp$ of m with respect to α; this is proven as in 14.3.8, but the details are simpler.

15.4.6. CONVEXITY. The figures in 15.3 seem to indicate that, for certain values of r and s, the images of quadrics of type $\mathrm{II}(r, s)$ and $\mathrm{III}(r, s)$ are frontiers of convex subsets of X. In fact, we have the following result:

15.4.7. PROPOSITION. *The image of a proper real affine quadric is the frontier of a convex set if and only if the quadric is of type* $\mathrm{II}(0, n)$ *or* $\mathrm{III}(n - 1, 0)$. *For quadrics of type* $\mathrm{II}(n - 1, 1)$ *each of the two connected components of the image* (cf. 15.4.3) *is the frontier of a convex set.*

Proof. For quadrics of type $\mathrm{III}(r, s)$ the condition follows from 11.8.11.2, since the images are graphs. The case $\mathrm{II}(0, n)$ is the unit sphere, which is the frontier of the closed unit ball $\sum_{i=1}^n x_i^2 \leq 1$. The case $\mathrm{II}(n - 1, 1)$ gives a convex set, since here the image is the union of the graphs of the two equations

$$x_n = \pm \sqrt{\sum_{i=1}^{n-1} x_i^2 + 1},$$

which represent convex functions (cf. 11.8.11.2). There remains to show that $\mathrm{II}(r, n - r)$ is never the frontier of a convex set for $2 \leq r \leq n - 1$. We again apply 11.8.11.2, but this time to a local expression of the image only: for example, take the point x with all coordinates zero except $x_n = 1$, near which the image is the graph of the function

$$x_n = \sqrt{1 + \sum_{i=1}^r x_i^2 - \sum_{i=r+1}^{n-1} x_i^2}.$$ □

15.4.8. NOTE. In an affine space X the frontiers of convex sets and images of quadrics share the property that almost every line of X (cf. 14.1.3.4) not contained in the set in question intersects it in at most two points. A very subtle result of Marchaud [MD] asserts that the converse is true. This and similar results from "finite geometry" can be found in [H–K] and [PL].

15.5. Polarity with respect to a proper affine quadric

Here K is again arbitrary.

15.5.1. If α is a proper quadric in X, the *polarity* with respect to α is the polarity (in $\tilde{X}$) with respect to $\tilde{\alpha}$. Whenever possible, we consider this map as a polarity in X: for example, when the polar hyperplane $m^\perp$ of $m \in X$ with respect to α is not ∞_X, the intersection $m^\perp \cap X$ is a hyperplane in X, which is still called the *polar hyperplane* of m and denoted by $m^\perp$, by abuse of notation.

15.5.2. THE CENTER. It is to be expected that the polar hyperplanes of points in ∞_X and the pole of ∞_X play some kind of role in the affine setting. Indeed, let $c = \infty_{\tilde{X}}^\perp$ be the pole of the hyperplane ∞_X of $\tilde{X}$. Two cases are possible: either $c \in \infty_X$ or $c \in X$. Assume first that $c \in X$, and apply 14.7.4 to $H = \infty_X$ and $m = c$; taking 6.4.2 into account, we see that the reflection $-\operatorname{Id}_{X_c}$ through the point c belongs to the group of $\tilde{\alpha}$. In particular, it leaves $\operatorname{im}(\alpha)$ invariant, that is, c is a center of symmetry for $\operatorname{im}(\alpha)$. This observation can be sharpened as follows:

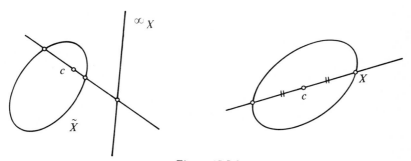

Figure 15.5.2

15.5.3. PROPOSITION AND DEFINITION. *If α is a proper affine quadric, the following conditions are equivalent:* (i) α *is of type* II *(cf. 15.2.2);* (ii) $\infty_{\tilde{X}}^\perp \notin X$; *and* (iii) *the hyperplane* ∞_X *is not tangent to* α.

If any of these conditions is satisfied, the point $c = \infty_{\tilde{X}}^\perp$ *is the center of symmetry of* $\operatorname{im}(\alpha)$, *called the center of* α; *we say that* α *is a central quadric. Quadrics of type* III *are called paraboloids.*

Proof. The equivalence between (ii) and (iii) is merely a rephrasing of 14.5.2.1. To show that (i) and (ii) are equivalent, we can use the equations of the center, given in 15.5.5.2 below: for the quadrics of type II, the homogeneous coordinates $(x_1, \ldots, x_n, t)$ of $\infty_X^\perp$ are $(0, \ldots, 0, 1)$, and for type III they can be taken to be $(0, \ldots, 0, 1, 0)$. The first of these points belongs to X, the second to ∞_X. $\square$

15.5.4. NOTE. One can work one's way backwards and use polarity to give a geometric proof for 15.2. This, however, is as complicated in the degenerate case as the explicit calculation given in 15.2, and the explicit calculation has the advantage of, allied with 13.4.8, yielding a complete reduction. In applying this reduction in practice, it is best to first calculate the center, if one exists, using 15.5.5.2, and then to use 15.2. See 15.7.4 and 15.7.10.

15.5.5. EXPRESSIONS. Let α be given by the equation q of 15.1.6.1, in a fixed affine frame. The polarity in $\tilde{X}$ with respect to α, expressed in the associated homogeneous coordinates $(x_1, \ldots, x_n, t)$, can be computed as follows (apply 13.1.3.6 and 14.5.3):

15.5.5.1. The polar hyperplane of $(\xi_1, \ldots, \xi_n, \theta)$ has equation

$$\sum_{i,j} a_{ij}\xi_i x_j + \sum_i b_i(\theta x_i + t\xi_i) + c\theta t = 0.$$

For example, in order to find the pole of ∞_X, we observe that the pole belongs to the polar hyperplanes of the n points $(1, 0, \ldots, 0)$ through $(0, \ldots, 0, 1, 0)$, which span ∞_X; this yields the equations for the center:

15.5.5.2 $$\frac{1}{2}\frac{\partial q}{\partial x_i} = \sum_i a_{ij}x_j + b_j = 0, \qquad i = 1, \ldots, n.$$

15.5.6. DIAMETERS. Consider a point $a \in \infty_X$ and its polar hyperplane $a^\perp$, and assume $a^\perp \neq \infty_X$. We see from 14.7.4 and 6.4.2 that the image of α is left globally invariant by the affine reflection through the affine hyperplane $a^\perp \cap X$ (also called $a^\perp$) and parallel to the lines whose point at infinity is a.

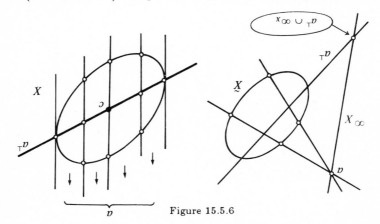

Figure 15.5.6

15.5.7. DEFINITION. *A diameter of a subset F of an affine space X is any hyperplane H for which that there exists a line direction $\vec{D}$ such that the reflection through H and parallel to $\vec{D}$ leaves F invariant.*

15.5.8. From 15.5.8 we see that the image of a proper affine quadric α has diameters. In fact, α has as many diameters as there are points $a \in \infty_X$ with $a^\perp \neq \infty_X$; but $a^\perp = \infty_X$ is equivalent to $a = \infty_X^\perp$, so there are only two possible cases:

15.5.8.1. First case. If α is a quadric with center c, the image $\text{im}(\alpha)$ has symmetries in all line directions, hence in all hyperplane directions, since the correspondence $a \mapsto a^\perp \cap \infty_X$ is the polarity with respect to $\vec{\alpha}$, which in this case is non-degenerate (cf. 15.1.3). Furthermore, all corresponding symmetry hyperplanes contain the center c.

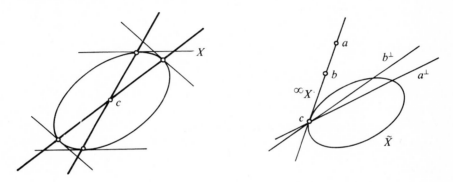

Figure 15.5.8.1

15.5.8.2. Second case. If α is a paraboloid, let $c = \infty_{\tilde{X}}^\perp \in \infty_X$ be its contact point with the plane at infinity. All line directions, except c, are directions of symmetry. Diameters cannot lie in any hyperplane direction; in fact, they all contain the direction of c. In the case of the plane, they are all parallel.

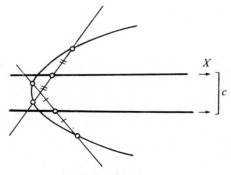

Figure 15.5.8.2

It is natural to ask whether the properties above characterize images of quadrics. The answer is neat for $K = \mathbf{R}$:

15.5.9. THEOREM. *Let B be a compact subset of a finite-dimensional, real affine Euclidean space. Assume that, for any hyperplane direction, B admits a diameter parallel to that direction. Then B is the union of homothetic quadrics with same center.*

Proof. Apply 2.7.5.9 and 2.7.5.10 to the stabilizer $G = \mathrm{GA}_B(X)$ of B in the affine group of X to conclude that there exists a Euclidean structure on X invariant under G. Then 8.2.9 shows that the affine symmetries that exist by assumption are orthogonal reflections for this Euclidean structure; since there are such reflections in all directions, 8.2.12 implies that $G = \mathrm{Is}_c(X)$, where c is the center of B. Finally, for all $x \in B$, the subset B must contain the orbit of x under G, which is the sphere with center c and containing x. $\square$

Theorem 15.5.9 can be given a geometric proof, based on 11.8.10.7.

15.5.10. NOTES. The reader can see that we have left some questions unanswered. More can be asked—for example, about the diameter of plane algebraic curves (see [LB2, 150]).

One application of 15.5.9 is to Minkowski geometry. Minkowski geometry is simply the metric structure of a normed finite-dimensional real vector space, with metric given by $d(x,y) = \|x-y\|$. Does such a geometry admit involutive isometries? By 15.5.9, if the answer is affirmative for all directions, the space is necessarily Euclidean. For the study of Minkowski metrics, see the very pleasant reference [B–K, 133 ff.]

Diameters are related to the locus of the midpoints of the chords of $\mathrm{im}(\alpha)$ parallel to a given direction; see 15.7.5.

Two diameters H and H' of a proper quadric α are called *conjugate* if their directions $\vec{H}$ and $\vec{H}'$ satisfy $(\vec{H})^\perp \perp (\vec{H}')^\perp$ with respect to $\vec{\alpha}$. In the case of conics, we obtain simple figures, which the reader should become familiar with: see 15.7.7, and also 15.6.4.

It is possible (cf. 14.5.6) to define polarity even when α is not proper; see in 15.7.4 an interpretation for 15.2 in the light of this operation.

15.5.11. ASYMPTOTIC CONE AND ASYMPTOTES. It is to be expected that the points of $\mathrm{im}(\alpha) \cap \infty_X$ also play a role here, as well as the tangents to α at such points. The *asymptotic cone* of α is the cone with vertex c (the center of our quadric) and circumscribed around α (cf. 14.5.3). (This is an interesting object only when $c \in X$.) When $n = 2$, the asymptotic cone consists of two lines or of the point c alone (the *asymptotes* of α.) For example, for $n = 2$ and $K = \mathbf{R}$, asymptotes exist only if α is a hyperbola (figure 15.5.11.1). For $n = 3$ and $K = \mathbf{R}$, the asymptotic cone exists only for the two types of hyperboloid (figure 15.5.11.2). See 15.7.11.

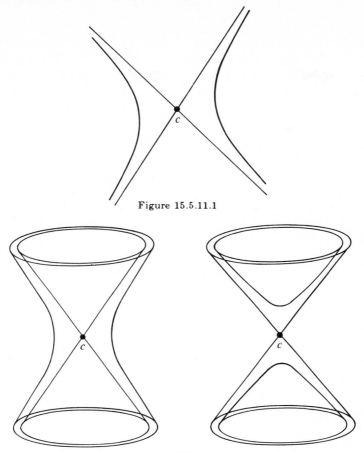

Figure 15.5.11.1

Figure 15.5.11.2

15.6. Euclidean affine quadrics

15.6.1. CLASSIFICATION. Assume that X is a Euclidean affine space; our goal will be to classify proper affine quadrics under the action of $\mathrm{Is}(X)$, not of $\mathrm{GA}(X)$. Applying 13.5.5 and the technique used in 15.3, we see that there is always an orthonormal frame in which α has one of the equations below:

$$\sum_{i=1}^{r} a_i x_i^2 - \sum_{i=r+1}^{n} a_i x_1^2 + 1 \quad \text{or} \quad \sum_{i=1}^{r} a_i x_i^2 - \sum_{i=r+1}^{n-1} a_i x_i^2 + 2x_n,$$

where $a_i > 0$ for all i. Equations of this kind are called *reduced*; see 15.7.10. To complete the classification, that is, to give only one equation for each orbit, it is enough to require the conditions $a_i \le a_j$ for all $i, j = 1, \ldots, r$ and all $i, j = r + 1, \ldots, n - 1$ or n.

15.6.2. We shall study Euclidean conics in detail in chapter 17; as for the arbitrary-dimensional case, we limit ourselves to the invariants below. For more results, see [EE].

15.6.3. An *ellipsoid* in the Euclidean space X is a quadric of type $II(0, n)$, that is, of the form

$$\mathcal{E} = \mathcal{E}(q) = \{ x \in X \mid q(x) = 1 \},$$

where $q \in Q(X_a)$ is a positive definite quadratic form and $a \in X$ is a point (cf. 15.3.2). A *conjugate set* for $\mathcal{E}$ is a basis $\{m_i\}_{i=1,\dots,n}$ of X_a in which q has the from $\sum_i x_i^2$, that is, an orthonormal basis for q (cf. 13.3.1). We now assume, in addition, that X is a Euclidean space, but whose Euclidean structure has nothing to do with q; as in 8.11.5, we denote by $\mathrm{Gram}(\cdot, \dots, \cdot)$ the Gram determinant of a set of vectors of X_a. We have the following result:

15.6.4. THEOREM OF APPOLONIUS. *An ellipsoid in the Euclidean space X has n associated scalars $A(\mathcal{E}, K)$ $(k = 1, \dots, n)$ such that*

$$A(\mathcal{E}, k) = \sum_{i_1 < \cdots < i_k} \mathrm{Gram}(m_{i_1}, \dots, m_{i_k})$$

for any set $\{m_i\}$ conjugate for $\mathcal{E}$.

15.6.5. COMPUTATION OF THE INVARIANTS. By 13.5.5, there exists an orthonormal basis $\{e_i\}$ for the Euclidean structure $\| \cdot \|^2$ that is orthogonal for q; write $q = \sum_i \lambda_i x_i^2$ in this basis. The set $\{m_i\}$, where $m_i = e_i / \sqrt{\lambda_i}$, is conjugate for $\mathcal{E} = \mathcal{E}(q)$; we clearly have

$$\mathrm{Gram}(m_{i_1}, \dots, m_{i_k}) = \|m_{i_1}\|^2 \times \cdots \times \|m_{i_k}\|^2 = \lambda_{i_1}^{-1} \cdots \lambda_{i_k}^{-1},$$

whence

$$A(\mathcal{E}, k) = \sum_{i_1 < \cdots < i_k} \lambda_{i_1}^{-1} \cdots \lambda_{i_k}^{-1}$$

for $k = 1, \dots, n$.

15.6.6. EXAMPLES. For n arbitrary, 8.11.6 shows that $A(\mathcal{E}, n)$ is the volume of the parallelepiped having a as a vertex and the m_i as edges (cf. 9.12.4.2); this shows that this volume is constant (see also 11.8.9.4). This result does not require a Euclidean structure on X (cf. 2.7.4); it is merely a geometric consequence of the fact that $|\det f| = 1$ for $f \in O(q)$ (cf. 13.6.2).

For n arbitrary and $k = 1$, we get $A(\mathcal{E}, n) = \sum_{i=1}^{n} \|m_i\|^2$. Thus, for an ellipse with equation

$$\frac{x^2}{a^2} + \frac{y^2}{b^2} - 1 = 0$$

in an orthonormal basis, we have $\|m_1\|^2 + \|m_2\|^2 = a^2 + b^2$ and the shaded area in figure 15.6.6 is equal to ab. Those where the original results of Appolonius.

For $n = 3$ and $k = 2$, we have (cf. 8.11.8)

$$A(\mathcal{E}, 2) = \|m_1 \times m_2\|^2 + \|m_2 \times m_3\|^2 + \|m_3 \times m_1\|^2,$$

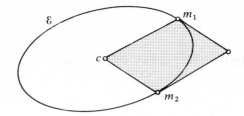

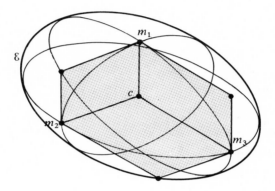

Figure 15.6.6

that is, the sum of the squares of the areas of three adjacent faces of the parallelepiped built on $\{m_1, m_2, m_3\}$ is a constant.

15.6.7. PROOF OF 15.6.4. Fix a conjugate set $\{m_i\}$ for $\mathcal{E} = q^{-1}(1)$, and let B be the matrix of the quadratic form $\|\cdot\|^2$ in this basis $\{m_i\}$. For another conjugate set $\{m_i'\}$, the matrix B' of $\|\cdot\|^2$ will be $b' = {}^t\!SBS$, with ${}^t\!SS = I$, by 13.1.3.8 and 13.6.5. This shows that $B' = S^{-1}BS$ is similar to B, and we have

$$\det(B' + \lambda I) = \sum_{k=0}^{n} \lambda_k A(\mathcal{E}, k) = \det(B + \lambda I),$$

where the $A(\mathcal{E}, k)$ depend only on B, that is, on $\mathcal{E}$. But, from linear algebra,

$$\det(B' + \lambda I) = \sum_{k=0}^{n} \lambda^k \left(\sum_{i_1 < \cdots < i_k} M'_{i_1 \cdots i_k} \right),$$

where $M'_{i_1 \cdots i_k}$ denotes the matrix with entries a'_{ij} for $i, j = i_1, \ldots, i_k$. By the definition of B' in 13.1.3.6, we have $b'_{ij} = (m_i' \mid m_j')$, so that

$$M'_{i_1 \cdots i_k} = \mathrm{Gram}(m_{i_1}, \ldots, m_{i_k}). \qquad \square$$

15.6.8. SYMMETRIES. The expressions in 15.6.1 show that the image of a proper quadric in a Euclidean affine space is invariant under certain reflections. Quadrics of type II are fixed under reflections through the n coordinate hyperplanes, $x_i \mapsto -x_i$ with the remaining coordinates fixed; quadrics of type III are fixed under the first $n-1$ of those reflections, but not under $x_n \mapsto -x_n$.

15.7. Exercises

15.7.1. Study versions of the Nullstellensatz for affine quadrics.

15.7.2. Study degenerate affine quadrics for $n = 2$ and 3 and $K = \mathbf{R}$ and $\mathbf{C}$.

15.7.3. Study the center and diameters of degenerate quadrics.

15.7.4. Define polarity for arbitrary affine quadrics. Show that, in the general case, the three types of quadrics are characterized by:

$$\text{type I}: \infty_X^\perp \cap \operatorname{im}(\alpha) \neq \emptyset;$$
$$\text{type II}: \infty_X^\perp \cap X \neq \emptyset \quad \text{and} \quad \infty_X^\perp \cap \operatorname{im}(\alpha) = \emptyset;$$
$$\text{type III}: \infty_X^\perp \cap X = \emptyset.$$

Give a geometric demonstration for 15.2.2.

15.7.5. Study the locus of the midpoints of chords of an affine quadric parallel to a given direction.

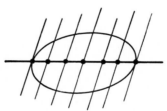

Figure 15.7.5

* **15.7.6.** Let $C = \operatorname{im}(\alpha)$ be the non-empty image of a proper plane conic, let m be a point on the plane, and $\langle m, a \rangle$ and $\langle m, b \rangle$ two distinct tangents to C at a and b, passing through m. Show that the line $D = \langle m, (a + b)/2 \rangle$ is a diameter of α, and that the tangents to α at the points of $C \cap D$ are parallel to $\langle a, b \rangle$. When α is a parabola, show that D always intersects C, and the intersection is the midpoint of m and $(a + b)/2$. Deduce a geometric construction for a sequence of points on an arc of parabola, given two points and the tangents at these points.

Observing that the area of the triangle $\{m, a', b'\}$ in figure 15.1 is one-fourth of the area of $\{m, a, b\}$, deduce that the shaded area is two-thirds of the area of $\{m, a, b\}$ (this limiting process is due to Archimedes). Prove the same fact using calculus.

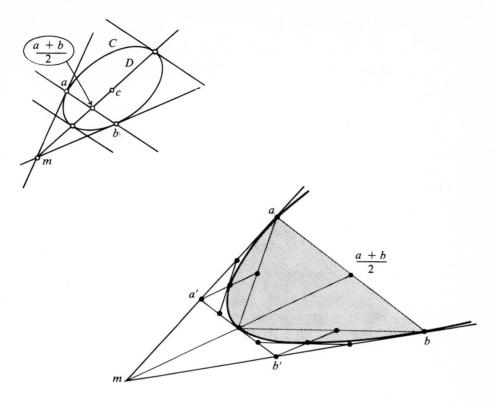

Figure 15.7.6

15.7.7. For a given ellipse and an arbitrary point on it, show that there exists a parallelogram circumscribed around the ellipse so that the contact points are at the middle of each side, and one of them is the given point on the ellipse. For the points α and β on figure 15.7.7, show that one always has $\overrightarrow{c\alpha} = \sqrt{2c\beta}$. For the points α and γ on the same figure, show that the directions $\overrightarrow{c\alpha}$ and $\overrightarrow{c\gamma}$ are conjugate with respect to $\bar{\alpha}$.

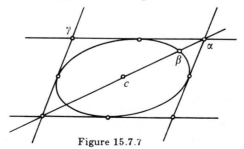

Figure 15.7.7

15.7.8. Prove 15.6.1 in detail.

* **15.7.9.** Let Q be an ellipsoid in a three-dimensional Euclidean affine space, and x a fixed point not in Q. Take three orthogonal lines D, E, F through x, and let the intersection points of Q with D be a, b, with E be c, d and with F be e, f. Show that the sum

$$\frac{1}{\overline{xa} \cdot \overline{xb}} + \frac{1}{\overline{xc} \cdot \overline{xd}} + \frac{1}{\overline{xe} \cdot \overline{xf}}$$

is constant. Give examples and generalize.

Now consider three lines D, E, F through x whose directions are pairwise conjugate relative to Q, and let the intersection points of Q with D be a, b, with E be c, d and with F be e, f. Show that the sum

$$\overline{xa} \cdot \overline{xb} + \overline{xc} \cdot \overline{xd} + \overline{xe} \cdot \overline{xf}$$

is constant.

15.7.10. For each of the quadrics in $\mathbf{R}^2$ and $\mathbf{R}^3$ whose equations are given below, find: an affine orthonormal frame in which the equation of the quadric has one of the forms given in 15.6.1; the equation in the new frame; the kind of quadric; the axes; and the asymptotes, if they exist. (When the equation involves parameters, discuss the results in terms of them.)

$$yz + zx + xy + \lambda(x + y + z) + \mu = 0$$
$$4x^2 + 3y^2 + 9z^2 + 8xz + 4xy + 4y + 8z + \lambda = 0$$
$$x^2 + (2\lambda^2 + 1)(y^2 + z^2) - 2(yz + zx + xy) + 2\lambda^3 - 3\lambda^2 + 1 = 0$$
$$(\mu z - \nu y)^2 + (\nu x - \lambda z)^2 + (\lambda y - \mu x)^2 = 1$$
$$4x^2 + 9y^2 + z^2 + 6yz + 4zx + 12xy + 2x + 3y + \lambda z + \mu = 0$$
$$x^2 + 2\lambda xy + y^2 - ax - by = 0$$
$$x^2 + \lambda(\lambda + 1)xy + \lambda^2 y(y - a) = 0$$
$$\alpha x^2 - 2xy + \beta y^2 - 2\beta x - 2\alpha y = 0$$
$$(\alpha - 1)x^2 + 2\beta xy - (\alpha + 1)y^2 + 2\alpha x + 2\beta y - (\alpha + 1) = 0$$

15.7.11. Show that, for a hyperbola in a real affine plane, the notion of asymptotes introduced in 15.5.11 coincides with the notion as defined for parametrized curves.

15.7.12. Show that, if a strict parallelogram is inscribed in a conic, its center is the center of the conic.

15.7.13. THE ORTHOPTIC SPHERE. Let Q be a central quadric in a three-dimensional Euclidean affine space. Show that the locus of the vertices of all right-angled trihedra circumscribed around Q is a sphere, called the *orthoptic sphere* of Q. What is the locus when Q is a paraboloid?

Show that a sphere is harmonically circumscribed around Q (cf. 14.5.4.4) if and only if it is orthogonal to the orthoptic sphere.

Show that the orthoptic spheres of a tangential pencil of quadrics form a pencil of spheres.

Compare with 17.4.2.3, 17.6.1, 17.9.5 and 15.7.20.

15.7.14. Let Q be a quadric in a three-dimensional Euclidean affine space $\mathcal{Q}$. Study the *circular sections* of Q, that is, the planes that intersect Q in a circle, as well as the circles themselves. (Hint: use 14.1.3.7 and 17.4.2 to study what happens at infinity in the complexified projectivization of the affine space.) Study also the spheres tangent to Q at two distinct points, and their intersection with Q.

* **15.7.15.** Let Q be a quadric in a three-dimensional Euclidean affine space, and let m be a point. Show that the number of normals to Q that pass through m is "in general" equal to six. Show that the feet of all normals to Q from m are contained in a second-degree cone with vertex m and containing the center of Q and the parallels to the axes of Q which go through m. Compare with 17.5.5.6.

15.7.16. Give a detailed demonstration of the string method for obtaining quadrics in a three-dimensional affine space (cf. 17.2.2.5 and 17.6.4).

* **15.7.17.** HOMOFOCAL QUADRICS. We consider in $\mathbf{R}^3$ the family of quadrics $Q(\lambda)$ whose equations are

$$\frac{x^2}{a^2 + \lambda} + \frac{y^2}{b^2 + \lambda} + \frac{z^2}{c^2 + \lambda} - 1 = 0,$$

with $a > b > c$. Find how many quadrics $Q(\lambda)$ pass through a fixed point (x_0, y_0, z_0); show that if three quadrics $Q(\lambda)$ pass through a point, their tangent planes at that point are orthogonal (cf. 17.6.3.3).

15.7.18. Draw four points on a real affine plane, no three of which are collinear. Draw and fill in the region of the plane containing the points x such that the conic passing through x and the four original points is an ellipse.

15.7.19. Find all the Euclidean reflections (cf. 9.2.4) that leave invariant a given Euclidean affine quadric.

* **15.7.20.** Let $\mathcal{E}$ be an ellipsoid with center O in an n-dimensional Euclidean affine space. We consider the sets $\{a_i\}_{i=1,\ldots,n}$ of points of $\mathcal{E}$ such that the vectors $\overrightarrow{Oa_i}$ are orthogonal. Show that

$$\sum_{i=1}^{n} \frac{1}{(Oa_i)^2}$$

is a constant.

Use this fact to find the envelope of the hyperplanes containing the a_i (cf. 10.13.12). Find another proof for 15.7.13 using polarity with respect to a sphere centered at O. Compare with 15.7.9.

Chapter 16
Projective conics

Conics have been the subject of an immense amount of study, since the time of the Greeks and especially in the hands of nineteenth-century geometers. The number of results is proportionally large, and the reader is referred to [EE] for a systematic treatment of them. Here we have selected the more or less classical material, and, as an example of a difficult result, the Poncelet theorem on polygons inscribed in a conic and circumscribed around another (section 16.6). In our view, this is by far the most beautiful result about conics.

Conics enjoy two advantages over other quadrics, from the point of view of their study: first, their property of being parametrizable by a projective line, and the attending ideas of cross-ratio and homographies (sections 16.2 and 16.3). Second, there is a theory of the intersection of two conics, and this leads to Bezout's theorem and the detailed study of pencils of conics (sections 16.4 and 16.5).

In this whole chapter, $P = P(E)$ is a projective plane over a commutative field K with characteristic different from two. We set $P^* = P(E^*)$. A point $m \in P$ will often be identified with an associated set (x, y, z) of homogeneous coordinates. If $a \neq b \in P$ are distinct points, we denote by $ab = \langle a, b \rangle$ the projective line joining a and b (see also 16.1.2).

As a rule, we shall discuss a fixed conic $\alpha = \mathrm{PQ}(E)$, and we will always denote its image by $C = \mathrm{im}(\alpha)$, and one of its equations by q. The exception is section 16.7, where α and q refer to a conic in an affine plane X.

16.1. Notation and expressions

16.1.1. From 14.1.3.2 and 14.1.7.1 we know that, if α is not proper, its image consists of a point, a line or a pair of lines.

16.1.2. CONVENTION. When discussing a proper conic α (explicitly mentioned or implied from the context), and two points a and b on $\mathrm{im}(\alpha) = C$, we shall *denote* by ab the projective line $\langle a, b \rangle$ if $a \neq b$, and the tangent to α at $a = b$ otherwise (cf. 14.1.3.5). Whenever this convention must be invoked (for example, 16.2.2 and 16.2.11), we shall leave to the reader the verification that the result is equally valid in both cases; the general reason is 14.1.3.2.

16.1.3. EXPRESSIONS. Let (x, y, z) be a set of homogeneous coordinates in P. In general, we shall write a matrix A and an equation q of α in the form

$$q = ax^2 + a'y^2 + a''z^2 + 2byz + 2b'zx + 2b''xy,$$

16.1.3.1
$$A = \begin{pmatrix} a & b'' & b' \\ b'' & a' & b \\ b' & b & a'' \end{pmatrix}.$$

Let p, q, r, s be the projective base of P associated with the homogeneous coordinates x, y, z, that is, $p = (1, 0, 0)$, $q = (0, 1, 0)$, $r = (0, 0, 1)$ and $s = (1, 1, 1)$ in those coordinates (cf. 4.4).

16.1.3.2. A triangle $\{p, q, r\}$ is self-polar with respect to α if and only if $b = b' = b'' = 0$ (cf. 14.5.4). A triangle $\{p, q, r\}$ is inscribed in C if and only if $a = a' = a'' = 0$; further, $s \in C$ if and only if $b + b' + b'' = 0$. A necessary and sufficient condition for q and r to belong to C and for pq and pr to be tangent to C is that the equation of α be of the form $ax^2 + 2byz$; if this is the case, the condition $s \in C$ is equivalent to the equation of α having the form $a(x^2 - yz)$ (use 14.5.3).

All of this allows us to sharpen 14.1.6 and 14.2.4 in the case of conics:

16.1.4. PROPOSITION. *Given five points of P, no three of which are collinear, there exists a unique conic containing the five. For K arbitrary, let α and β be conics such that $\mathrm{im}(\alpha)$ is neither empty nor reduced to a point; then $\mathrm{im}(\alpha) = \mathrm{im}(\beta)$ implies $\alpha = \beta$.*

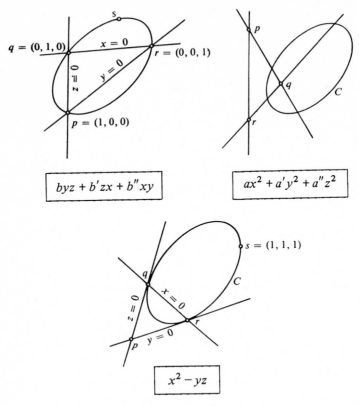

Figure 16.1.3

Proof. By proposition 4.6.8 we can take a projective base formed by the first four of the given points, and call them p, q, r, s; in this base, the equation of C is of the form $by(z - x) + b'(z - y)$, by 16.1.3.2. Let $(x_0, y_0, z_0) = t$ be the fifth given point. Saying that t does not lie on any of the lines pq, qr, rp, ps, qs and rs implies that x_0, y_0 and z_0 are all distinct and non-zero, so that

$$by_0(z_0 - x_0) + b'x_0(z_0 - y_0) = 0$$

has a unique solution (b, b') in $\tilde{K}$, with $bb' \neq 0$ and $b + b' \neq 0$; this shows that the conic $byz + b'zx + b''xy$ is proper, since $\det A = bb'b'' \neq 0$.

Now consider α and β with $\text{im}(\alpha) = \text{im}(\beta)$. Suppose first that α is degenerate. Then $\alpha = \beta$ from 14.1.3, 14.1.7.1 and the assumptions. If α is proper, let $k = \#K$. If $k > 3$ and $C \neq \emptyset$, we have $\#C \geq 5$; in fact, if $a \in C$, there are $k + 1 > 4$ lines containing a, and all but the tangent intersect C at a point distinct from a. This proves that $\alpha = \beta$ in view of the first part of the theorem. There remains the case $\#K = 3$. By the reasoning above, C contains at least four points. If the four are taken as a projective base, the previous paragraph shows that the equation of α has to be $b(xy + yz + zx)$ since α is proper.   □

16.2. Good parametrizations, cross-ratios and Pascal's theorem

16.2.1. The discussion above leads to the idea of associating to each line D containing $m \in C$ the point $D \cap C$. To be more precise: for a fixed $m \in P$, *denote* by m^* the line in P^* consisting of the lines in P that go through m (cf. 4.1.3.5). For α proper and $m \in C$, *define* the map

16.2.2 $$\pi_m : C = \operatorname{im}(\alpha) \ni n \mapsto mn \in m^*$$

(recall the convention introduced in 16.1.2).

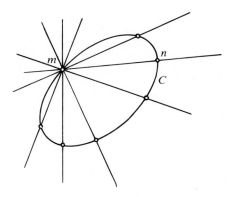

Figure 16.2.1

16.2.3. PROPOSITION. *For every $m \in C$ the map π_m is a bijection. For every m and $n \in C$ the composition $\pi_n \circ \pi_m^{-1} : m^* \to n^*$ is a homography. Conversely, if m and n are distinct points in P and $f : m^* \to n^*$ is a homography, there exists a conic α in P such that*

$$\operatorname{im}(\alpha) = \{\, D \cap f(D) \mid D \in m^* \,\}.$$

This conic contains m and n, and is degenerate if and only if $f(mn) = mn$, in which case it is composed of mn and another line.

Proof. Bijectivity follows from 14.1.3.2 and 14.1.3.5. If $m, n \in C$, take a projective frame in which $m = (0, 1, 0)$, $n = (0, 0, 1)$ and α has equation $x^2 - yz$; this is always possible by 16.1.3.2. Then m^* consists of lines D with equation $\lambda x + \mu z = 0$, and n^* of lines D' with equation $\lambda' x + \mu' y = 0$; thus $D \cap D' \in C$ is equivalent to $\lambda \lambda' = \mu \mu'$, which indeed indicates a homography between $(\lambda, \mu) \in \tilde{K}$ and $(\lambda', \mu') \in \tilde{K}$.

To prove the converse, suppose first that $f(mn) = mn$; then our assertions have been demonstrated in 6.5.9. Otherwise, apply 16.1.4 to m, n and three distinct points of the form $D \cap f(D)$, where $D \neq mn$. $\qquad \square$

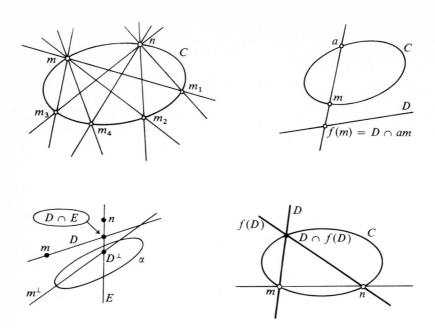

Figure 16.2.3

16.2.4. COROLLARY. *If $K = \mathbf{C}$ and α is proper, $\operatorname{im}(\alpha)$ is homeomorphic to $P^1(\mathbf{C})$; if $K = \mathbf{R}$ and α is proper and has non-empty image, $\operatorname{im}(\alpha)$ is homeomorphic to $P^1(\mathbf{R})$.*

Proof. In the frame chosen for the previous proof, the map π_m^{-1} has the form

$$\tilde{K} \ni (\lambda, \mu) \mapsto (\lambda\mu, -\mu^2, -\lambda^2),$$

and that map is continuous if $K = \mathbf{R}$ or $\mathbf{C}$. □

16.2.5. PROPOSITION. *Let α be a proper quadric with non-empty image C, and m_i $(i = 1, 2, 3, 4)$ four points of C of which no more than two are equal; then the cross-ratio $[mm_i]$ of the four lines mm_i (cf. 6.5 and 16.1.2) does not depend on the point $m \in C$. This number is called the cross-ratio of the four points m_i of C, and is denoted by $[m_i]$ or $[m_i]_C$. If exactly three of the m_i are identical, we set $[m_i] = -1$ by convention (cf. 6.4.1, harmonic division).*

Proof. This all follows from 16.1.4 and 6.1.4. □

16.2.6. Let α be proper, let $a \in C$ be a point and D a line such that $a \notin D$. Then, as can be seen by applying 6.5, the map

$$C \ni m \mapsto am \cap D \in D$$

is a bijection, and preserves cross-ratios. One example of this is stereographic projection, for $n = 1$ and $K = \mathbf{R}$ or $n = 2$ and $K = \mathbf{C}$ (see 18.1.4, 18.10.7 and 20.6).

16.2.7. EXAMPLES

16.2.7.1. Let α be proper and $m \neq n$ points in P. The set of points of the form $D \cap E$, where D and E are orthogonal lines containing m and n, respectively, is the image of a conic, and contains m and n. In fact, by 14.5, we have $E = nD^{\perp}$; but $m^* \ni D \mapsto D^{\perp} \in m^{\perp}$ is a homography, and so is $m^{\perp} \ni x \mapsto nx \in n^*$, so the second part of 16.2.3 completes the proof.

$\qquad$ An example of such a locus is the orthoptic circle of a Euclidean conic; see 17.4.2.3.

16.2.7.2. By duality, we deduce from 16.2.3 that if $f : D \to E$ is a homography between two lines of P, the line $\langle m, f(m) \rangle$, for m ranging over D, is the envelope of a tangential conic of P tangent to D and E. A very practical case of this result has been encountered in 9.6.7.1.

16.2.7.3. From 14.6.1 we deduce that if α is a proper conic with non-empty image we can define the *cross-ratio* of four tangents to α in either of two equivalent ways: as the cross-ratio of the contact points, or as the cross-ratio of the four points where an arbitrary tangent intersects the four given tangents.

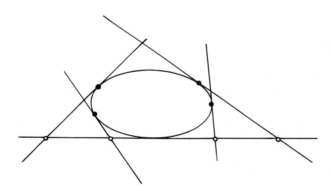

Figure 16.2.7.3

16.2.7.4. A nice application, involving the angle subtended by an arc, is given in 17.4.2.2.

16.2.8. REMARK. By 16.1.3.2, every proper conic with non-empty image has an equation of the form $y^2 - xz$; but then the map

$$K^2 \ni (\lambda, \mu) \mapsto (\lambda^2, \lambda\mu, \mu^2) \in E$$

passes to the quotient, and yields a map $f : \tilde{K} \to C$ which is bijective and preserves cross-ratios, being, in fact, one of the maps of the proof of 16.2.3. More generally, we have the following

16.2.9. PROPOSITION. *A good parametrization of a proper conic with non-empty image C is a bijection $f : \tilde{K} \to C$ that preserves cross-ratios. For*

*every good parametrization f there exists a map $g \in P_2^\bullet(K^2; E)$ (cf. 3.3.1)
such that the diagram*

$$
\begin{array}{ccc}
K^2 & \xrightarrow{\ g\ } & E \\
\downarrow{\scriptstyle p} & & \downarrow{\scriptstyle p} \\
\tilde{K} & \xrightarrow{\ f\ } & P
\end{array}
$$

*commutes, that is, such that $f \circ p = p \circ g$; such a map is unique up to a
scalar in K^*. Conversely, take $g \in P_2^\bullet(K^2; E)$ and let $f : \tilde{K} \to P$ be the
map obtained by taking quotients; if f is injective, its image $f(\tilde{K}) = C$ is the
image of a proper conic and f is a good parametrization. For any two good
parametrizations $f, g : \tilde{K} \to C$ the composition $g^{-1} \circ f$ is a homography of $\tilde{K}$.*

Proof. The last assertion follows from the definition and 6.1.4. The existence
of g follows from the particular $g : (\lambda, \mu) \mapsto (\lambda^2, \lambda\mu, \mu^2)$ already met; uniqueness is trivial, following from the definition of $P_2^\bullet$. To prove the converse,
set

$$
\begin{aligned}
x &= u\lambda^2 + u'\lambda\mu + u''\mu^2 \ , \\
y &= v\lambda^2 + v'\lambda\mu + v''\mu^2 \ , \\
z &= w\lambda^2 + w'\lambda\mu + w''\mu^2 \ ;
\end{aligned}
$$

we start by showing that the determinant $\begin{vmatrix} u & u' & u'' \\ v & v' & v'' \\ w & w' & w'' \end{vmatrix}$ is non-zero. For
otherwise we would obtain a map of the form

$$
(\lambda, \mu) \mapsto \left(a\lambda^2 + b\lambda\mu + c\mu^2, a'\lambda^2 + b'\lambda\mu + c'\mu^2\right),
$$

and we leave it to the reader to show that the quotient map $\tilde{K} \to \tilde{K}$ can
never be injective if $\#K \geq 3$.

We deduce from this that there exist linearly independent vectors ϕ, η, ς
in E^* such that $\lambda^2 = \phi(x, y, z)$, $\lambda\mu = \eta(x, y, z)$ and $\mu^2 = \varsigma(x, y, z)$. This
implies that the image of $f(\tilde{K})$ is the conic with equation $\eta - \phi\varsigma = 0$, and
this conic is proper because ϕ, η and ς are linearly independent. Up to a
homography of $\tilde{K}$ and one of P, our map is of the form

$$
(\lambda, \mu) \mapsto (\lambda^2, \lambda\mu, \mu^2),
$$

which is certainly a good parametrization by 16.2.8. $\square$

16.2.10. NOTE. This is a particular case of a general theorem on unicursal
curves, due to Lüroth (and valid only when K is algebraically closed). See
[WK, 149-151], for example.

16.2.11. PASCAL'S THEOREM. *Let C be the image of a proper conic, and a,
b, c, d, e and f points on C such that there are at most three pairs of identical
points among the six (cf. 16.1.2). Then the points $ab \cap de$, $bc \cap ef$ and $cd \cap fa$
are collinear.*

Proof. Set $x = bc \cap ed$, $y = cd \cap ef$ $z = ab \cap de$, and $t = af \cap dc$; by 16.2.5
and 6.5.2 we have

$$
[z, x, d, e] = [ba, bc, bd, be] = [fa, fc, fd, fe] = [t, c, d, y],
$$

and zt, xc, ey are concurrent by 6.5.8.   □

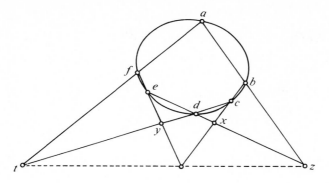

Figure 16.2.11

16.2.12. Observe that Pappus' theorem is the counterpart of Pascal's in the case that C consists of two lines, cf. 5.4.1. Apparently, nobody knows how Pascal proved his theorem for circles, but it is known that he passed from circles to conics by projection (cf. 17.1.5). For other proofs and properties of the "mystic hexagon", see 16.8.3, 16.8.4 and 16.8.5. Pascal's theorem admits an obvious converse, due to the fact that a line intersects a conic in at most two points. We will use this converse in 16.3.3 and 16.7.3.

From 14.6 we deduce the following fact:

16.2.13. COROLLARY (BRIANCHON). *If a hexagon is circumscribed to a proper conic, its three diagonals are concurrent.*   □

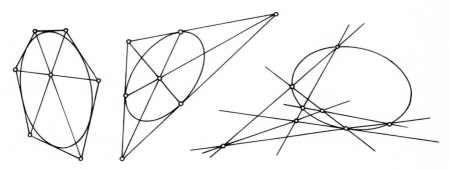

Figure 16.2.13

The results in 16.2.11 and 16.2.13 allow for degenerate cases (cf. 16.1.2), two of which are shown in figure 16.2.13 (with maximal degeneracy in the middle diagram).

16.2.14. If α is proper and $(m_i)_{i=1,2,3,4} \subset C$, we have $\left[f(m_i)\right]_{f(C)} = [m_i]_C$ for any $f \in GP(P)$.

16.3. Homographies and group of a conic. Applications

> In this section α denotes a proper conic with non-empty image C.

16.3.1. DEFINITION. *A homography of α, or of C, is a bijection f of C that preserves cross-ratios. A homography f is called an involution if $f^2 = \mathrm{Id}_C$ and $f \neq \mathrm{Id}_C$ (cf. 6.7). The group of homographies of α is denoted by $\mathrm{GP}(\alpha)$ or $\mathrm{GP}(C)$.*

16.3.2. By 16.2, a necessary and sufficient condition of f to be a homography of C is that

$$\pi_a \circ f \circ \pi_a^{-1} \in \mathrm{GP}(a^*),$$

where $a \in C$ is arbitrary; in fact, any good parametrization will do instead of π_a. The involutions of C correspond to involutions in the sense of 6.7. There exists a unique homography transforming three given points of C into three other points. Thus $\mathrm{GP}(C)$ is always isomorphic (but not canonically so) to $\mathrm{GP}(\tilde{K})$ (this group is also *written* $\mathrm{GP}(1; K)$).

Observe that there is a completely different way, suggested by 16.2.14, to obtain homographies of α: just take the restriction to C of any $f \in \mathrm{PO}(\alpha)$ (cf. 14.7). We will soon see (16.3.8) that all homographies are obtained in this way. To do this we need to relate homographies of C to the whole space P.

16.3.3. THEOREM. *For $f \in \mathrm{GP}(C)$, there exists a line Δ of P such that, for any two points $m \neq n$ on C and for $m' = f(m)$, the condition $n' = f(n)$ is equivalent to $m'n \cap mn' \in \Delta$. We say that Δ is the (homography) axis of f. The fixed points of f are the points of $\Delta \cap C$.*

Proof. (Recall 16.1.2 if necessary.) Let $(a_i)_{i=1.2.3.4.5}$ be points in C, and set $f_i' = f(a_i)$ and $s_{ij} = a_i a_j' \cap a_i' a_j$. Assume a_1, a_2 and a_3 are distinct and let Δ be the line $s_{12}s_{13}$. By 6.5.8 and 16.2.5 the points s_{14} and s_{15} belong to Δ. By 16.2.11 (and its converse, cf. 16.2.12), the point s_{45} belongs to Δ (and conversely). $\square$

16.3.4. REMARK. If K is algebraically closed, we can prove 16.3.3 directly from 6.5.8, without having to resort to 16.2.11. In fact, $\Delta = s_{12}s_{13}$ has to intersect C, and the intersections are the fixed points of f; thus Δ is characterized as the line joining these two points (cf. 16.1.2). Now it is 16.2.11 that follows trivially from 16.3.3! Thus we can use an algebraic closure of K and the extension of α to that closure to prove 16.2.11 in the general case. Notice that for fields that are not algebraically closed, f may not have fixed points even when $\mathrm{im}(\alpha) \neq \emptyset$ (see 16.3.5 below).

16.3.5. EXAMPLE. Rotations of a circle are homographies; this follows from 8.8.7.2, and shall be mentioned again in 17.4.2.

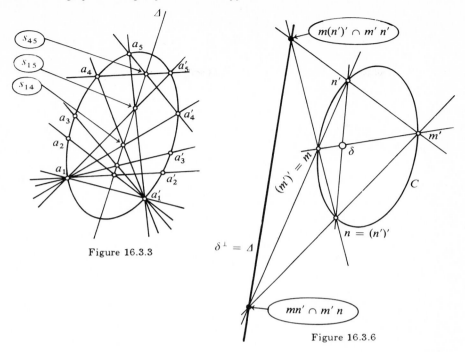

Figure 16.3.3

Figure 16.3.6

16.3.6. COROLLARY (FRÉGIER). *The bijection $f \in \mathrm{GP}(C)$ is an involution if and only if there exists $\delta \in P \setminus C$ such that, for any $m \in C$, we have*

$$\{m, f(m)\} = C \cap \delta m$$

(cf. 16.1.2). The point δ is called the Frégier point of f, and coincides with the pole $\Delta^{\perp}$ of the axis of f.

Proof. Necessity comes from 14.5.2.6 (cf. figure 14.5.2); sufficiency follows from 6.7.4, which, allied to 16.3.3, shows that the condition $\{m, f(m)\} = C \cap \delta m$ does determine an involution, for δ as described. $\qquad \square$

16.3.6.1. Question. Even if $f \in \mathrm{GP}(C)$ is not an involution, we can still ask what happens to the line $\langle m, f(m) \rangle$ as m runs over C. See the answer in 16.8.8.

16.3.7. EXAMPLE. If C is a circle in a Euclidean plane with center δ and $a \in C$ is arbitrary, the involution of C determined by δ takes each point into the diametrically opposed one, and the involution of a^{*} determined by π_a takes each line into the perpendicular one. We shall return to these points in 16.3.10.2 and 17.5.

We can now prove the result previewed in 16.3.2:

16.3.8. PROPOSITION. *The restriction map $\mathrm{PO}(\alpha) \ni g \mapsto g|_C \in \mathrm{GP}(C)$ is bijective; in particular, $\mathrm{GP}(C)$ is naturally isomorphic to $\mathrm{PO}(\alpha)$.*

Proof. We know that $g|_C$ is indeed in $\mathrm{GP}(C)$, by 16.2.14. The map is injective because C always contains at least four points (exactly four if $\#K = 3$) that form a projective base (see proof of 16.1.4). To prove surjectivity, we use 6.7.3 to reduce the problem the the case of an involution; but an involution, with Frégier point δ and axis $\Delta = \delta^{\perp}$, is the restriction of a map $g \in \mathrm{PO}(\alpha)$, by 14.7.4. $\square$

16.3.9. COROLLARY. *If K is algebraically closed, we have, whenever α is proper (or, equivalently, q is non-degenerate):*

$$\mathrm{GP}(K^2) = \mathrm{GP}(\tilde{K}) = \mathrm{GP}(1; K) \cong \mathrm{PO}(q) \cong O^+(K^3; x^2 + y^2 + z^2).$$

If $K = \mathbf{R}$ and $q = x^2 + y^2 - z^2$, we have

$$\mathrm{GP}(\mathbf{R}^2) = \mathrm{GP}(\tilde{\mathbf{R}}) = \mathrm{GP}(1; \mathbf{R}) \cong \mathrm{PO}(q) \cong O^+(2, 1). \qquad \square$$

Thus we recover isomorphisms between certain "classical groups": see [DE1, chapter IV].

16.3.10. APPLICATIONS.

16.3.10.1. Since the homography axis is known whenever we have three pairs of homologous points, 16.3.3 gives a geometric construction for the image of an arbitrary point under a homography. In the case of an involution, two pairs of homologous points are enough, since they determine the Frégier point. In particular, the double points of an involution are the contact points of the tangents to C issued form the Frégier point, and the pair of homologous points common to two involutions is the intersection of C with the line joining the two Frégier points.

By 16.2.6, the constructions above also solve the same problems on a line D, or for a pencil a^* of lines (cf. 16.2.3); see figure 16.3.10.1.1, where we use a Euclidean affine plane X containing D and an auxiliary circle Γ of X.

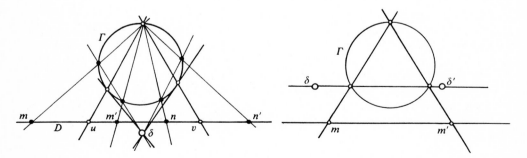

Figure 16.3.10.1.1   Figure 16.3.10.1.2

16.3.10.2. Finding the axes of an ellipse. Let $\mathcal{E}$ be an ellipse in an affine Euclidean plane X, with center c, and assume known two conjugate diameters cu and cv. (This situation arises relatively often in practice, as in descriptive geometry or in 16.7.3.) By 15.7.7, the pairs of vectors $\overrightarrow{cu}, \overrightarrow{cv}$ and $\overrightarrow{cp}, \overrightarrow{cq}$ in figure 16.3.10.2.1 are both conjugate with respect to $\vec{\alpha}$, so their directions are in involution on the projective line c^*: in fact, c^* is naturally isomorphic to ∞_X, and polarity on ∞_X with respect to $\vec{\alpha}$ is an involution, since it is involutive, homographic and distinct from the identity.

Finding the axes of $\mathcal{E}$ is the same as finding orthogonal directions that are conjugate with respect to $\vec{\alpha}$. This can be achieved by transferring the involution from c^* to a circle Γ containing c (figure 16.3.10.2.2); the axes correspond to diametrically opposite points of Γ that are in involution. See also 17.9.22.

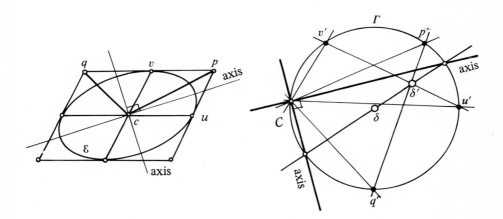

Figure 16.3.10.2.1   Figure 16.3.10.2.2

16.3.10.3. Castillon's problem. Consider, in a Euclidean plane, a circle Γ and n points $a_i \notin \Gamma$ $(i = 1, \ldots, n)$. The problem is to find a polygon $\{m_i\}_{i=1,\ldots,n}$ inscribed in Γ and such that $a_i \in m_i m_{i+1}$ $(i = 1, \ldots, n)$. The problem is equally easy to solve for a conic α with non-empty image C in a projective plane P, and points $a_i \in P \setminus C$. The idea is that m_1 is a fixed point of the homography $g \in \mathrm{GP}(C)$ defined by $f = \phi_n \circ \cdots \circ \phi_1$, where ϕ_i is the involution of C having Frégier point a_i. The explicit construction is carried out by forming $\alpha' = f(\alpha)$, $\beta' = f(\beta)$ and $\gamma' = f(\gamma)$, where α, β and γ are arbitrary distinct points on C, and then writing the axis Δ of f as

$$\Delta = \langle \alpha\beta' \cap \alpha'\beta, \, \beta\gamma' \cap \beta'\gamma \rangle.$$

The options for m_1 are the points where Δ and C intersect. See figure 16.3.10.3 and 10.11.4.

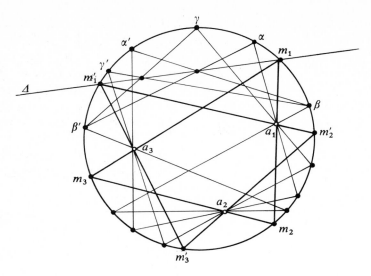

Figure 16.3.10.3

16.4. Intersection of two conics. Bezout's theorem

From now till the end of the chapter, we assume that $\#K \geq 5$.

16.4.1. Let $\phi \in P_n^\bullet(K^2)$ be a homogeneous polynomial over K^2, of degree $n \leq 4$. If $(a, b) \in K^2$ is such that $\phi(a, b) = 0$, we have $\phi(ka, kb) = 0$ for all $k \in K$, so we can define a *root* of ϕ in $\tilde{K} = P(K^2)$ as any point $m \in \tilde{K}$ such that $m = p(a, b)$ with $\phi(a, b) = 0$. Given a point $(a, b) \neq (0, 0)$, the only linear form ω over K^2 satisfying $\omega(a, b) = 0$ is, up to a scalar in K^*,

$$\omega : (\lambda, \mu) \mapsto b\lambda - a\mu.$$

In view of this, the following lemma makes sense:

16.4.2. LEMMA. *Consider a non-zero polynomial $\phi \in P_n^\bullet(K^2)$, with $n \leq 4$.*
  i) *For m to be a root of ϕ it is necessary and sufficient that there exist $\psi \in P_{n-1}^\bullet(K^2)$ such that $\phi = (b\lambda - a\mu)\psi$. The order of m is, by definition, the integer ω such that $\phi = (b\lambda - a\mu)^\omega \psi$, where $\psi \in P_{n-\omega}^\bullet(K^2)$ and m is not a root of ψ. We shall denote by $R(\phi) = \{(m_i, \omega_i)\}_{i=1,\dots,k}$ the set of distinct roots m_i of ϕ in $\tilde{K}$, where ω_i is the order of m_i. If $\phi' \in P_n^\bullet(K^2) \setminus 0$, writing $R(\phi) = R(\phi')$ will mean that ϕ and ϕ' have the same roots and each root has the same order.*
  ii) *For $m_i = p(a_i, b_i)$, we have $\phi = \prod_{i=1}^k (b_i\lambda - a_i\mu)^{\omega_i} \psi$, where ψ is in $P_{n-\sum_{i=1}^k \omega_i}^\bullet(K^2) \setminus 0$ and has no roots; in particular, either $\sum_{i=1}^k \omega_i = n$ or $\sum_{i=1}^k \omega_i \leq n - 2$.*
  iii) *If $R(\phi) = R(\phi')$ and $\sum_{i=1}^k \omega_i = n$, we have $\phi' = k\phi$, where $k \in K^*$.*

iv) *If $f \in \mathrm{GP}(K^2)$, we have $R(\phi \circ f) = \left\{ f^{-1}(m_i), \omega_i \right\}_{i=1,\ldots,k}$.*

v) *If K is algebraically closed, we have $\sum_{i=1}^{k} \omega_i = n$ for every ϕ.*

Proof. The proof consists in reducing the problem to usual polynomials over K and with values in K; this is achieved by taking $\lambda/\mu = x$ as the unknown. In doing this, we may be dropping a root $p(1,0) = \infty \in \tilde{K}$; but, since $\phi \neq 0$, there is only a finite number of roots, and we can apply a homography of $\tilde{K}$ so that ∞ is no longer a root. This done, all the properties above are classical in $K[X]$, since the cardinality of K is greater than the degree of the polynomials in question. $\square$

16.4.3. INTERSECTION OF A CONIC WITH A PROPER CONIC. Let α be a proper conic with non-empty image C, and α' a second conic, not necessarily proper, with equation q'. Let f be a good parametrization for α, and g the associated map as in 16.2.9; then, if $m = p(x) \in C \cap C'$, where C' is the image of α', we have $q'(x) = 0$. But $m \in C$, so there exists $(a, b) \in K^2$ with $x = (a, b)$; hence $q'\bigl(g(a, b)\bigr) = 0$, that is, $(q' \circ g)(a, b) = 0$. The map $q' \circ g : K^2 \to K$ clearly lies in $\mathcal{P}_4^{\bullet}(K^2)$, but we calculate it explicitly for later reference. Since $g(\lambda, \mu) = (\lambda^2, \lambda\mu, \mu^2)$ and q' is given by

$$q' = ax^2 + a'y^2 + a''z^2 + 2byz + 2b'zx + 2b''xy,$$

we have

16.4.4 $q' \circ g = a\lambda^4 + 2b''\lambda^3\mu + (a' + 2b')\lambda^2\mu^2 + 2b\lambda\mu^3 + a''\mu^4.$

Observe first that $q' \circ g = 0$ implies $q' = a'(y^2 - xz) = a'q$, whence $\alpha' = \alpha$. This, incidentally, gives another proof for the last statement in 16.1.4. From now on we assume that $\alpha' \neq \alpha$, which implies $q' \circ q \neq 0$ and enables us to use 16.4.2.

For $m \in C \cap C'$, the order of $f^{-1}(m)$ as a root of $q' \circ g$ depends only on α and α', not on q' or on the good parametrization f chosen for C; this follows from 16.4.2 (iv) and the last assertion in 16.2.9. This leads to the

16.4.5. DEFINITION. *If $m \in C \cap C'$, the order of $f^{-1}(m)$ is called the order of m. We shall write $C \square C' = \left\{ (m_i, \omega_i) \right\}_{i=1,\ldots,k}$, where m_i runs over $C \cap C'$ and ω_i is the order of m_i. We say that C' is tangent to (resp. osculates, superosculates) C at m if the order of m is ≥ 2 (resp. ≥ 3, $= 4$). If $m \notin C \cap C'$, we say that the order of m is zero.*

16.4.6. REMARKS. That $\#(C \cap C')$ is finite follows from 16.4.2 and 16.4.3.

Watch out for the fact that $C \square C'$ only makes sense if C is proper. We shall see in 16.4.7.4 that, if C' is also proper, $C' \square C = C \square C'$.

The nomenclature introduced in 16.4.5 will be justified in 16.4.7.3 and 16.4.12.1.

16.4.7. INTERPRETATION. Since the order does not depend on the choice of a good parametrization, we can suppose that the data for q and q' are as in

16.4.4. Let us find the order of the point $m = p(1,0,0)$. Since $m = f(p(1,0))$, the order of m is the order of $(1,0)$ as a root of 16.4.4; thus

16.4.7.1 m has order
$$\begin{cases} \geq 1 & \text{if and only if } a = 0; \\ \geq 2 & \text{if and only if } a = b'' = 0; \\ \geq 3 & \text{if and only if } a = b'' = a' + 2b' = 0; \\ = 4 & \text{if and only if } a = b'' = a' + 2b' = b = 0. \end{cases}$$

Order ≥ 1. The condition for $m \in C'$ is that $a = 0$.

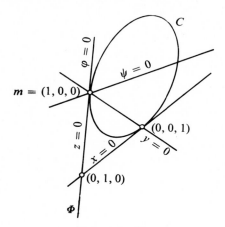

Figure 16.4.7

Order ≥ 2.   By 14.5.3 and the remarks following 13.1.3.7, the polar of (x_0, y_0, z_0) with respect to α' has equation

16.4.7.2
$$ax_0 x + a' y_0 y + a'' z_0 z + b(y_0 z + z_0 y) + b'(z_0 x + x_0 z) + b''(x_0 y + y_0 x) = 0;$$

in particular, the polar $m^\perp$ of $m = (1,0,0)$ with respect to α' is

$$ax + b'z + b''y = 0,$$

whence $m \in C'$. The condition $b'' = 0$ expresses the fact that the line $z = 0$, which is already tangent to α at m, is also a tangent to α' at m (watch out—C' can be degenerate).

Order ≥ 3. In this case we can write

$$q' = a'(y^2 - xz) + z(2by + a''z),$$

that is, $q' = kq + \phi\psi$, where $k \in K$ and ϕ, ψ are non-zero linear forms, ϕ being an equation for the tangent Φ to α at m. Conversely, if $q' = kq + \phi\psi$ with ϕ an equation for Φ, the order of m is ≥ 3 (note that the line with equation ψ automatically contains the point m).

Order $= 4$. Here we can write $q' = kq + \phi^2$, where ϕ is an equation for the tangent to α at m, and conversely.

The following proposition sums up the situation:

16.4.7.3. Proposition. *Given a proper conic* $\alpha \in PQ(P)$ *with equation* q *and non-empty image* C, *and an arbitrary conic* α' *with image* C' *distinct from* C, *consider a point* $m \in C \cap C'$ *and an equation* ϕ *for the tangent* Φ *to* α *at* m. *Then*

  i) m *has order* ≥ 2 *if and only if* Φ *is tangent to* α';
  ii) m *has order* ≥ 3 *if and only if every equation of* α' *is of the form* $q' = kq + \phi\psi$, *with* $\psi \in E^* \setminus 0$ *and* $k \in K$ $\big($*and* $m \in \phi^{-1}(0)\big)$;
  iii) m *has order* ≥ 4 *if and only if every equation of* α' *is of the form* $q' = kq + \phi^2$, *with* $k \in K$.

16.4.7.4. Corollary. *If* C *and* C' *are both proper,* $C \,\square\, C' = C' \,\square\, C$.

Proof. By inspection of the table of possibilities for $C \,\square\, C'$ (use 16.4.2).

16.4.7.5. Table

type	$C \,\square\, C'$	$\sum_{i=1}^{k} \omega_i$	$\#(C \cap C')$	points with order ≥ 2	points with order 4
	$(a,1)$	1	1	0	0
	$(a,1),(b,1)$	2	2	0	0
	$(a,2)$	2	1	1	0
I	$(a,1),(b,1),(c,1),(d,1)$	4	4	0	0
II	$(a,2),(b,1),(c,1)$	4	3	1	0
III	$(a,2),(b,2)$	4	2	2	0
IV	$(a,3),(b,1)$	4	2	1	0
V	$(a,4)$	4	1	1	1

By 16.4.7.3 the integers in the last three columns are symmetric in C and C', assuming both conics are proper; but they are enough to differentiate between any two lines of the table. □

16.4.8. THEOREM. *Consider a proper conic* $\alpha \in PQ(P)$ *with non-empty image* C, *and arbitrary conics* α' *and* α'' *with images* C' *and* C'' *distinct from* C. *Set* $C \,\square\, C' = \big\{(m_i, \omega_i)\big\}_{i=1....k}$. *Then*

  i) $\sum_{i=1}^{k} \omega_i$ *can have the values 0, 1, 2 or 4;*
  ii) *if* $\sum_{i=1}^{k} \omega_i = 4$, *the condition* $C \,\square\, C' = C \,\square\, C''$ *is equivalent to* C *being in the pencil determined by* α *and* α' *(cf. 14.2.7);*
  iii) *if* $F : P \rightarrow P'$ *is a homography between* P *and another projective plane* P', *we have* $f(C) \,\square\, f(C') = f(C \,\square\, C') = \big\{(f(m_i), \omega_i)\big\}_{i=1....k}$;
  iv) *if* K *is algebraically closed, we necessarily have* $\sum_{i=1}^{k} \omega_i = 4$.

Proof. Parts (i) and (iv) follow from the discussion above and 16.4.2 (v). Part (iii) comes from definition 14.1.3.8. To prove (ii), let q, q', q'' be equations for $\alpha, \alpha', \alpha''$, let g be a good parametrization for α, and set $\phi' = q' \circ g$ and $\phi'' = q'' \circ g$. In the notation of 16.4.2, we have $R(\phi') = R(\phi'')$ if $C \,\square\, C' = C \,\square\, C''$; then 16.4.2 (iii) implies that $\phi'' = k\phi'$ for some $k \in K^*$, whence $(q'' - kq') \circ g = 0$, so that $q'' - kq' = hq$ for $h \in K$, by the remarks following 16.4.4. This is the same as saying that α'', with equation $q'' = kq' + hq$, belongs to the pencil determined by α and α' (cf. 14.2.7.1). The converse is trivial. □

16.4.9. NOTES. Part (iv) of 16.4.8 is a particular case of Bezout's theorem, which holds for any pair of algebraic curves in P. The difficulty with the more general version is, first, to define the order (or "multiplicity") of an intersection point, and then to show that the sum of the multiplicities is equal to the product of the degrees of the curves (here $2 \times 2 = 4$). See [FN, 112] for a recent exposition of this theorem, and also [WK, 111] for a more elementary exposition, (which, unfortunately, does not use commutative algebra systematically enough).

Another approach, more geometric than the one we've followed, is mentioned in 16.4.11.2.

16.4.10. EXAMPLES OF INTERSECTIONS. DEGENERATE CONICS IN A PENCIL. Here we give examples for the five types of intersections shown in Table 16.4.7.5, and at the same time we study systematically the degenerate conics in the pencil determined by α and α'.

16.4.10.1. Four distinct points (type I). Let a, b, c, d be the four points, and ϕ, ψ, ξ, η equations for the lines ab, cd, bc, da, respectively. The equation for the pencil F of conics passing through these four points is $\left\{ k\phi\psi + h\xi\eta \mid (k, h) \in \tilde{K} \right\}$ (cf. 16.2.7.2). Clearly this pencil contains three degenerate conics, namely, the pairs of lines $\{ab, cd\}$, $\{ac, db\}$, $\{ad, bc\}$; by 14.2.7.5 there are no others. By the construction indicated in figure 14.5.2, the triangle

$$\{p = ab \cap cd, q = ac \cap db, r = ad \cap bc\}$$

is self-polar with respect to all the conics in $\mathcal{F}$. It is the only triangle with this property: in the projective base $\{p, q, r, s\}$, where s belongs to all the conics in $\mathcal{F}$, the equations of the conics are of the form $k(x^2 - z^2) + h(y^2 - z^2)$, for $(k, h) \in \tilde{K}$. Thus, for some curves, the coefficients of x^2, y^2 and z^2 are distinct, so there is a unique orthogonal basis, and consequently a unique self-polar triangle, with respect to such curves (cf. 14.5.4).

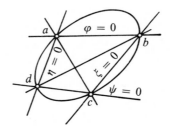

Figure 16.4.10.1.1

A pleasant consequence of the uniqueness of this self-polar triangle is that, if α and α' also have four distinct common tangents, their six intersection points all lie on the sides of the triangle $\{p, q, r\}$ (figure 16.4.10.1.2). To see this, apply the duality relation introduced in 14.6: the common tangents to α and α' are the points common to α^* and α'^*, so the three lines AA', BB'

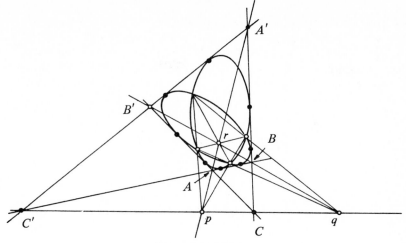

Figure 16.4.10.1.2

and CC' are pairwise conjugate with respect to α and α' and the triangle they form is self-polar with respect to the two conics.

16.4.10.2. One tangency point (type II). Take a projective frame whose first three points are the intersection points a, b, c (where a is the tangency point). Assume that the equation of the common tangent Φ at a is $x + y$ (cf. 16.4.7.3); then $\mathcal{F}$ consists of conics with equation

$$\{\, ky(x + z) + hxz \mid (k, h) \in \tilde{K} \,\}.$$

Two degenerate conics are given by $y(x+z) = 0$ and $xz = 0$, corresponding to $\{ab, ac\}$ and $\{\Phi, bc\}$. There are no others because $\det A = k^2 h/4$ (cf. 14.1.4.2).

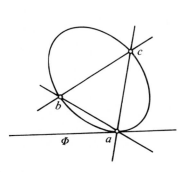

Figure 16.4.10.2

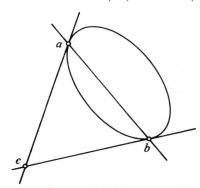

Figure 16.4.10.3

16.4.10.3. Bitangent conics (type III). By 16.1.3.2, the conics in Φ have equation $kx^2 + hyz$ in a base $\{a, b, c\}$, where a and b are the tangency points and c is the intersection of the two common tangents. The degenerate conics in $\mathcal{F}$ are $\{ca, cb\}$ and the double line ab; there are no others because $\det A = -kh^2/4$.

16.4.10.4. Osculating conics (type IV). By 16.4.7 we can assume the conics in $\mathcal{F}$ to be given by $k(y^2 - zx) + hyz$; all of them osculate $y^2 - xz$ at a and go through b. The pencil has only one degenerate conic, whose image is $\{ab, \Phi\}$. Geometric characterizations of the osculation are given in 16.4.12 for the real case and in 16.4.13 in general, this latter using homographies.

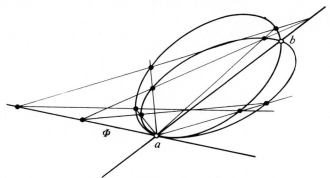

Figure 16.4.10.4

16.4.10.5. Superosculating conics (type V). By 16.4.7, $\mathcal{F}$ can be described by $k(y^2 - zx) + hz^2$. There is a unique degenerate conic, having for equation z^2 and for image the tangent Φ (a double line). A construction of the conics in $\mathcal{F}$ starting from one of them is given in figure 16.4.10.5.1. A geometric characterization of superosculating conics is difficult; in the Euclidean case, they must have same curvature, but this is not enough, unless the conics have a common (Euclidean) symmetry axis going through a, in which case osculation implies superosculation. Figure 16.4.10.5.2 represents the osculating (hence superosculating) circles to the vertices of an ellipse, and gives the construction to find their centers. For arbitrary K, see a characterization in 16.4.13. We shall meet superosculating conics again in 19.6.8.3.

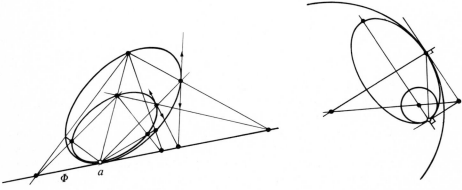

Figure 16.4.10.5.1   Figure 16.4.10.5.2

16.4.11. NOTES

16.4.11.1. The actual search for intersection points of two conics amounts to solving an equation of the fourth degree (16.4.4, for example); however, if one knows a degenerate conic in the pencil $\mathcal{F}$ determined by the conics, the problem is reduced to solving two equations of the second degree. Finding the degenerate conics of $\mathcal{F}$ amounts to solving a third-degree equation (cf. 14.2.7.5); since any fourth-degree equation can be cast in the form 16.4.4, we conclude that fourth-degree equations can be reduced to third-degree ones (and consequently solved by radicals).

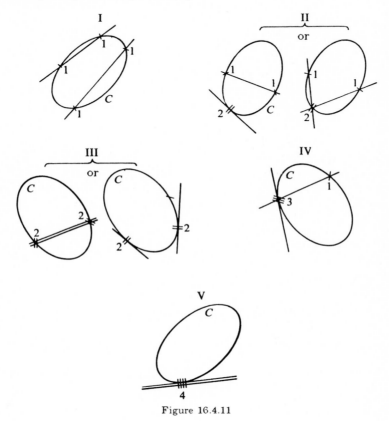

Figure 16.4.11

16.4.11.2. The study of the five types in 16.4.10 shows that $C \square C'$ can be calculated by taking a degenerate conic C'' of the pencil $\mathcal{F}$ and adding up the orders of the intersection points of C'' and C, according to the following rules: If there is a tangency, the order of the point is two; if a point lies on both lines that make up C'', count its order twice; and if C'' is a double line, count all orders twice. (See figure 16.4.11.) In order to define $C \square C'$ in this geometric way, it is necessary to show that the result does not depend on the choice of C''; see [SG, chapter XII] for details. This method also assumes that K is algebraically closed, otherwise there is no guarantee that $\mathcal{F}$ has degenerate conics, or that a conic intersects a line.

16.4.11.3. Contact theory. It is possible to construct a contact theory for differentiable curves. A contact of order at least two implies tangency; orders at least three and four correspond to our notions of osculation and superosculation (for conics and $K = \mathbf{R}$.) For this theory, see [LF1, 74 ff.] or [DE4, volume III, p. 34, problem 9].

16.4.12. OSCULATING CURVES AND DIFFERENTIAL GEOMETRY. Let P be a real projective plane, D a line in P and $X = P \setminus D$ the associated real affine plane (cf. 5.1.3). Consider two proper conics C, C' in P, tangent at $m \notin D$. By 16.7.2, C and C' can be considered as C^∞ regular geometric arcs (cf. [B–G, chapter 8] for this notion and the following ones). Finally, put a Euclidean structure on X.

16.4.12.1. Proposition. *The curves C and C' are osculating at m if and only if the associated geometric arcs in X have same curvature.*

Proof. We can assume that C and C' are parametrized by good parametrizations and that the image of $(0, 1)$ is m under both parametrizations. Further, by taking the tangent at m as one of the coordinate axes, the parametrizations take the form $t \mapsto \big(f(t), g(t)\big)$ for C and $t \mapsto \big(u(t), v(t)\big)$ for C', with $g'(0) = v'(0) = 0$. If q is an equation for C, definition 16.4.5 and the standard algebraic criterion for a polynomial to have a triple root imply that C osculates C' if

$$\frac{d^2\big(q(u(t), v(t))\big)}{dt^2}(0) = q_{xx}(m)\big(f'(0)\big)^2 + q_y(m)g''(0) = 0.$$

But we also have $q\big(f(t), g(t)\big) = 0$ for all t, whence

$$q_{xx}(m)\big(u'(0)\big)^2 + q_y(m)v''(0) = 0.$$

Given that $g'(0) = v'(0)$, the two relations above imply that the curvatures of C and C', which are given by $\dfrac{|g''(0)|}{\big(f'(0)\big)^2}$ and $\dfrac{|v''(0)|}{\big(u'(0)\big)^2}$, are equal. This also shows the converse. □

16.4.13. HOMOGRAPHIES AND OSCULATING CONICS. Here K is again arbitrary. Given a line D of P and a point $a \in D$, there exist homographies f of P fixing D pointwise and leaving invariant any line S that passes through a. Such a homography is called an *elation* with axis D and center a. An elation is determined by giving D and two points $\big(m, f(m)\big)$ collinear with a. It has no fixed points outside D.

To prove the above assertions, it is enough to send D to infinity (cf. 5.4); we see then that the restriction of f to the affine plane $F \setminus D$ must be a translation by a vector in the direction of a.

In homogeneous coordinates in which the equation of D is $z = 0$ and $a = p\big((1,0,0)\big)$, the matrix of f will be of the form

$$\begin{pmatrix} 1 & 0 & \alpha \\ 0 & 1 & 0 \\ 0 & 0 & 1 \end{pmatrix}.$$

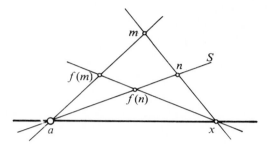

Figure 16.4.13

16.4.13.1. Proposition. *Let C and C' be proper conics, having the same tangent T at the point $a \in C \cap C'$. A necessary and sufficient condition for C and C' to osculate at a is that there be an elation f with center a and axis D such that $f(C) = C'$. Furthermore, C and C' are superosculating if and only if $D = T$.*

Notice that this statement justifies figures 16.4.10.4 and 16.4.10.5.1.

Proof. Let f be an elation with center a and axis D; if $D \neq T$, let b be a second point in $D \cap C$. Then $f(C)$ is a conic tangent to T at a and containing b, by 16.4.14; on the other hand, C and $f(C)$ have no point in common other than a and b. This means the intersection of C and $f(C)$ is of type IV, by table 16.4.7.5. If $D = T$ and $z = 0$ is an equation for D, we can write

$$q = a'y^2 + a''z^2 + 2byz + 2b'zx$$

(cf. 16.4.7.1); but we have seen that f is of the form $(x, y, z) \mapsto (x + \alpha z, y, z)$, so f^{-1} is given by $(x, y, z) \mapsto (x - \alpha z, y, z)$, and the equation of $f(C)$ will be given by

$$q' = a'y^2 + a''z^2 + 2byz + 2b'zx - 2b'\alpha z^2$$

(cf. 14.1.3.8), and $f(C)$ is clearly superosculating to C, by 16.4.7.3.

Conversely, let C and C' be osculating at a, and call m and m' the second intersections of C and C' with a line going through a and such that $m \neq m'$. Let $D = ab$, where $b \in C \cap C'$ is distinct from a unless C and C' are superosculating. Then the lifting f with axis D and center a and taking m to m' yields a conic $C'' = f(C)$ which, according the the above and 16.4.8 (ii), belongs to the pencil determined by C and C'. But $f(m) = m' \in C \cap C'$, whence $C'' = C'$. $\qquad\square$

16.5. Pencils of conics

Pencils of conics are assumed non-degenerate in this section (cf. 14.2.7.5).

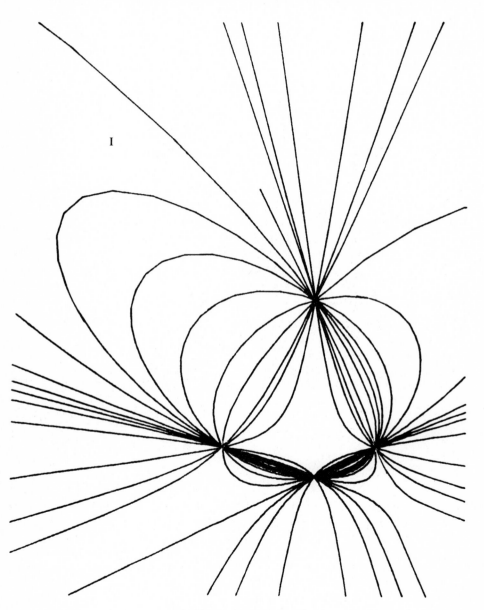

Figure 16.5.1.1

16.5.1. PROPOSITION. *Let P be a projective plane, α a proper conic in P with non-empty image C, and m_i ($i = 1, \ldots, k$) points of C with integer weights $\omega_i \geq 1$ such that $\sum_{i=1}^{k} \omega_i = 4$. Then the set*

$$\mathcal{F} = \left\{ \alpha \in \mathrm{PQ}(P) \,\middle|\, C' = \mathrm{im}(\alpha'), C \,\square\, C' = \{(m_i, \omega_i)\}_{i=1,\ldots,k} \right\}$$

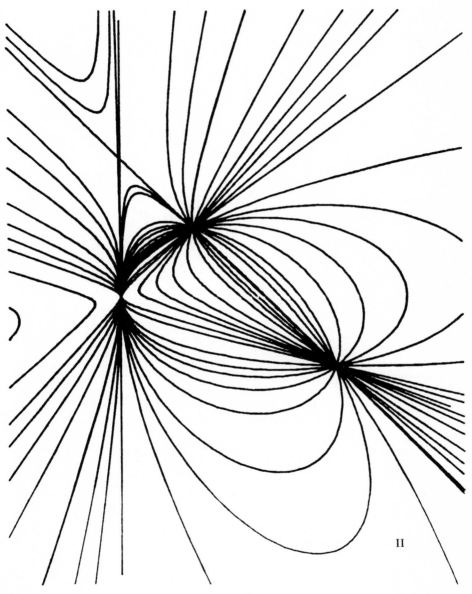

II

Figure 16.5.1.2

A pencil of this type is called full, and we associate to it a type according to 16.4.7.5. The set $\left\{(m_i, \omega_i)\right\}_{i=1,\ldots,k}$ depends only on $\mathcal{F}$, and not on the proper conic α we started with; in particular, the type of $\mathcal{F}$ does not depend on α. Given two full pencils $\mathcal{F}$ and $\mathcal{F}'$, a necessary condition for the existence of a homography $f \in \mathrm{GP}(P)$ taking $\mathcal{F}$ to $\mathcal{F}'$ is that the two pencils have the same type. This condition is also sufficient if the pencils are of type I, II or III, or if K is algebraically closed or equal to $\mathbf{R}$.

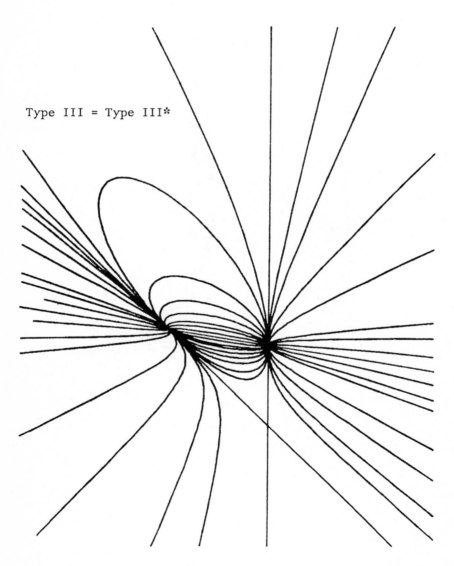

Type III = Type III*

Figure 16.5.1.3

Proof. To show that $\mathcal{F}$ is a pencil, it is enough to invoke the examples in 16.4.10, which exhibit conics C' with $C \square C' = \{(m_i, \omega_i)\}$, and to apply 16.4.8. From 16.4.7.4 we deduce that the type does not depend on C, only on $\mathcal{F}$. That $\mathcal{F}$ and $f(\mathcal{F})$ have same type follows from 16.4.8 (iii). For the converse when the type is I, II or III we use the fact that the pencil is well-determined by the data $\{a, b, c, d\}$, $\{a, b, c, \Phi\}$ or $\{a, b, c\}$, respectively, and

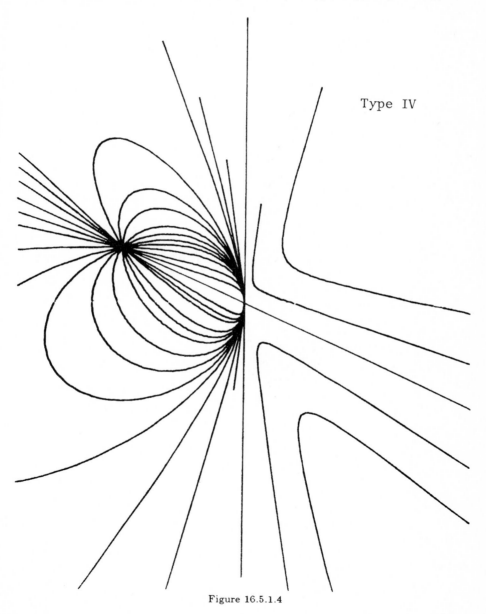

Type IV

Figure 16.5.1.4

that $GP(P)$ acts transitively on the set of such data (cf. 4.5.10). For types
IV and V this reasoning doesn't work, because then it is necessary to have a
conic in order to determine a pencil, and two conics aren't always conjugate
under $GP(P)$ (cf. 13.1.4.5). If K is algebraically closed or $K = \mathbf{R}$, we use
14.1.5.3 to reduce everything to the case of a fixed conic C. Then type IV
is characterized by two points in C, and type V by one point; by 16.3.8, the
group of C acts transitively on triples of points in C.   □

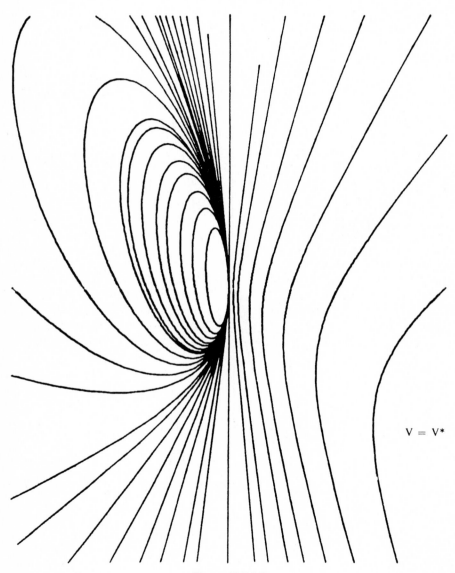

V = V*

Figure 16.5.1.5

16.5.2. NOTES. We have given in 16.4.10 explicit equations for all types of full pencils, and located the degenerate conics in such pencils.

For a discussion of non-full pencils in the real case, see 16.8.15. One technique of attack is to complexify $\mathcal{F}$ into a pencil $\mathcal{F}^C$ of conics in P^C (cf. 7.3 and 7.6). The results obtained in P^C are interpreted for conics in P, if possible. See examples of this technique in 17.5.

16.5.3. USES OF LINEARITY.

16.5.3.1. Let $\mathcal{F}$ be a pencil of conics in P and $m \in P$ a point. There exists $m' \in P$ such that m and m' are conjugate with respect to every conic in $\mathcal{F}$. To see this, use the extension of polarity given in 14.5.6: let q, q' be the equations of two conics in $\mathcal{F}$, with polar forms Π and Π'. The polar of m with respect to q (resp. q') is $\Pi(x, \cdot) = 0$ (resp. $\Pi'(x, \cdot) = 0$), where $m = p(x)$; thus any point $m' = p(x')$ such that $\Pi(x, x') = \Pi'(x, x') = 0$ has the desired property, since the polar form of $kq = kq'$ is $k\Pi = k\Pi'$.

16.5.3.2. We see that when $\mathcal{F}$ is of type I, the point m is unique as long as m does not lie on a side of the common self-polar triangle. The map $m \mapsto m'$ from a subset of P into itself is called a *quadratic transformation*, and plays important roles, cf., for example, [DQ, 153], [ML, 52] and [FN, 171 ff.]

16.5.3.3. It follows from 16.5.3.1 that, for $m \in P$, the polar of M with respect to any conic in $\mathcal{F}$ always contains m'. If the polar is always a line (this is the case if m does not belong to the radical of one of the degenerate conics in $\mathcal{F}$), that line, read in m'^*, is in homography with the pair $(k, h) \in \check{K}$ that defines the conic $kq + hq'$ in $\mathcal{F}$. Take a line D in P not intersecting any radical of a degenerate conic in $\mathcal{F}$, and fix two points $m, n \in D$; then the pole of D with respect to $kq + hq'$ is the intersection of the polars of m and n with respect to $kq + hq'$. The remarks above and 16.2.3 show that the set of poles of D is a conic in P, containing m' and n'. In fact, that conic is the image of D under the quadratic transformation associated with $\mathcal{F}$. Thus that transformation takes lines into conics, whence its name (see also 16.8.24).

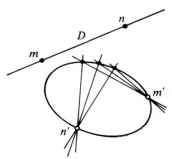

Figure 16.5.3.3

16.5.4. DESARGUES'S THEOREM. We recall corollary 14.2.8.3, since we shall be needing it several times in what follows. Let $\mathcal{F}$ be a non-degenerate pencil of conics. A line D is called *good* with respect to $\mathcal{F}$ if D contains no point in the base of $\mathcal{F}$ (cf. 14.2.7.3) and is not contained in the image of a (degenerate) conic in $\mathcal{F}$. Desargues's theorem says that *if D is a good line with respect to $\mathcal{F}$, there exists an involution f of D such that, for any $\alpha \in \mathcal{F}$, the intersection $\mathrm{im}(\alpha) \cap D$ is of the form $\{m, f(m)\}$ for some $m \in D$.*

16.5.5. APPLICATIONS.

16.5.5.1. The eleven-point conic. Let $\{a, b, c, d\}$ be a projective base; set $p = ab \cap cd$, $q = ac \cap db$, $r = ad \cap bc$ and let D be a line not containing any of these seven points. There exists a conic containing the following eleven points:

- p, q, r;
- the harmonic conjugate u' of $u = ab \cap D$ with respect to a and b, and the five corresponding points associated with ac, ad, bc, bd and cd;
- the two double points α and β (if they exist) of the Desargues involution on D, whose homologous points are $\{ab \cap D, cd \cap D\}$, $\{ac \cap D, db \cap D\}$ and $\{ad \cap D, bc \cap D\}$.

To show this, introduce the pencil of conics containing a, b, c, d and observe that the eleven points in question are on the image of D under the quadratic transformation associated with this pencil.

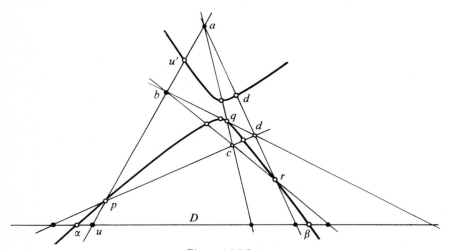

Figure 16.5.5.1

16.5.5.2. Conics in $\mathcal{F}$ tangent to a line. If $\mathcal{F}$ is a pencil of conics and D is a good line with respect to $\mathcal{F}$ (cf. 16.5.4), there exist either two or zero conics in $\mathcal{F}$ that are tangent to D. If they exist, the contact points are the fixed points of the involution determined by $\mathcal{F}$ on D. An explicit construction derives from 16.3.10.1; in particular, one can use the degenerate conics of $\mathcal{F}$.

16.5.5.3. The great Poncelet theorem for $n = 3$. Let $\mathcal{T}$ and $\mathcal{T}'$ be triangles inscribed in the image C of a proper conic; there exists a conic Γ tangent to all six sides of $\mathcal{T}$ and $\mathcal{T}'$. Corollary: if the conics C and Γ are such

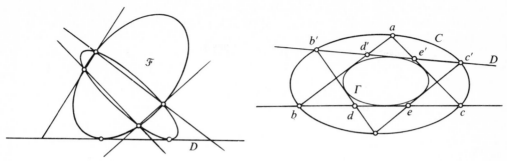

Figure 16.5.5.2   Figure 16.5.5.3

that there exists one triangle T inscribed in C and circumscribed around Γ, there exist many other such triangles, one for each point $a \in C$ from which one can draw two distinct tangents to Γ. (Thus, if K is algebraically closed, we can choose a arbitrarily in $C \setminus \Gamma$, and if $K = \mathbf{R}$, we can choose a arbitrarily in a neighborhood of each vertex of T.)

To demonstrate the corollary, we want to apply 16.2.7.2 to the lines D and E in figure 16.5.5.3, and this, by 6.1.4, amounts to showing that $[b, d, e, c] = [d', b', c', e']$. By 16.2.5, both these cross-ratios are equal to $[b, b', c', c]_C$.

We have already encountered this question in 10.10.4 and 10.13.3; we will discuss it at length in section 16.6.

16.5.6. DUALITY.

16.5.6.1. We now place ourselves in the framework introduced in 14.6. A *tangential pencil* of conics is a pencil of conics in P^*, that is, a line in $PQ(P^*)$. A tangential pencil is called full if it is full as a pencil in P^* (in the sense of 16.5.1); full pencils form five types, referred to as I*, II*, III*, IV* and V*. The three figures below show the situation in P, not in P^*, else they would be identical to the corresponding figures 16.5.1, by definition! In P, degenerate conics in a pencil of type I* are represented by pairs of points $\{a, a'\}$, $\{b, b'\}$ and $\{c, c'\}$; the reader should consider the other four cases.

16.5.6.2. In fact, the reason we haven't drawn pencils of type III* and V* is that they look the same as pencils of type III and V! To explain why this is so, let's consider the following question: let $\mathcal{F}$ be a pencil in P. To each non-degenerate conic $\alpha \in \mathcal{F}$, 14.6.1 associates a conic $\alpha^* \in PQ(E^*)$. What is the set of such α^*?

In an arbitrary set of coordinates, let A and B be matrices for two conics defining $\mathcal{F}$. By 14.6.1, the matrices of the desired conics α^* will be of the form $(\lambda A + \mu B)^{-1}$, for all λ, μ such that $\det(\lambda A + \mu B) \neq 0$. The rule for matrix inversion shows that the coefficients of $(\lambda A + \mu B)^{-1}$ are degree-two homogeneous polynomials in λ and μ, multiplied by $(\det(\lambda A + \mu B))^{-1}$; we can omit this common factor and still have the same conics. Thus the set of conics α^* is contained in the image of a map $f \in P_2^{\bullet}(K^2; Q(E^*))$. This map is of the form $\lambda^2 U + \lambda\mu V + \mu^2 W$, where U, V and W are vectors in $Q(E^*)$; thus its image lies in $KU + KV + KW$. By taking quotients and applying the proof of 16.2.9, we see that the image is a conic or a straight line in a projective plane inside $PQ(E^*)$. Explicit inversion of matrices of types I through V, in the coordinate system described in 16.4.10, shows that the image is a proper conic for types I, II and IV and a line for types III and V.

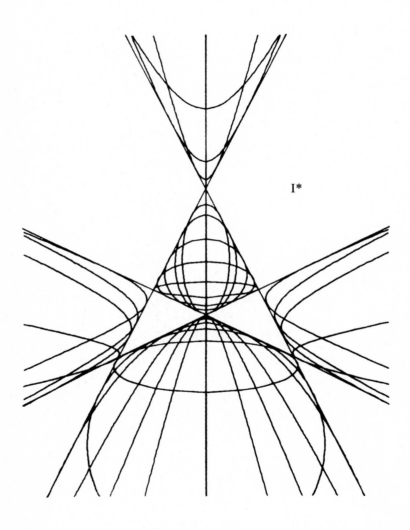

Figure 16.5.6.1

Thus pencils of type III and V are also tangential pencils (though degenerate conics in this situation are not the same as for punctual pencils), and pencils of type I, II and IV yield conics in $PQ(E^*)$. This latter result is a further proof of 16.5.3.3. We can also say that the map $* : PPQ(E) \rightarrow PPQ(E^*)$ introduced in 14.6.3.1 can be naturally extended to certain lines in $PQ(E)$.

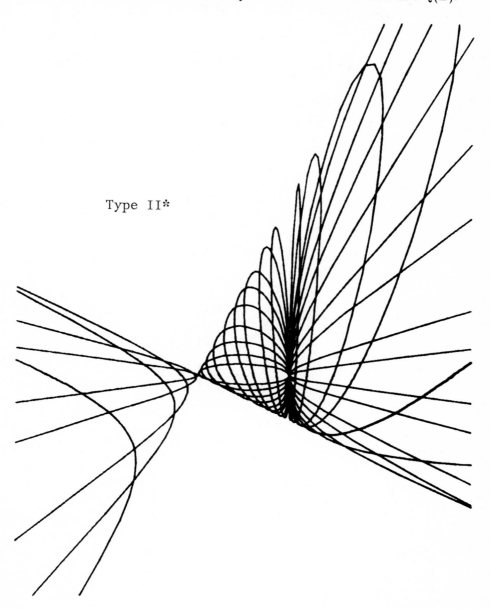

Type II*

Figure 16.5.6.2

16.5.6.3. Plücker's theorem. Interpreting 16.5.4 in P^*, we see that, if $\mathcal{F}^*$ is a pencil of conics in P_* and m is a point of P that does not lie on any common tangent to $\mathcal{F}^*$ and is not a degenerate point of $\mathcal{F}^*$, the tangents drawn from m to the conics of $\mathcal{F}^*$ form an involution of m^* (cf. 16.2.1).

For instance (cf. 16.5.5.2), every appropriate point in the plane lies in two or zero conics of a tangential pencil (cf. figures 16.5.6).

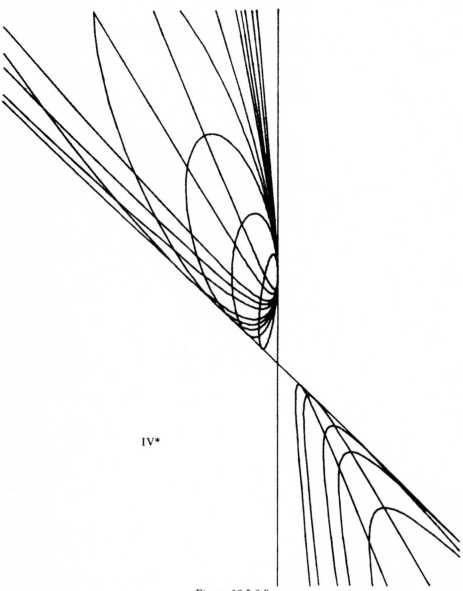

IV*

Figure 16.5.6.3

16.6. The great Poncelet theorem

16.6.1. Our aim is to show that if two conics C and Γ are such that there is an n-sided polygon inscribed in C and circumscribed around Γ, there exist many other such polygons. All known proofs of this result are rather long and recondite. The one given here is no exception; it is due to Hart, and follows the exposition in [LB2, 116–120]. More powerful demonstrations, based on an idea of Cayley, involve a cubic associated with the pencil determined by C and Γ; the idea is to endow that cubic with a group structure (cf. [FN, 124], for example), and to use a parametrization of it by elliptic functions (cf. [LG1, 12], for example). This affords a deep explanation for Poncelet's theorem. For a recent exposition, see [GF, section I(d)].

Another elegant demonstration for Poncelet's theorem, due to Chasles, is based on the notion of a "multivalued algebraic correspondence", in this case the $(2,2)$ correspondence; we mention it here because the notion of a correspondence has recently undergone important developments. In any case, a careful statement and proof of corollary 16.6.11, for instance, seem to be necessarily long, as the reader can verify by looking them up [DE5, chapter IV, § 12] and [DQ, 158–159].

16.6.2. Contrary to the common demonstrations which are restricted to $K = \mathbf{C}$, or, more exactly, K algebraically closed, we assume here that K is an arbitrary field with at least five elements, as stated in 16.4. Our building blocks will be: every point is contained in a conic of a given pencil; every line containing a point on a conic intersects that conic again; and the Desargues involution (cf. 16.5.4).

16.6.3. The idea of the proof is an induction on the number n of sides of the polygon, but it only works in the more general context of polygons inscribed in C and whose sides are tangent to n conics $\Gamma_1, \ldots, \Gamma_n$ in the pencil containing C and Γ; the original statement of the theorem is the particular case $\Gamma_1 = \cdots = \Gamma_n$. This is a good example of a result that gets easier to prove when generalized.

From now one we fix a pencil $\mathcal{F}$ of conics in a projective plane, and a proper conic $\alpha \in \mathcal{F}$ with non-empty image C. All conics we shall consider will be proper and have non-empty image, and, by 16.1.4, will be identified with their images. All points considered will lie in the complement of the base of $\mathcal{F}$, and all lines will be good with respect to $\mathcal{F}$ (cf. 16.5.4 and 14.2.7.3).

16.6.4. PROPOSITION. *Let $a, b, c \in C$ be distinct points, and $\Gamma, \Gamma' \in \mathcal{F}$ such that Γ is tangent to ab at α and Γ' is tangent to ac at β. Set $\Delta = \alpha\beta$. There exists $d \in C$ such that Γ is tangent to cd at $\Delta \cap cd$ and Γ' is tangent to bd at $\Delta \cap bd$; moreover, there exists $\Gamma'' \in \mathcal{F}$ tangent to ad at $\Delta \cap ad$ and to bc at $\Delta \cap bc$.*

Let $a, b, c, d \in C$ be distinct points, and $\Gamma \in \mathcal{F}$ tangent to ab at α and to cd at γ; there exists $\Gamma' \in \mathcal{F}$ tangent to ac at $\alpha\gamma \cap ac$ (and so, by the first part of the proposition, also tangent to bd at $\alpha\gamma \cap ac$).

Proof. Let γ be the point of $\Gamma \cap \Delta$ other than α, and let $\mathcal{G}$ be the pencil defined by Γ and the degenerate conic with image $\{ab\} \cup \{c\gamma\}$. The involutions determined by $\mathcal{F}$ and $\mathcal{G}$ on ac are the same, namely, the involution associated to the pencil of quadrics in ac determined by the quadric with image $\{a, c\}$ and the quadric $\Gamma \cap ac$ (which is well-defined, cf. 14.1.3.3, though the set $\Gamma \cap ac$ may be empty).

On the other hand, the involution determined by $\mathcal{G}$ on ac has $\Gamma' \cap ac = \{\beta, \beta\}$ as a fixed point; thus, there exists $\Sigma \in \mathcal{G}$ such that $\Sigma \cap ac = \{\beta, \beta\}$. But since $\Sigma \cap \Delta = \{\alpha, \alpha\}$, the conic Σ must be degenerate with image $\alpha\beta$, and thus all conics in $\mathcal{G}$ are tangent to $c\gamma$ at γ. In particular, the tangent to Γ at γ contains c. The desired point d is then $d = c\gamma \cap C$.

In fact, the same reasoning, with c, b, d in lieu of a, b, c, shows that Γ' is tangent to bd at $bd \cap \Gamma$; applied to the point $\delta = ad \cap \Delta$ and the conic Γ'' in $\mathcal{F}$ that goes through δ, it shows that Γ'' is tangent to ad at δ; and finally that Γ'' is tangent to bc at $\Delta \cap bc$. This reasoning also proves the second part of the proposition.   □

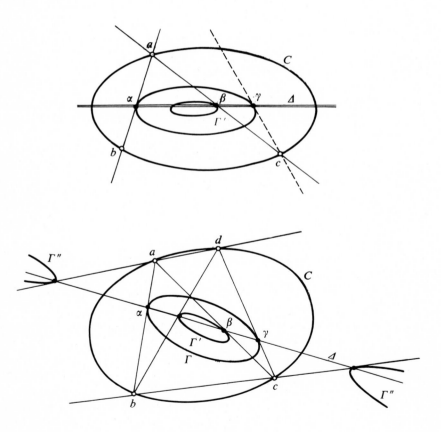

Figure 16.6.4

16.6.5. COROLLARY AND DEFINITION. *Let $a, b, c \in C$ be distinct points, and $\Gamma', \Gamma'' \in \mathcal{F}$ such that Γ' is tangent to ab at γ and Γ'' is tangent to ac at β. There exist exactly two conics $\Gamma_1, \Gamma_2 \in \mathcal{F}$ tangent to bc. The tangency point α_1 of Γ_1 is collinear with β and γ, and the tangent point α_2 of Γ_2 is such that $a\alpha_2, b\beta, c\gamma$ are concurrent. A sextuple $(a, b, c, \Gamma, \Gamma', \Gamma'')$, where $a, b, c \in C$ are distinct and $\Gamma, \Gamma', \Gamma'' \in \mathcal{F}$ are tangent to bc, ca, ab at α, β, γ, respectively, is called positive if $a\alpha, b\beta$ and $c\gamma$ are concurrent.*

Proof. The existence of Γ_1 follows from 16.6.4; the existence of Γ_2 from that of Γ_1 and from 16.5.5.2. That $a\alpha_2, b\beta$ and $c\gamma$ are concurrent is a consequence of 6.7.2 and 6.4.5. $\qquad\square$

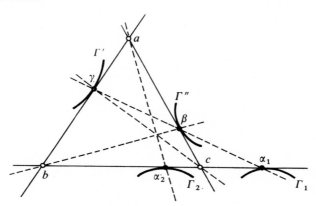

Figure 16.6.5

16.6.6. Consider a $2n$-tuple $(a_1, \ldots, a_n, \Gamma_1, \ldots, \Gamma_n)$, where the points $a_i \in C$ are distinct and Γ_i is tangent to $a_i a_{i+1}$ for all $i = 1, \ldots, n$ (with the obvious convention $a_{n+1} = a_1$). We use corollary 16.6.5 to define a new set of conics $\Gamma'_i \in \mathcal{F}$ ($i = 1, \ldots, n - 3$), as follows: We first require that the sextuple $(a_1, a_2, a_3, \Gamma_2, \Gamma'_1, \Gamma_1)$ be positive; then, by induction, we require that the sextuple $(a_1, a_{i+1}, a_{i+2}, \Gamma_{i+1}, \Gamma'_i, \Gamma'_{i-1})$ be positive for $i = 2, \ldots, n - 3$. At the end of the run we're left with the sextuple $(a_1, a_{n-1}, a_n, \Gamma_{n-1}, \Gamma_n, \Gamma'_{n-3})$, and if this sextuple is positive we say that the $2n$-tuple $(a_1, \ldots, a_n, \Gamma_1, \ldots, \Gamma_n)$ is *positive*. (See figure 16.6.6.)

With these preliminaries, the statement of the great Poncelet theorem is the following:

16.6.7. THEOREM. *Let $(a_1, \ldots, a_n, \Gamma_1, \ldots, \Gamma_n)$ be a positive $2n$-tuple and $b', b'' \in C$ two distinct points such that $b'b''$ is tangent to Γ_1. Then there exists a positive $2n$-tuple $(b_1, \ldots, b_n, \Gamma_1, \ldots, \Gamma_n)$ such that $b_1 = b'$ and $b_2 = b''$.*

16.6.7.1. The case $n = 3$. The idea is use 16.6.4 to make successive transfers. Let α_1, α_2 and α_3 be the contact points of Γ_1, Γ_2 and Γ_3 with $a_1 a_2$, $a_2 a_3$ and $a_3 a_1$, respectively, and let α' be the contact point of $b'b''$ with Γ_1. The last part of 16.6.4 shows that there exists $\Sigma \in \mathcal{F}$ tangent to $a_1 b'$ at $\beta' = a_1 b' \cap \alpha_1 \alpha'$ and to $b'' a_2$ at $\beta'' = b'' a_2 \cap \alpha_1 \alpha'$. By the first part of 16.6.4,

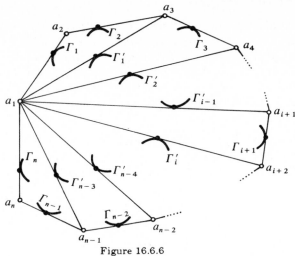

Figure 16.6.6

there exists $b''' \in C$ such that Γ_2 is also tangent to $b''b''''$ at $\alpha'' = b''b''' \cap \beta''\alpha_2$ and Σ is also tangent to a_3b''' at $\beta''' = a_3b''' \cap \alpha_2\beta''$. Again the last part of 16.6.4 yields $\Theta \in \mathcal{F}$ tangent to a_3a_1 at $\lambda = a_3a_1 \cap \beta'''\beta'$ and to $b'''b'$ at $\mu = b'''b' \cap \beta'''\beta'$. We will conclude by showing that $\Theta = \Gamma_3$.

The lines a_1b', a_2b'' and a_3b''' are not concurrent, since they're all tangent to Σ. Applying 5.5.1 to the pairs of triangles $\{\{a_1, a_2, a_3\}, \{\beta', \beta'', \beta'''\}\}$ and $\{\{\beta', \beta'', \beta'''\}, \{b', b'', b'''\}\}$, we see that neither $\alpha_1, \alpha_2, \lambda$ nor α', α'', μ are collinear. Since, by assumption, $\alpha_1, \alpha_2, \alpha_3$ are not collinear, 16.6.5 shows that $\lambda = \alpha_3$, whence $\Theta = \Gamma_3$; but since α', α'', μ are not collinear, the sextuple $(b', b'', b''', \alpha', \alpha'', \mu)$ is positive.

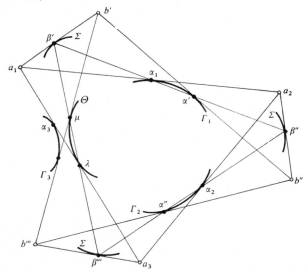

Figure 16.6.7

16.6.7.2. Reduction to $n - 1$. The induction step is trivial with the help of 16.6.7.1 and the construction in 16.6.6. From $(a_1, a_2, a_3, \Gamma_2, \Gamma'_1, \Gamma_1)$, which is positive by definition, we deduce a positive sextuple $(b_1 = b', b_2 = b'', b_3, \Gamma_2, \Gamma'_1, \Gamma_1)$; then, from $(a_1, a_3, a_4, \Gamma_3, \Gamma'_2, \Gamma'_1)$, another positive sextuple $(b_1, b_3, b_4, \Gamma_3, \Gamma'_2, \Gamma'_1)$; and so on. The $2n$-tuple $(b_1, b_2, b_3, \ldots, b_n, \Gamma_1, \ldots, \Gamma_n)$ we end up with is positive by definition. $\square$

16.6.8. NOTE. If K is arbitrary, it is not true that, given $b' \in C$, there exists a $b'' \in C$ distinct from b' and such that $b'b''$ is tangent to Γ_1 (see figure 16.6.12, for example). However, such a tangent exists if K is algebraically closed. If $K = \mathbf{R}$, one sees by continuity that the tangent exists if b' is taken close enough to a_1. In those two cases we conclude that if there exists one positive $2n$-tuple, there exist uncountably many of them.

16.6.9. We now have to return to the original problem that motivated 16.6.7, namely, finding $2n$-tuples of the form $(a_1, \ldots, a_n, \Gamma, \ldots, \Gamma)$, which correspond to polygons inscribed in C and circumscribed around Γ. Observe that the condition $\Gamma_1, \ldots, \Gamma_n \in \mathcal{F}$ is automatically satisfied now for the pencil determined by C and Γ. There remains to show the lemma:

16.6.10. LEMMA. *A $2n$-tuple $(a_1, \ldots, a_n, \Gamma, \ldots, \Gamma)$ is always positive.*

Proof. By 16.6.5, there exists a positive $2n$-tuple $(a_1, \ldots, a_n, \Gamma, \ldots, \Gamma, \Gamma')$; applying 16.6.7 to this $2n$-tuple with $b' = a_2$ and $b'' = a_3$, we deduce a positive $2n$-tuple which can be no other than $(a_2, \ldots, a_n, a_1, \Gamma, \ldots, \Gamma, \Gamma')$, since a point can only be contained in two tangents to a conic. Thus $a_1 a_2$ is also tangent to Γ', and so are $a_2 a_3, \ldots, a_{n-2} a_{n-1}$. Since Γ and Γ' share n distinct tangents $a_i a_{i+1}$ $(i = 1, \ldots, n)$, we have $\Gamma = \Gamma'$ as long as $n \geq 5$ (cf. 16.1.4). This leaves the cases $n = 3$ and $n = 4$. The case $n = 3$ has already been solved (figure 16.2.13, and compare 16.8.2). For $n = 4$ a polarity argument shows the existence of collinear points as in figure 16.6.10, implying exactly that $(a_1, a_2, a_3, \Gamma, \Gamma', \Gamma)$ and $(a_1, a_3, a_4, \Gamma, \Gamma, \Gamma)$ are positive. $\square$

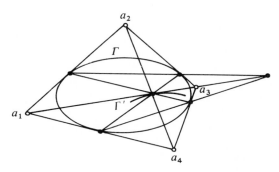

Figure 16.6.10

16.6.11. COROLLARY. *Let $(a_1, \ldots, a_n)$ be an n-sided polygon inscribed in a conic C and circumscribed around the conic Γ. For any two points*

$b', b'' \in C$ *such that* $b' \neq b''$ *and* $b'b''$ *is tangent to* Γ, *there exists a polygon* $(b', b'', b_3, \ldots, b_n)$ *inscribed in* C *and circumscribed around* Γ. *In particular, if* $K = \mathbf{C}$ *or* $\mathbf{R}$, *there exist uncountably many such polygons.* $\square$

For the existence of such polygons, see 16.6.12.4 and 17.6.7.

16.6.12. REMARKS.

16.6.12.1. See in 17.6.5 the interesting case that C and Γ are homofocal.

16.6.12.2. If (a_i) is inscribed in C and circumscribed around Γ, 16.6.7 shows that each line $a_i a_{i+2}$ $(i = 1, \ldots, n)$ is tangent to the same conic Γ_2 of the pencil $\mathcal{F}$ determined by C and Γ (argue as in the proof of 16.6.10). Similarly, each $a_i a_{i+3}$ $(i = 1, \ldots, n)$ is tangent to the same conic $\Gamma_3 \in \mathcal{F}$, and so on. By analogy with regular polygons and concentric circles, one may suspect that the geometry admits a symmetry group analogous to the rotation group of the circles. The deep reason why Poncelet's theorem is difficult is that such a group exists, but does not correspond to any subgroup of $\mathcal{S}_X$. See 16.6.1.

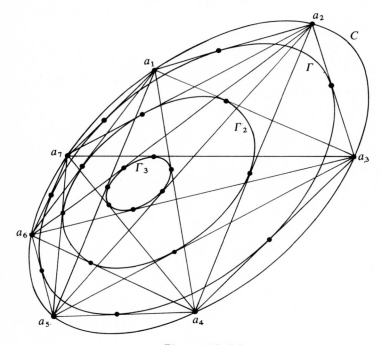

Figure 16.6.12.2

16.6.12.3. Poncelet polygons can give rise to interesting degenerate cases (i.e., some of the a_i coincide). For example, in figure 16.6.12.3, the quadrilateral (a_1, a_2, a_3, a_4) degenerates into (b_1, b_2, b_1, b_4) when two of the vertices are common to C and Γ. Notice also the quadrilateral having the asymptotes of Γ as two of its sides. See also [LB2, 142].

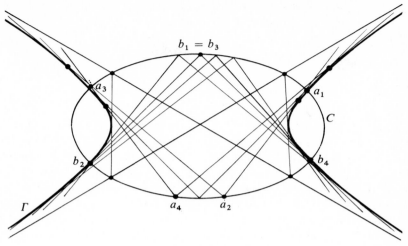

Figure 16.6.12.3

16.6.12.4. Cayley found the explicit condition on the equations of C and Γ for the existence of an n-sided polygon inscribed in C and circumscribed around Γ. If A and B are matrices representing C and Γ, consider the formal development of the function $\sqrt{\det(A + \lambda B)}$ in λ:

$$\sqrt{\det(A + \lambda B)} = \sum_{i=0}^{\infty} \sigma_i \lambda^i.$$

The desired condition is then

$$\begin{vmatrix} \sigma_3 & \sigma_4 & \cdots & \sigma_{p+1} \\ \sigma_4 & \sigma_5 & \cdots & \sigma_{p+2} \\ \vdots & \vdots & \ddots & \vdots \\ \sigma_{p+1} & \sigma_{p+2} & \cdots & \sigma_{2p-1} \end{vmatrix} = 0 \qquad \text{if } n = 2p,$$

$$\begin{vmatrix} \sigma_2 & \sigma_3 & \cdots & \sigma_{p+1} \\ \sigma_3 & \sigma_4 & \cdots & \sigma_{p+2} \\ \vdots & \vdots & \ddots & \vdots \\ \sigma_{p+1} & \sigma_{p+2} & \cdots & \sigma_{2p} \end{vmatrix} = 0 \qquad \text{if } n = 2p + 1.$$

The reader should write out the first two or three conditions when A and B are diagonal (this case happens, for example, when C and Γ are homofocal; see 17.6), and also for the case of two circles (compare with 10.13.3). For the proof of Cayley's formulas, see [LB2], or the more recent reference [GR–HA1].

16.6.12.5. On the subject of Poncelet's theorem, see also [B–H–H].

16.7. Affine conics

16.7.1. EQUATIONS. We preserve the notations introduced in 15.1, special-ized to the case of an affine plane X. An equation for α will be written

16.7.1.1 $ax^2 + 2bxy + cy^2 + 2dx + 2ey + f;$

the conic α will be proper if $\begin{vmatrix} a & b & d \\ b & c & e \\ d & e & f \end{vmatrix} = 0.$ The points at infinity of α

(cf. 15.1.3.2) are given by $ax^2 + 2bxy + cy^2 = 0$, or $cm^2 + 2bm + a = 0$, where we have *denoted* by m the *slope* of the corresponding line. (This a very convenient usage, but one must watch out for the case $m = \infty$, cf. 5.2.4, for example.) If $K = \mathbf{R}$ and the proper conic α has non-empty image C, it follows from 15.3.3.2 that C is an ellipse if $ac - b^2 > 0$, a parabola if $ac - b^2 = 0$ and a hyperbola if $ac - b^2 < 0$. The center of 16.7.1.1 is given by the equations $ax + by + d = 0$ and $bx + cy + e = 0$, by 15.5.5.2. To find the asymptotes one can use 14.5.3.

16.7.2. GOOD PARAMETRIZATIONS. By 16.2.6, good parametrizations can be obtained by considering a revolving line D through a point $a \in C$ and taking the slope m of D as a parameter (in $\tilde{K} = K \cup \infty$) for the second point in $C \cap D$. By 16.2.9 the result is always of the form

$$x = \frac{um^2 + u'm + u''}{wm^2 + w'm + w''}, \qquad y = \frac{vm^2 + v'm + v''}{wm^2 + w'm + w''}.$$

A particular case consists in taking $a \in \infty_X \cap C$, when C is a parabola or a hyperbola; this is the same as intersecting C with a family of parallel lines. We recover the well-known formulas $x \mapsto \alpha x^2 + \beta x + \gamma$ and $x \mapsto \dfrac{\alpha x + \beta}{\gamma x + \delta}$, which is a way to show that the graphs of these functions are parabolas and hyperbolas, respectively. See also 17.7.1 and 17.8.2.

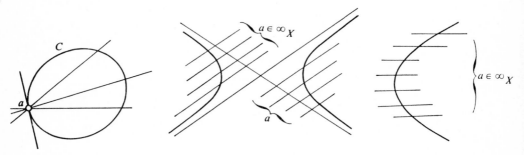

Figure 16.7.2

16.7.3. CONSTRUCTION OF THE CONIC PASSING THROUGH FIVE POINTS. Figure 16.2.11 shows how to find the intersection of a conic C, defined by five

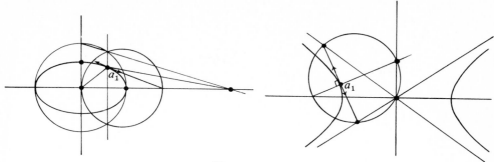

Figure 16.7.3.1

of its points, with a line D passing through one of the five. Figure 16.2.13 shows how to find the tangent to C at a known point. From 15.5.6 we can deduce the diameters of C, and thus the center, as the intersection of two diameters (figure 16.7.3.1).

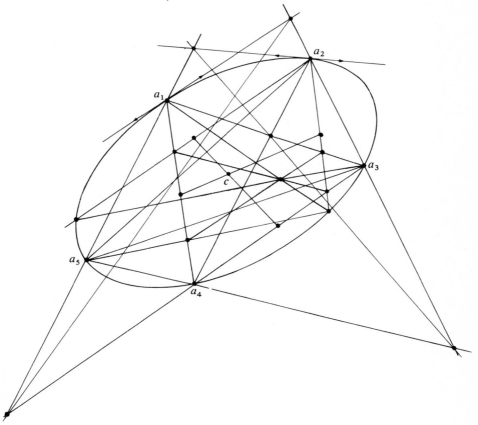

Figure 16.7.3.2

Having the center, two points and their tangents, one can determine
conjugate directions (cf. 15.6.3 and 15.7.5), hence the directions of the axes,
following the construction in 16.3.10.2. (This construction is good for any
Euclidean structure, though in practice one is interested in the Euclidean
structure of one's piece of paper and the length unit of one's ruler.) Given
the axes (cf. 15.6.8), plus one point and its tangent, the determination of the
vertices, etc., follows trivially from 17.7 and 17.8 (figure 16.7.3.2).

If the conic is a parabola, two points and their tangents suffice (see 16.7.4
and figure 16.7.3.3).

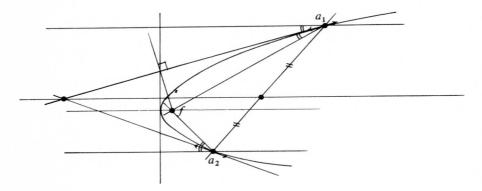

Figure 16.7.3.3

16.7.4. THE PARABOLA TANGENT TO FOUR LINES. From 16.1.4 and
a duality argument it follows that there exists a unique conic tangent to
$\infty_X \subset \hat{X}$ (that is, a parabola) and four given lines D_i in X, no two of which
are parallel and no three of which are concurrent. In order to determine it
using a Euclidean geometric construction, we first find its contact points with
D_1 and D_2 as in figure 16.7.4 (which is dual to figure 16.2.13); then 15.5.8
gives the direction of the axis, and 17.9.18.1 takes care of the rest.

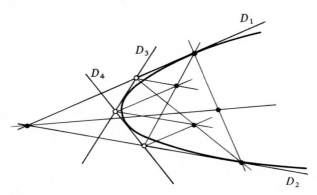

Figure 16.7.4

16.7.5. THE NINE-POINT CONIC. This is the affine version of 16.5.5.1. Given $a, b, c, d \in X$, there is a conic going through p, q, r and the midpoints of ab, ac, ad, bc, bd and cd; its points at infinity are the fixed points of the involution of ∞_X taking $\overrightarrow{ab}$ to $\overrightarrow{cd}$, $\overrightarrow{ad}$ to $\overrightarrow{bc}$ and $\overrightarrow{ac}$ to $\overrightarrow{db}$. This makes it easy to construct geometrically (cf. 16.3.10). Further, its center is $(a+b+c+d)/4$, as follows from 3.4.10 and 15.7.7. The nine-point circle is a particular case in which d is the orthocenter of $\{a, b, c\}$, cf. 10.11.3 and 17.5.4.

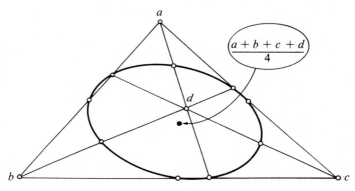

Figure 16.7.5

16.8. Exercises

16.8.1. Show that, given four points a, b, c, d in the image C of a proper conic, we have $[a, b, c, d] = -1$ if and only if $ab \perp cd$.

* **16.8.2.** Show that if $\{a, b, c\}$ is a triangle circumscribed around C, and α, β, γ are the tangency points, then the segments $a\alpha, b\beta, c\gamma$ are concurrent (cf. figure 16.12.3). Use only analytic geometry in your proof.

16.8.3. THE PASCAL LINES. Let C be a proper conic and a_i $(i = 1, \ldots, 6)$ points on C. The points a_i, taken in an arbitrary order, determine by 16.2.11 a line, called the *Pascal line* of the sextuple. Show that the same six points determine at most sixty Pascal lines; find an example where all sixty are distinct. Show that some triples of Pascal lines are concurrent, and that we obtain in this way twenty points, forming ten pairs of points conjugate with respect to C.

16.8.4. Let $C \subset P$ be the non-empty image of a proper conic, and embed P in a three-dimensional projective space $\hat{P}$. Show that there exists a quadric with neutral equation (cf. 14.4) whose image Q satisfies $C = Q \cap P$. Let a, b, c, d be four points on C, let X and X' be the generating lines in Ξ (cf. 14.4) containing a and b, and T and T' the lines in Θ containing c and d. Show that the points $ab \cap cd$, $X \cap T'$ and $X' \cap T$ are collinear. Deduce from this Pascal's theorem. Use this technique to do the previous exercise.

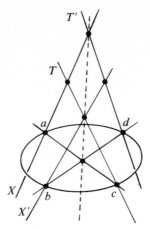

Figure 16.8.4

* **16.8.5.** Prove Pascal's theorem using calculus, by taking a projective base formed by four of the six points on the conic.

* **16.8.6.** Let C be the non-empty image of a proper conic, and let p, q, r be such that C is tangent to pq at q and to pr at r. Show that, for any $m, n \in C$, the following holds:

$$[q, r, m, n]^2_C = [pq, pr, pm, pn].$$

See an application for this in 17.4.2.2.

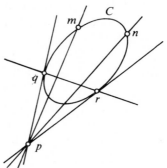

Figure 16.8.6

* **16.8.7.** Show that two involutions of a proper conic whose image is non-empty commute if and only if their Frégier points are conjugate.

* **16.8.8.** Let f be a non-involutive homography (different from the identity) of a proper conic with non-empty image C. Show that the set of lines $\langle m, f(m) \rangle$, for m ranging through C, is the set of tangents to a proper conic which is bitangent to C. Prove a converse statement. Deduce a simple proof for the results in 16.6 in the case of two bitangent conics.

16.8.9. Define and study the cross-ratio of four conics in a pencil.

* **16.8.10.** Prove rigorously that there are regions of the plane which do not intersect any of the conics of a given tangential pencil (figures 16.5.6).

16.8.11. Justify the constructions in figures 16.4.10.5 through 16.4.10.7.

16.8.12. Carry out the explicit calculation of dual pencils in 16.5.6.2.

16.8.13. What happens to the lines joining every p-th vertex of a $2p$-sided polygon inscribed in C and circumscribed around Γ?

16.8.14. Show that a tangential pencil of type IV* admits an osculation point for all its conics.

16.8.15. CLASSIFICATION OF PENCILS OF REAL CONICS. Classify (not necessarily non-degenerate) pencils of conics for $K = \mathbf{R}$ (cf. [LY, 259]).

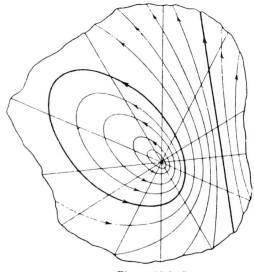

Figure 16.8.15

* **16.8.16.** Study the intersection of $xz - y^2 = 0$ and $xy - z^2 = 0$ over the field K with three elements.

16.8.17. Let K be the field with seven elements. Show that $xz - y^2 = 0$ and

$$2x^2 + z^2 - 2xy - yz = 0$$

intersect in a single point and that the pencil determined by them has only one degenerate conic. Show that $x^2 + 4yz = 0$ and $z^2 + xy = 0$ intersect in a point and their pencil has no degenerate conic.

16.8.18. Deduce 16.1.4 from 16.2.3.

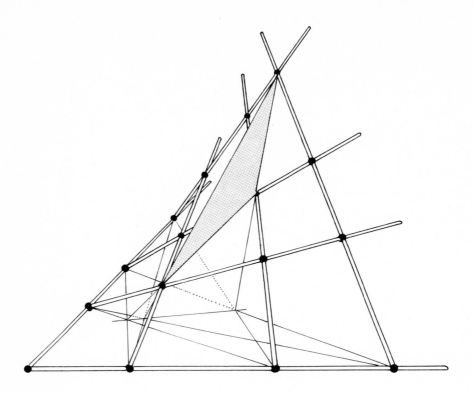

Figure 16.8.19

16.8.19. Prove Pappus' theorem using the technique of 16.8.4.

16.8.20. Given six tangents D_i $(i = 1, \ldots, 6)$ to a conic, with equations ϕ_i, show that the polynomials ϕ_i^2 are linearly dependent in $Q(E)$ (cf. 13.1.3.3).

16.8.21. What are the images of degenerate conics in $\mathcal{F}$ under the map $\mathcal{F} \to PQ(E^*)$ introduced in 16.5.6.2?

16.8.22. Find an explicit construction for the parabolas passing through four given points in an affine plane.

16.8.23. Are there pairs of conics having four common points but no four common tangents?

16.8.24. Calculate a simple quadratic transformation explicitly, in homogeneous coordinates.

16.8.25. Under what conditions is the conic

$$ax^2 + a'y^2 + a''z^2 + 2byz + 2b'zx + 2b''xy = 0$$

tangent to the three sides of the triangle $x = 0$, $y = 0$, $z = 0$? Can such a conic always be reduced (in the appropriate sense) to the form

$$x^2 + y^2 + z^2 - 2yz - 2zx - 2xy = 0?$$

Show that the curve with equation

$$x^8 + y^8 + z^8 - 2y^4 z^4 - 2z^4 x^4 - 2x^4 y^4 = 0$$

can be decomposed into four pairwise bitangent conics. Study the family of their lines of bitangency.

Chapter 17
Euclidean conics

Here we use the results from the preceding chapters to study conics in a Euclidean plane: ellipses, parabolas and hyperbolas. Except for circles, conics can be defined as the loci of points whose distances to a fixed point (the focus) and to a fixed line (the directrix) bear a constant ratio. They can also be defined, with the exception of parabolas, as the loci of points whose distances to two fixed points have a constant sum or difference.

These elementary results, proved in section 17.2, have been known since antiquity, but it was not until the nineteenth century that they were unified in a deep-reaching, encompassing theory. This development is mentioned in section 17.4. The techniques introduced in 17.4 are used to deduce from the projective results in chapter 16 a number of results on Euclidean conics (sections 17.5 and 17.6). The great Poncelet theorem is applied to the study of polygons of maximal perimeter inscribed in an ellipse; the perhaps surprising conclusion is that there are infinitely many such polygons.

Finally, sections 17.7 and 17.8 state, without proof, a number of statements that apply to ellipses alone, hyperbolas alone or equilateral hyperbolas alone. The reader can have his fun using one or the other of the techniques accumulated along the last few chapters in the proof of any of these statements.

For additional results on Euclidean conics, see [EE], [DQ] and [ML].

Throughout this chapter X denotes a Euclidean affine plane, $\tilde{X}$ its projective completion, $\tilde{X}^C$ the complexification of $\tilde{X}$ (cf. 7.6), and I, J the cyclic points of X (cf. 9.5.5.1). Unless we mention otherwise, C denotes the non-empty image of a proper conic α in X; by 16.1.4, it's all right to identify α and C.

17.1. Descartes's principle

17.1.1. By 15.6 and definition 15.3.3.2 an orthonormal frame can be found in which the equation of C is

17.1.2
$$\begin{cases} \dfrac{x^2}{a^2} + \dfrac{y^2}{b^2} - 1 & (a \geq b) & \text{(ellipse)}, \\[2mm] \dfrac{x^2}{a^2} - \dfrac{y^2}{b^2} - 1 & & \text{(hyperbola)}, \\[2mm] y^2 - 2px & & \text{(parabola)}. \end{cases}$$

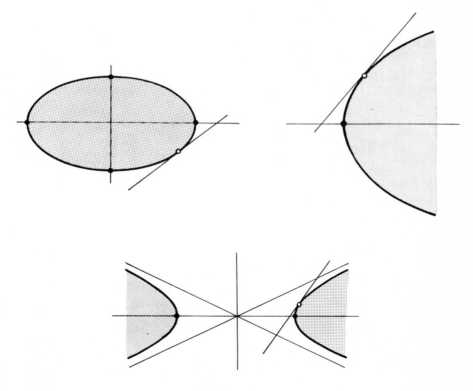

Figure 17.1.2

17.1.3. Notice that $a = b$ for the ellipse gives a circle. For a hyperbola it gives, by definition, an *equilateral hyperbola*. Equilateral hyperbolas have perpendicular asymptotes, and, just like circles, they have extra properties that ordinary hyperbolas don't share (see 17.8.3, for example).

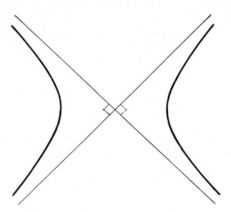

Figure 17.1.3

A point of a conic lying on an axis is called a *vertex*. Ellipses have four vertices, hyperbolas have two and parabolas, one.

17.1.4. TANGENTS AND NORMALS. CONVEXITY. By 15.4.6, ellipses and parabolas are always the frontiers of convex sets; so is each of the two connected components of a hyperbola.

The tangent to C at a point $m \in C$ can be defined in four equivalent ways: as the polar line of m (14.5.2.1), as the tangent space to the C^∞ submanifold C (14.3.8), as the tangent to the parametrized curve C (16.2.9), and as the unique supporting line at m for the convex set bounded by the connected component of C containing m (11.6.4).

The *normal* to C at $m \in C$ is the line going through m and perpendicular to the tangent to C at m.

17.1.5. THE WORD "CONIC". Conics have been known since the fourth century B.C., but it was Appolonius (second century B.C.) who first called them "conic sections", having proved that they are exactly the curves that one obtains by intersecting a cone of revolution with a plane (see figures 17.3.1 and 17.3.2). The analytic proof of this fact is left to the reader; see 17.3 for a geometric proof.

17.1.6. ANALYTIC GEOMETRY. The first result of algebraic geometry ever proved was that any second-degree equation yields a conic, in the sense of 17.1.5 or 17.1.2; it is due to Descartes (1637). Appolonius only knew equations 17.1.2. Descartes, the creator of analytic geometry, was the first to define curves of degree n, and illustrated the power of his technique by solving a problem open since the time of the Greeks: given four lines D_i $(i = 1, 2, 3, 4)$ in the plane and $k \in \mathbf{R}_+^*$, find the set

$$\{\, m \in X \mid d(m, D_1)d(m, D_2) = kd(m, D_3)d(m, D_4) \,\}.$$

Using analytic geometry, we take one equation ϕ_i for each D_i, and our set is given by $|\phi_1 \phi_2| = k|\phi_3 \phi_4|$, which is the equation of two conics $\phi_1 \phi_2 = \pm k \phi_3 \phi_4$ (incidentally, Descartes left one of them out). Notice that, as k changes, the conics form a pencil (cf. 16.4.10). Compare with [I–R, 367, exercise 350].

17.1.7. See in 17.9.1 another application of this technique.

17.2. Metrical properties (elementary)

17.2.1. MONOFOCAL PROPERTIES

17.2.1.1. Proposition. *Non-empty images of proper conics in X, with the exception of circles, are exactly the subsets of X of the form*

$$\{\, m \in X \mid fm = e \cdot d(m, D) \,\},$$

where $D \subset X$ is a line, $f \notin D$ a point in X and $e \in \mathbf{R}_+^*$.

Proof. Let $x = h$ be the equation of D and $(c, 0)$ the coordinates of f. Our subset is defined by

$$(x - c)^2 + y^2 = e^2 (x - h)^2.$$

If $e \neq 1$, this can be reduced to one of the equations in 17.1.2 by setting $h = c/e^2$:

17.2.1.2
$$\frac{x^2}{(c/e)^2} + \frac{y^2}{(1 - e^2)(c/e)^2} - 1 = 0.$$

If $e = 1$, we take $h = -c$ and find

17.2.1.3
$$y^2 = -4cx.$$

Conversely, an equation of the form 17.1.2 can be written in the form 17.2.1.2 as follows:

$$e = 1 \qquad p = -2c, \quad \text{for a parabola},$$

$$c = \sqrt{a^2 - b^2}, \quad e = c/a, \quad h = c/e^2 \quad \text{for an ellipse (if } a > b),$$

$$c = \sqrt{a^2 + b^2}, \quad e = c/a, \quad h = c/e^2 \quad \text{for a hyperbola.} \qquad \square$$

17.2.1.4. The number e is called the *eccentricity* of C; an ellipse (resp. parabola, hyperbola) is characterized by $e < 1$ (resp. $e = 1$, $e > 1$). A point f as above is called a *focus* of C, and D the associated *directrix*. The preceding study shows that ellipses and hyperbolas have two foci, lying on the x-axis (which is called the *focal axis*, or, specifically, the *major axis* for ellipses and the *transverse axis* for hyperbolas). A parabola only has one focus, lying on the x-axis. (See figure 17.2.1.4.)

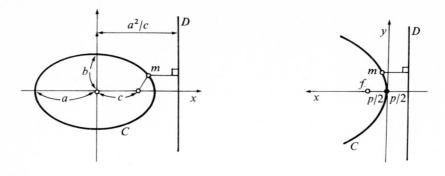

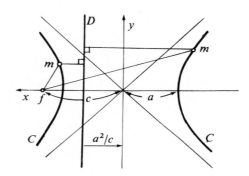

Figure 17.2.1.4

17.2.1.5. The corresponding convex regions (cf. 17.1.4) are given by

$$\{ m \mid mf \le e \cdot d(m, D) \}.$$

17.2.1.6. Consider $m, n \in C$ and $p = D \cap mn$ (work in $\tilde{X}$ if necessary). Then pf is a bisector of $\{fm, fn\}$ (cf. 8.7.7.4).

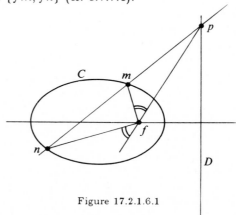

Figure 17.2.1.6.1

This follows from 10.3.8, because

$$\frac{pm}{pn} = \frac{d(m, D)}{d(n, D)} = \frac{fm}{fn}.$$

Passing to the limit shows that the tangent T to C at m intersects D at a point p such that $\overline{fm, fp} = \pi/2$. This gives a construction for the tangent (figure 17.2.1.6.2) and also shows that, if S is a line through f, its conjugate S' (cf. 15.5.1) through f satisfies $S' \perp S$ (figure 17.2.1.6.3). This last property characterizes foci (cf. 17.4.3), and implies the so-called little Poncelet theorem: if ma and mb are tangent to C at a and b and f is a focus of C, the line fm is the bisector of fa and fb. In fact, set $p = ab \cap D$; polarity shows that $p^\perp$ goes through m and f, so that $[fa, fb, fp, fm = -1]$, and the assertion follows from 8.7.7.5. See also 17.6.3.6.

Figure 17.2.1.6.2

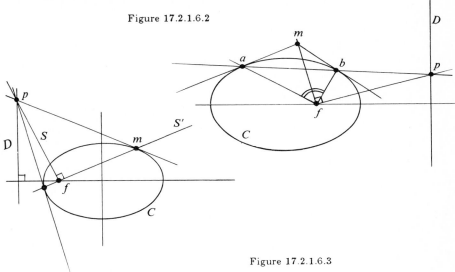

Figure 17.2.1.6.3

17.2.1.7. Motion of planets. The trajectory of a planet is given, in polar coordinates, by an equation $\rho = \rho(\theta)$ satisfying

$$\frac{1}{\rho} + \frac{d^2}{d\theta^2}\frac{1}{\rho} = \text{constant}$$

(Binet's formula); this is a result of Newton's laws of motion and gravitation (cf. [SY, 175], for example). Integrating we get $1/\rho = \alpha \cos\theta + \beta \sin\theta + \gamma$, or, upon changing the polar axis,

$$\rho(\theta) = \frac{1}{\alpha \cos\theta + \gamma}.$$

Back to cartesian coordinates we have $x^2 + y^2 = k(x + h)^2$, the equation of a conic with one focus at the origin. Thus the trajectories of planets, comets and asteroids are all conics with the sun at one focus.

17.2.2. BIFOCAL PROPERTIES.

17.2.2.1. Let f, f' and D, D' be the two foci and associated directrices of a conic C (that is not a parabola or a circle). By 17.1.4 we have, for every $m \in C$:

$$d(m, D) + d(m, D') = d(D, D') \quad \text{for an ellipse,}$$
$$|d(m, D) - d(m, D')| = d(D, D') \quad \text{for a hyperbola.}$$

By 17.2.1.1, this means that $mf + mf' = 2a$ for an ellipse, and $|mf - mf'| = 2a$ for a hyperbola, for every $m \in C$. This makes the next proposition plausible, though it does not yet prove it:

17.2.2.2. Proposition. *Non-empty images of proper conics in X, with the exception of parabolas, are exactly the subsets of X of the form $\{ m \in X \mid mf + mf' = 2a \}$ (resp. $\{ m \in X \mid |mf - mf'| = 2a \}$), where f and f' run through all pairs of points in X and $2a > ff'$ (resp. $2a < ff'$).*

Proof. The case of circles corresponds to $f = f'$ and is immediate. We discuss ellipses, and leave hyperbolas to the reader.

Take axes in which $f = (c, 0)$, $f' = (-c, 0)$; for $m = (x, y)$, we have

$$(mf)^2 + (mf')^2 = 2(x^2 + y^2 + c^2)$$

and $(mf')^2 - (mf)^2 = 4cx$ (these formulas also follow from 9.7.6). Setting $mf + mf' = 2\alpha > 0$, we have $mf' - mf = 2cx/\alpha$, then $mf' = \alpha + cx/\alpha$ and $mf = \alpha - cx/\alpha$. Thus we always have

$$\left(\alpha + \frac{cx}{\alpha} \right) + \left(\alpha - \frac{cx}{\alpha} \right) = 2(x^2 + y^2 + c^2),$$

or

$$(\alpha^2 - c^2) \left(\frac{x^2}{\alpha^2} - 1 \right) + y^2 = 0;$$

for $\alpha = a$, we recover the ellipse $x^2/a^2 + y^2/b^2 - 1 = 0$, where $b^2 = a^2 - c^2$.

Conversely, from $(\alpha^2 - c^2) \left(\dfrac{x^2}{\alpha^2} - 1 \right) + y^2 = 0$ we deduce that

$$(c^2 x^2 - a^2 \alpha^2)(a^2 - \alpha^2) = 0,$$

the case $a \neq \alpha$ being excluded for it implies $x = a\alpha/c$ and the contradiction $mf' - mf = 2a > ff'$. ◻

17.2.2.3. The corresponding convex regions (cf. 17.1.4) are $\{ m \in X \mid mf + mf' \leq 2a \}$ for the ellipse and $\{ m \in X \mid mf - mf' \leq 2a \}$ and $\{ m \in X \mid mf' - mf \leq 2a \}$ for the hyperbola.

17.2.2.4. Tangents. From 9.10 and 17.1.4 we deduce that the tangent to C at m is the interior (resp. exterior) bisector of $\overrightarrow{mf}$ and $\overrightarrow{mf'}$ for a hyperbola (resp. ellipse). Conversely, 9.10 shows that every curve whose tangent has

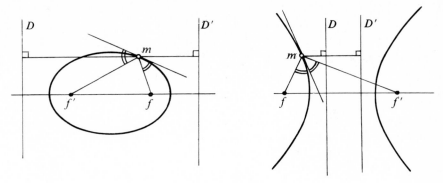

Figure 17.2.2.4

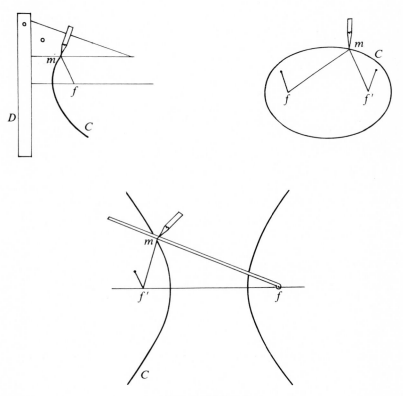

Figure 17.2.2.5.1

this property , or such that $mf \pm mf'$ is constant, is a subset of a conic with foci f, f'. This renders 17.2.2.2 even more reasonable. See also 17.6.3.5.

17.2.2.5. String construction. Figure 17.2.2.5.1 shows how to generate conics using a piece of string; see in 17.7.1 another mechanical generation for the ellipse. Figure 17.2.2.5.2 shows a similar construction for quadrics in dimension three; details are given in [H–C, 26] and 15.7.16. See also [SAL, section 421b], and the reference given in 17.6.4.

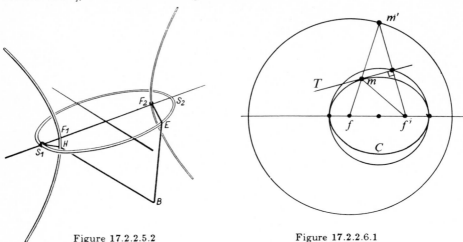

Figure 17.2.2.5.2   Figure 17.2.2.6.1

17.2.2.6. The pedal curve of a focus. Assume C is not a parabola. Using 17.2.2.4, 17.2.2.2 and taking the reflection $m' = \sigma_T m$ of m through the tangent T to C at m, we deduce that the pedal curve of C with respect to the focus f (cf. 9.6.8) is the circle concentric with C and bitangent to C at both ends of the focal axis (the *principal circle.*) If C is a parabola, we use 17.2.1.1 and 17.2.1.6 to show that the tangent T to C at m is the bisector of mf and the parallel to the axis that contains m; the pedal curve is the tangent to C at the vertex.

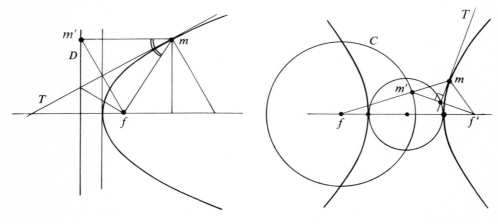

Figure 17.2.2.6.2   Figure 17.2.2.6.3

17.3. Metrical properties (spatial)

It is natural to try to show, in a purely geometric way, the identity between conics as defined by Appolonius (sections of cones of revolution) and as defined by their monofocal or bifocal properties. This is done in figures 17.3.1 and 17.3.2, due to the Belgians Dandelin and Quételet.

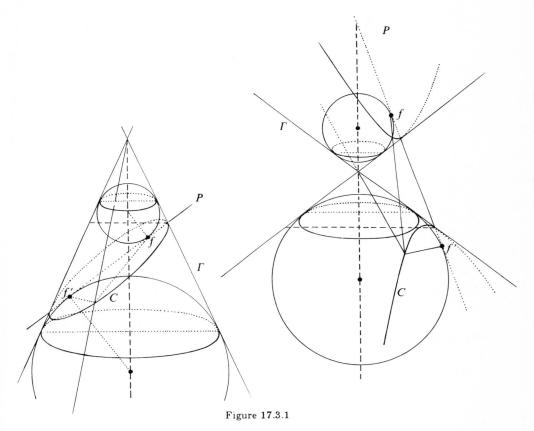

Figure 17.3.1

Figure 17.3.1 displays the bifocal properties and figure 17.3.2 the mono-focal properties. The key ideas in the proof of figure 17.3.2 are the following: (i) two segments of tangents to a sphere drawn from the same point have the same length; (ii) there exists at least one sphere Σ inscribed in the cone Γ and tangent to the intersecting plane P; (iii) if Q is the plane containing the circle where Σ touches Γ, points on the cone are characterized by $d(m, Q)/d(m, \Sigma \cap Q) = $ constant; (iv) if $D = P \cap Q$, points in P satisfy $d(m, Q)/d(m, D) = $ constant. Notice that the condition in (iii) is necessary and sufficient, otherwise we'd only be able to conclude that $C = P \cap \Gamma$ is contained in a conic with directrix D and focus f (the point where Σ touches P). The reader can fill in the details of this proof as an exercise, or look them

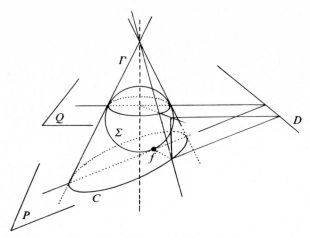

Figure 17.3.2

up in [D–C1, chapter VIII]. He can also recover the properties of tangents (17.2.1.6 and 17.2.2.4). See also 17.6.3.4.

17.4. Metrical properties (projective)

17.4.1. By 8.8.7 and 9.5.5.2 we know how to interpret Euclidean properties of X projectively. As required, we introduce the projective completion $\tilde{X}$ of X, the complexification $\tilde{X}^C$ of $\tilde{X}$, and the cyclic points I, J of X.

By 7.3, 7.6, 16.7 and 16.1.4, there corresponds to every proper conic α with non-empty image (or to every image C of such a conic) a unique projective conic in $\tilde{X}^C$, *denoted* by $\overline{\alpha}$, and a unique subset $\overline{C}$ of $\tilde{X}^C$, the image of $\overline{\alpha}$. We will make systematic use of this extension in our study of Euclidean conics from now on; this may appear artificial and involved, but will pay off eventually. There are two big advantages to this method, as often when a mathematical problem is studied in more generality: it affords a profound and unified understanding of the properties of circles and conics, and it yields many new properties, some of which would be very hard to prove in an elementary way.

17.4.2. THE CIRCLE AS A CONIC. From 14.1.3.2 and 8.8.6.1 we deduce the following

17.4.2.1. Criterion. The non-empty image C of a proper conic is a circle if and only if $I, J \in \overline{C}$ (if and only if $I \in \overline{C}$, since the involution σ of $\tilde{X}^C$, cf. 7.5.1 and 8.8.6.1, fixes $\overline{C}$ and takes I into J). More generally, if α is a conic in X, not necessarily proper or with non-empty image, saying that $I \in \overline{\text{im}(\alpha)}$ is the same as saying that every equation q of α is for the form $\vec{\alpha} = k\| \cdot \|^2$ in the Euclidean norm of $\vec{X}$. Such conics are called *generalized circles,* and will be studied in chapter 20.

17.4.2.2. This criterion explains the property stated in 10.9.4: if a and b are two points on a circle C, the oriented angle between lines $\overline{xa}, \overline{xb}$ is constant for $x \in C$, and conversely. In fact, the cross-ratio $[xa, xb, xI, xJ]$ in $\overline{C} \subset \hat{X}^C$ is constant, and Laguerre's formula (8.8.7.4) proves the equivalence. As to the fact that $2\widehat{\overline{xa}, \overline{xb}} = \overrightarrow{\widehat{\omega a, \omega b}}$ if ω is the center of C, it follows from 16.8.6.

17.4.2.3. Application: the orthoptic circle. Here we look for the set of points m such that the two tangents ma, mb to a conic C are orthogonal. By 8.8.7.4 this condition is equivalent to $[ma, mb, mI, mJ] = -1$ in $\hat{X}^C$, and by 14.2.5.6 this is equivalent to $mI \perp mJ$ (with respect to $\overline{C}$ in $\hat{X}^C$. Finally, by 16.2.7.1, this is equivalent to m belonging to a certain conic $\overline{S} \in \hat{X}^C$ that passes through I and J. This conic is proper if and only if IJ is not self-conjugate (that is, if C is not a parabola) and I, J are not conjugate with respect to $\tilde{C}^C$ (that is, if C is not an equilateral hyperbola). Returning to X, we see that m should lie in $\overline{S} \cap X$, a generalized circle (or a line if C is a parabola). There is no reason to suppose that the whole circle will be represented, since the tangents dropped from m to C may not exist. A complete discussion goes as follows: for an ellipse, we get a full circle, with radius $a^2 + b^2$; for a parabola, the whole directrix; for a hyperbola, the empty set if the asymptotes form an obtuse angle, a single point (the center) if it's an equilateral hyperbola, and a circle minus four points if the asymptotes form an acute angle. The reader can prove all of this in an elementary way, using 14.5.3 or geometrically.

    Notice that, for a tangential conic degenerated into two distinct points a and b (cf. 14.6.3.2), the orthoptic circle is the circle with diameter ab.

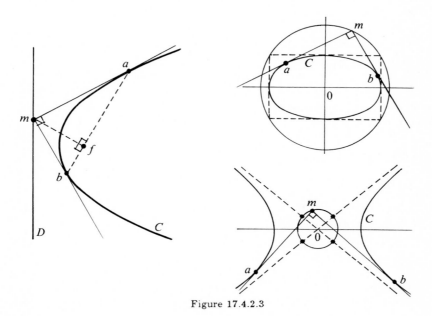

Figure 17.4.2.3

17.4.3. The foci and the cyclic points

17.4.3.1. We have remarked in 17.2.1.6 that the involution taking a line through a focus f of C into its conjugate with respect to C does, in fact, take each line into its orthogonal. Its fixed points are the tangents to C from f, by the result dual to 14.5.2.6; but, at the same time, they should be fI and fJ (in $\hat{X}^C$), by Laguerre's formula. The conclusion is that fI and fJ are tangent to $\overline{C}$.

17.4.3.2. Conversely, one can ignore all of the above, and *define* the foci of a conic C in X as the points f such that the lines fI and fJ are tangent to $\overline{C}$. Since $\sigma(I) = J$ and $\sigma(\overline{C}) = \overline{C}$ (cf. 17.4.2.1), there are only three possible cases: either $I, J \in \overline{C}$ and f is unique and coincides with the center, or C is a parabola and f is also unique, or there are four tangents to $\overline{C}$ emanating from I and J, and they intersect in four distinct points. In the latter case the set of four tangents is globally invariant under σ, so two of the points, say f, f', satisfy $\sigma(f) = f$ and $\sigma(f') = f'$, implying $f, f' \in X$, whereas the other two, say g, g', satisfy $\sigma(g) = g'$, so that $g, g' \notin X$, but $gg' \subset X$. Polarity shows that the lines ff' and gg' are orthogonal, since

$$[I, J, ff' \cap IJ, gg' \cap IJ] = -1,$$

and that $ff' \cap gg'$ is the pole of ∞_X, hence the center of C. Thus ff' and gg' are the axes of C, and we recover the arrangement described in 17.1.2.4.

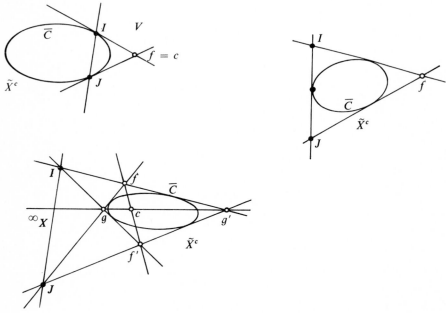

Figure 17.4.3.2

17.4.3.3. There still remains to see that a focus f and its polar $D = f^\perp$ with respect to C satisfy the metric definition given in 17.2.1.1. Let $m, n \in C$ be points, and set $p = mn \cap D$ (in $\tilde{X}$); since, in $\tilde{X}^C$, the tangents to $\overline{C}$ emanating from f are fI and fJ, polarity shows that, if u is the harmonic conjugate of P with respect to m and n, we have $[fI, fJ, fp, fu] = -1$. Thus, by 8.8.7, the lines fp and fu are orthogonal in X and fp is the bisector of fm and fn; the discussion in 17.2.1.6 implies that $fm/fn = d(m, D)/d(n, D)$, and this for arbitrary m, n, q.e.d.

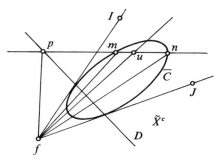

Figure 17.4.3.3

17.4.3.4. Property 17.2.2.2 will be explained in 17.6.3.5.

17.4.3.5. Application: parabolas and the Simson line. *If the three sides of a triangle $\{a, b, c\}$ are tangent to a parabola, the focus f of this parabola belongs to the circle circumscribed around $\{a, b, c\}$.* This is a direct consequence of 16.5.5.3, 17.4.2.1 and 17.4.3.2, applied to the six points a, b, c, f, I, J in $\tilde{X}^C$.

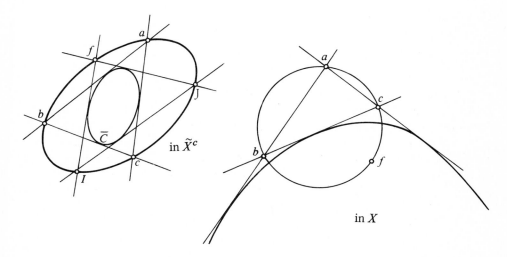

Figure 17.4.3.5.1

By 17.2.2.6, the projections of f on the sides ab, bc and ca are collinear, since they all lie on the tangent to C at the vertex. Conversely, we can define a parabola by its focus and the tangent at the vertex if the three projections are collinear. This shows that a necessary and sufficient condition for the projections of a point on the sides of $\{a, b, c\}$ to be collinear is that the point belong to the circle circumscribed around $\{a, b, c\}$; we have recovered 10.4.5.4.

Here is a corollary of this result and 16.7.4: given four lines on the plane, the circles circumscribed around each of the four triangles determined by the lines are concurrent, and their intersection is the focus of the parabola tangent to the four lines. See also exercise 10.13.19.

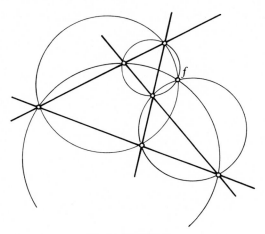

Figure 17.4.3.5.2

17.4.3.6. For a practical method for finding the foci, see 17.9.24.

17.5. Pencils of conics and the cyclic points

17.5.1. We shall understand by a *pencil of conics* in X a non-degenerate pencil of conics in $\tilde{X}$ (cf. 14.2.7.5). We will be studying the points at infinity of conics in X, that is, the restriction of the pencil $\mathcal{F}$ to ∞_X; when necessary we complexify and study the pencil $\mathcal{F}^C$ in $\tilde{X}^C$ (cf. 16.5.2), so as to be able to use the cyclic points. The material below, as long as it involves only points at infinity of conics $\alpha \in \mathcal{F}$, is amenable to an elementary treatment using slopes and the notation introduced in 16.7.1; the reader is strongly encouraged to develop this analysis.

In the sequel, the expression "equilateral hyperbola" will include pairs of orthogonal lines.

17.5.2. Assume that ∞_X is good with respect to $\mathcal{F}$, and consider the behavior of the involution f of $\mathcal{F}^C$ in ∞_{X^C} with reference to the cyclic points I and J, taking into account that $\sigma(I) = J$, $\sigma(J) = I$ and that pairs of points matched by f must be invariant under σ (which is not an involution of ∞_{X^C}, cf. 7.5). There are three possible cases:

$$\text{case I}: f(I) = I \quad \text{and} \quad f(J) = J,$$
$$\text{case II}: f(I) = J \quad \text{and} \quad f(J) = I,$$
$$\text{case III}: \{f(I), f(J)\} \cap \{I, J\} = \emptyset.$$

Recalling 6.7.2, 8.8.7, 8.7.7.5, 17.4.2 and 17.1.3, we see that:

17.5.3. PROPOSITION. *Let $\mathcal{F}$ be such that ∞_X is good with respect to $\mathcal{F}$. In general, $\mathcal{F}$ contains a single equilateral hyperbola (case III); if it contains two, it contains only equilateral hyperbolas (case I). A necessary and sufficient condition for $\mathcal{F}$ to contain a generalized circle is that two conics in $\mathcal{F}$ have axes in the same directions, in which case all conics in $\mathcal{F}$ have axes in the same directions (case II).* $\qquad\square$

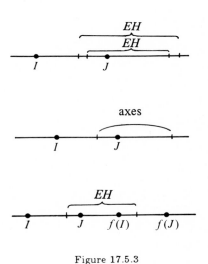

Figure 17.5.3

17.5.4. APPLICATIONS OF CASE I. Consider four lines D, D' and E, E' with $D \perp D'$ and $E \perp E'$ (in the Euclidean metric). The four define a pencil with base $\{D \cap E = a, D \cap E' = b, D' \cap E = c, D' \cap E' = d\}$, all of whose conics are equilateral hyperbolas. In particular, ad and bc are orthogonal by 17.5.3; this shows that the altitudes of a triangle are concurrent (cf. 10.2.5).

A similar reasoning shows that, for any three points $a, b, c \in C$, where C is an equilateral hyperbola, the orthocenter of $\{a, b, c\}$ also lies in C.

If we apply 16.5.5.1 to a triangle $\{a, b, c\}$ and its orthocenter d, we recover 10.11.3. The nine-point circle is the locus of the centers of equilateral hyperbolas containing a, b, c and d. The characterization of the center of the circle given in 10.11.3 follows from 16.7.5.

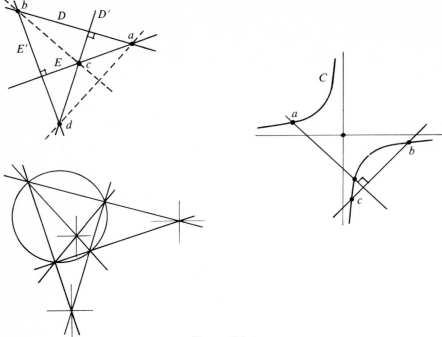

Figure 17.5.4

17.5.5. APPLICATIONS OF CASE II.

17.5.5.1. Let $\{D, D'\}$ and $\{E, E'\}$ be two pairs of lines; for the four points $D \cap E$, $D \cap E'$, $D' \cap E$ and $D' \cap E'$ to be cocyclic it is necessary and sufficient that the bisectors of the pairs $\{D, D'\}$ and $\{E, E'\}$ lie in the same directions.

17.5.5.2. Let $\mathcal{F}$ be a pencil of conics in X, one of which is a circle, and assume that ∞_X is good with respect to $\mathcal{F}$. Then the locus of the center of conics in $\mathcal{F}$ is an equilateral hyperbola of type II, whose asymptotes are the axes of the two parabolas contained in $\mathcal{F}$ (figure 17.5.5.2).

This follows from 16.5.5.1, except for the part about the asymptotes. To show that part, let u and v be the points in ∞_X corresponding to the directions of the axes. By the construction in 16.5.3.3, when $m \in H$ the lines um and vm are conjugate with respect to the conic in $\mathcal{F}$ having center m; when $m = v$ this means that the tangent to h at u is the axis of the parabola tangent to ∞_X at u.

17.5.5.3. When two parabolas C and C' in X form a full pencil $\mathcal{F}$ (cf. 16.5.1), a necessary and sufficient condition for their axes to be perpendicular is that $\mathcal{F}$ contain a circle; further, if $C \sqcap C' = \{(m_i, \omega_i)_{i=1\ldots k}\}$, the axes of the parabolas go through the center of mass

$$\frac{1}{4} \sum_{i=1}^{k} \omega_i m_i.$$

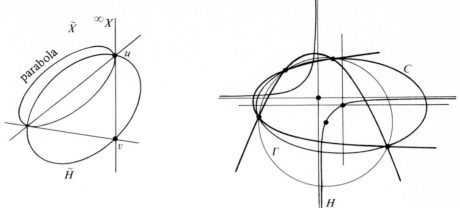

Figure 17.5.5.2

This follows from 17.5.5.2 and 16.7.5; we deduce an immediate construction for the parabolas passing through four distinct cocyclic points.

The center of mass statement can also be proven in elementary ways: one can use diameters and 17.5.5.1, or the formula for the sum of the roots of a fourth-degree equation if the coordinate system is chosen parallel to the axes of the parabolas.

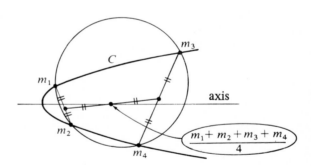

Figure 17.5.5.3

17.5.5.4. The osculating circle. *Let C be the non-empty image of a proper conic, and $m \in C$ a point. There exists a unique circle, denoted by $\Gamma_m C$, that osculates C at m in the sense of 16.4.5.* (It is also osculating in the differential geometry sense, cf. [B–G, 8.4.15].) *This circle will be called the osculating circle to C at m. Assume m is not a vertex, and let D be the line through D satisfying the condition that the bisectors of D and the tangent $T_m C$ to C at m are parallel to the axes. Then $\Gamma_m C$ is characterized as the circle tangent to C at m and containing the point m' where D intersects C again; also as the circle tangent to C and crossing from one side to the*

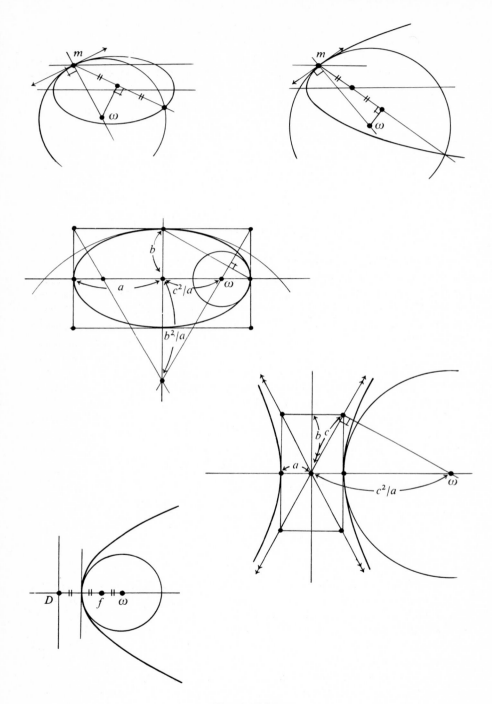

Figure 17.5.5.4

other of C in the neighborhood of m. If m is a vertex, $\Gamma_m C$ is automatically superosculating to C at m, and $C \cap \Gamma_m C = \{m\}$.

Proof. This follows from 17.5.3 and 16.4.10, together with symmetry arguments (for the superosculation) and continuity arguments (for the crossing). □

If m is not a vertex, the circle $\Gamma_m C$ is geometrically determined by the criterion above. In the case of a parabola, 17.5.5.3 shows that the axis contains the point $(3m + m')/4$, which gives an immediate construction for $\Gamma_m C$, not even requiring the determination of m'.

If m is a vertex, one must use the criterion stated in 16.4.7.3, and the fact that

$$\frac{x^2}{a^2} \pm \frac{y^2}{b^2} - 1 + k(x - a)^2$$

is the equation of a circle in order to find the superosculating circle to C at $(a, 0)$; this is done in figure 17.5.5.4. In 17.7.4 we shall encounter other constructions for the center of curvature ω at a point of C.

17.5.5.5. An example from mechanics. Take a string with one end fixed (call it ω) and a weight at the other end. The weight being at rest at the point a directly below ω, set it in motion by giving it some horizontal initial velocity. The weight moves along a vertical circle Γ of center ω while the string is taut; but at some point m of its trajectory, when the radial component of the weight becomes equal to the centrifugal "force" (this happens if the initial velocity was chosen within a certain range) the string slackens and the weight starts falling freely. Determine the initial speed so that the free-fall trajectory of the weight contains the starting point a.

The solution is found by remarking that the free-fall trajectory is a parabola with vertical axis, and that this parabola osculates the circle at m, since the velocity and acceleration of the weight must remain the same as it passes through m. The point m' where the parabola intersects the circle again is easily found: by 17.5.5.4 and 8.7.2.4, we have $\widehat{\overrightarrow{\omega a}, \overrightarrow{\omega m'}} = -3\widehat{\overrightarrow{\omega a}, \overrightarrow{\omega m}}$.

Plugging in $m' = a$, we get $3\widehat{\overrightarrow{\omega a}, \overrightarrow{\omega m}} = 0$, and $\widehat{\overrightarrow{\omega a}, \overrightarrow{\omega m}} = 0$ or $2\pi/3$ or $4\pi/3$.
For similar questions, see 17.9.2.

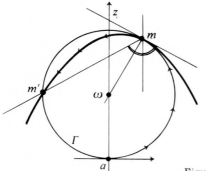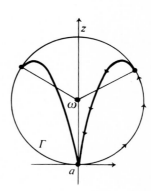

Figure 17.5.5.5.1

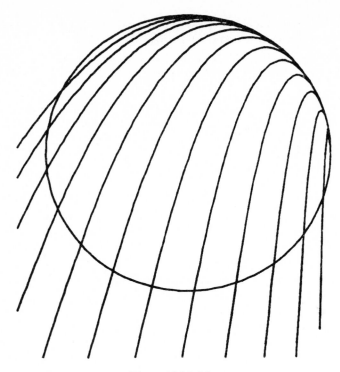

Figure 17.5.5.5.2

17.5.5.6. Normals to a conic from a point. Appolonius' hyperbola.

Let C be a non-empty proper conic and $a \in X$ a point. There exist at least one and at most four normals to C containing a; the feet of these normals are the intersections $C \cap A_a C$, where $A_a C$ is an equilateral hyperbola well defined by a and C, called the Appolonius' hyperbola of C with respect to a. The hyperbola $A_a C$ is geometrically determined by the following properties: its asymptotes are parallel to the axes of C; it contains a and the center of C (if C is a parabola, one of its asymptotes is the axis of C); its center is $\frac{1}{4} \sum_{i=1}^{k} \omega_i m_i$, where $\{(m_i, \omega_i)\}_{i=1,\ldots,k} = C \square \Gamma$ for any circle Γ intersecting C so that $\sum m_i = 4$; and it contains the points $u, v \in a^\perp$ such that au, av are orthogonal (in particular, if a lies outside C, the lines au, av are the bisectors of the tangents to C issued from a).

Proof. We show first that the locus of the centers of the conics in the pencil $\mathcal{F}$ determined by C and a generalized circle Γ centered at a depends only on a and C. In fact, 16.5.3.3 shows that m is the center of a conic in $\mathcal{F}$ if and only if its polars with respect to C and Γ are parallel. Now, the polar of a point m with respect to a circle Γ with center a is orthogonal to am, so its direction depends only on a, not on Γ.

This locus is a hyperbola, denoted by $A_a C$, as can be seen by applying 17.5.5.2 and 17.5.5.3 to an arbitrary Γ—the degenerate conic with image a

and equation $d^2(\cdot, a) = 0$, say. There remains to show that, for $m \in C$, the line am is perpendicular to the tangent T_mC to C at m if and only if $m \in A_aC$. Again, the tangent to C at m is the polar of m with respect to C, and it is parallel to the polar to m with respect to Γ if and only if $am \perp T_mC$ in the Euclidean sense.

We suggest that the reader draw a few hyperbolas A_aC for various choices of a and C; see also 17.9.18.2. Notice the following amusing corollary: If a circle Γ intersects a hyperbola in four distinct points, the center of mass of the points depends only on the center of the circle.

For a discussion of the number of normals, see 17.7.4.   □

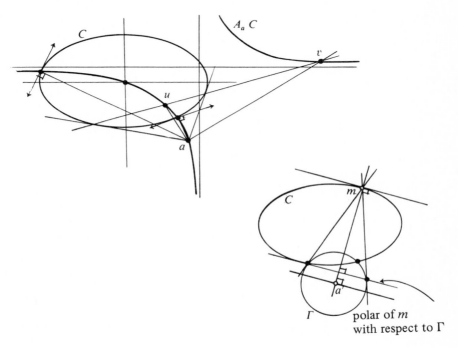

Figure 17.5.5.6

17.6. Tangential pencils of conics. Homofocal conics

Here we make systematic use of 16.5.6.3. In this section $\mathcal{F}^*$ is a *tangential pencil* in X, which is, by definition, a tangential pencil in $\tilde{X}$. In order to be able to use the cyclic points I and J of X, we introduce the complexification $\mathcal{F}^{*C}$ of $\mathcal{F}^*$, which is a tangential pencil in $\tilde{X}^C$.

17.6.1. ORTHOPTIC CIRCLES OF A TANGENTIAL PENCIL. Let $\mathcal{F}^*$ be a tangential pencil, and $\mathcal{F}^{*C}$ its complexification. Given two conics $\alpha^*, \beta^* \in \mathcal{F}^*$, let Γ and Σ be their orthoptic circles. Then $\overline{\Gamma}$ and $\overline{\Sigma}$ are conics in $\tilde{X}^C$; take $m \in \overline{\Gamma} \cap \overline{\Sigma}$. The tangents dropped to the conics in $\mathcal{F}^{*C}$ from m form an involution in m^* (cf. 16.5.6.3) whose two pairs associated to $\overline{\Gamma}$ and $\overline{\Sigma}$ are harmonically conjugate with respect to the pair $\{mI, mJ\}$. Thus all orthoptic circles Θ of conics of $\mathcal{F}^*$ satisfy $m \in \overline{\Theta}$; in other words, the circles $\overline{\Theta}$ form a pencil of conics in $\tilde{X}^C$, and the circles Θ for a pencil of circles in X, in the sense of 10.10.

17.6.2. APPLICATIONS.

17.6.2.1. *If a, b, c, d, e, f are the opposite vertices of a complete quadrilateral, the three circles with diameters $\{a, b\}$, $\{c, d\}$ and $\{e, f\}$ belong to a pencil. In particular, the points $(a+b)/2$, $(c+d)/2$ and $(e+f)/2$ are collinear.*

Proof. This follows from 17.6.1 and the end of 17.4.2.3, applied to the tangential pencil formed by conics tangent to the four lines ac, cb, bd, da.   □

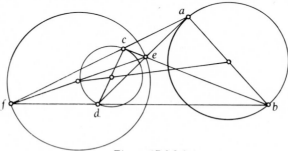

Figure 17.6.2.1

17.6.2.2. Apply 17.6.1 to the tangential pencil of conics in X that are tangent to the three sides ab, bc, ca of a given triangle and to the line at infinity ∞_X. Since the conics are parabolas, their orthoptic circles are their directrices; in particular, for the degenerate conic $\{a, bc \cap \infty_X\}$, the directrix is the line orthogonal to bc and containing a. This shows again that the altitudes of $\{a, b, c\}$ are concurrent, and also that the directrix of every parabola tangent to the sides of $\{a, b, c\}$ goes through the orthocenter of $\{a, b, c\}$. One consequence is that the orthocenters of the four triangles determined by four given lines are collinear, since they all lie on the directrix of the parabola tangent to the four lines (cf. 16.7.4 and 17.4.3.5).

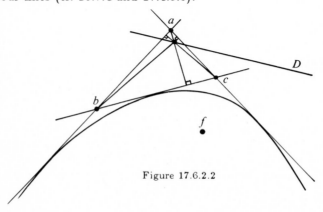

Figure 17.6.2.2

Another consequence is that if f is a point on the circle circumscribed around $\{a, b, c\}$, the reflections of f in the three sides of $\{a, b, c\}$ are collinear, lying on a line that goes through the orthocenter of $\{a, b, c\}$ (and is called the *line of the images*, cf. 10.13.16).

17.6.3. HOMOFOCAL CONICS.

17.6.3.1. Two conics are said to be *homofocal* if they share both foci. If $f, f' \in X$ are distinct points, the set of conics with foci at f and f' is a tangential pencil $\mathcal{F}^*$. To see this, work in $\tilde{X}^C$; by 17.4.3, a conic C lies in $\mathcal{F}^*$ if and only if $\overline{C}$ belongs to the tangential pencil of conics in $\tilde{X}^C$ that are tangent to the four lines fI, fJ, $f'I$ and $f'J$. Clearly, $\mathcal{F}^*$, the restriction of this tangential pencil to X, is also a tangential pencil.

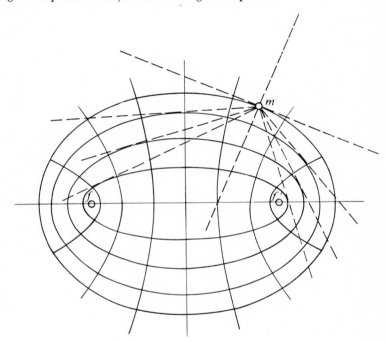

Figure 17.6.3.1

17.6.3.2. One can also prove this using calculus. If $f = (c, 0)$, $f' = (-c, 0)$, the conics in question are given by the equations

17.6.3.3
$$\frac{x^2}{c^2 + \lambda} + \frac{y^2}{\lambda} - 1 = 0, \qquad \lambda \in]-c^2, 0[\cup]0, +\infty[.$$

The envelope equation is given by the matrix

$$\begin{pmatrix} \frac{1}{c^2+\lambda} & 0 & 0 \\ 0 & \frac{1}{\lambda} & 0 \\ 0 & 0 & -1 \end{pmatrix}^{-1} = \begin{pmatrix} c^2 + \lambda & 0 & 0 \\ 0 & \lambda & 0 \\ 0 & 0 & -1 \end{pmatrix},$$

which is indeed linear in λ (cf. 14.6.1 and 15.1.6). Notice (figure 17.6.3.3) that the limiting *curve* is different for $\lambda \to 0^+$ and $\lambda \to 0^-$, whereas the limiting *conic* for $\lambda \to 0$ is, of course, well-defined, and its image is $\{f, f'\}$ (cf. 14.6.3). In the sense of 16.5.6.2, the limit as $\lambda \to 0$ is the double line $y = 0$.

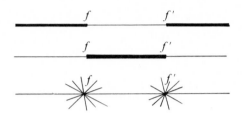

Figure 17.6.3.3

17.6.3.4. Tangents. Let $\mathcal{F}^*$ be the pencil of homofocal conics with foci f, f' and $m \notin ff'$ a point in X. By 16.5.6.3, the tangents drawn from m to conics in $\mathcal{F}^*$ form an involution in m^*, whose double (fixed) lines are the tangents at m to the conics in $\mathcal{F}^*$ which go through m. This involution contains the pair $\{mf, mf'\}$ and, when extended to $\check{X}^C$, also the pair $\{mI, mJ\}$. Thus there are double lines in X, namely, the bisectors of mf and mf' (use 8.7.7.5, 8.8.7 and 6.7.4). Consequently, there exist two conics of $\mathcal{F}^*$ going through m, one tangent to the interior bisector of $\{\overrightarrow{mf}, \overrightarrow{mf'}\}$ (a hyperbola), and the other tangent to the exterior bisector (an ellipse).

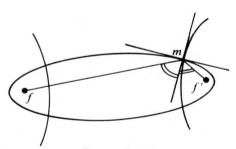

Figure 17.6.3.4

17.6.3.5. Observe that we have not used 17.2.2.4 here; on the contrary, we have derived it anew. In fact, the remark above, applied to a projective conic and allied to 17.4.3, shows that the tangent to a conic with foci f, f' is the bisector of $\{mf, mf'\}$. Applying 15.4.5 and 9.10.1, we obtain in this way a projective demonstration for 17.2.2.2.

17.6.3.6. The "little" Poncelet theorem. The above involution shows also that, if D and D' are the tangents dropped from m to a conic C with foci f, f', the pairs $\{mf, mf'\}$ and $\{D, D'\}$ have the same bisectors; these bisectors are also the tangents to the conics in $\mathcal{F}^*$ that go through m. This result is called the "second little Poncelet theorem" (cf. 17.1.2.6). It can be applied to demonstrate, among other things, the second part of 11.9.21.

17.6.4. ANOTHER STRING TOY. The construction in 17.2.2.5 was generalized by Graves: *given an ellipse C and a closed piece of string with length strictly bigger than the perimeter of C, the locus of a pencil used to pull the string taut around C is another ellipse C', homofocal with C.*

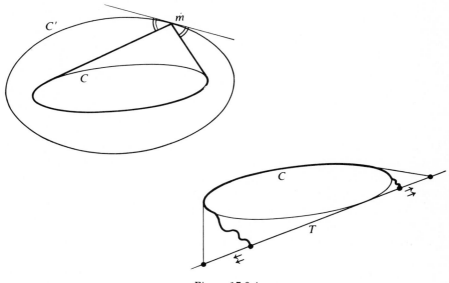

Figure 17.6.4

Proof. Let m be a point such that the string is taut, and C' the ellipse homofocal with C and containing x (cf. 17.6.3.4); then the string is taut for every point of C', by 9.10.4 and 17.6.3. To show the converse, observe that on each tangent T to C there exist exactly two points for which the string is taut (use continuity, for example), and T intersects C in two points (cf. 14.1.3.4). $\square$

For a generalization to quadrics, see [SD1], [SD2], or [GE]. See also 17.2.2.5 and [SAL, section 421b].

17.6.5. PONCELET REVISITED: MAXIMUM PERIMETER POLYGONS INSCRIBED IN AN ELLIPSE. Here we discuss a version of 16.6.11, due to Chasles and involving homofocal ellipses. This discussion does not purport to be a proof of 16.6.11, even in the particular case of homofocal ellipses, but just an illustration. See also 9.4.3 and 9.14.33.

17.6.6. THEOREM. *Let C be an ellipse and $n \geq 3$ an integer. Among the convex polygons with n distinct vertices inscribed in C, there exist some* *with maximum perimeter. In fact, there are infinitely many such maximum-perimeter polygons (MPP); one vertex of an MPP can be chosen arbitrarily on C. Furthermore, the sides of all n-vertex MPP's are tangent to the same ellipse C'_n, homofocal with C.*

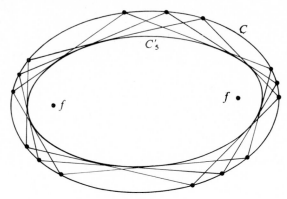

Figure 17.6.6.0

17.6.6.1. Notice first that the condition that the polygon is convex and has distinct vertices is essential. For example, if n is even, the degenerate polygon formed by alternately taking vertices at either end of the major axis of C has perimeter $2na$ (where C has equation $x^2/a^2 + y^2/b^2 - 1$), and a polygon with distinct vertices but close to the one just mentioned will have a perimeter as close to $2na$ as desired. The perimeter of an MPP is much smaller: for $n = 4$, for instance, it is given by $4\sqrt{a^2 + b^2}$ (see figure 17.6.6 and exercise 17.9.6).

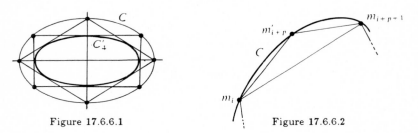

Figure 17.6.6.1   Figure 17.6.6.2

Proof.

17.6.6.2. Convex polygons inscribed in C and having n distinct vertices form a subset $\mathcal{P}$ of C^n; let P be a maximal perimeter polygon lying in the closure $\overline{\mathcal{P}}$ of $\mathcal{P}$ in C^n (cf. 9.4.2.2). We first show that $P \in \mathcal{P}$. Let the vertices of P be $(m_i)_{i=1,\dots,n}$, in that order. If we had $m_i = m_{i+1} = \cdots = m_{i+p}$ and $m_{i+p+1} \neq m_i$, we could replace m_{i+p} by a new vertex m'_{i+p} in the open interval between m_{i+p} and m_{i+p+1} to find a polygon with perimeter strictly greater than P. Having shown that all the vertices are distinct, we use 9.10.5 to conclude that the tangent $T_{m_i}C$ to C at m_i is the exterior bisector of $\{\overrightarrow{m_i m_{i-1}}, \overrightarrow{m_i m_{i+1}}\}$.

17.6.6.3. We can now use 17.6.3.6 to show that the sides of a given MPP
are tangent to an ellipse C', homofocal with C. By 16.6.11, there are other
polygons inscribed in C and circumscribed around the same ellipse C', and
one vertex of such a polygon can be chosen arbitrarily. Applying 9.10.1 to each
vertex of the new polygons, we conclude that they have the same perimeter
as the original MPP. There remains to see that every MPP is obtained in this
way. Let $m \in C$ be a vertex of an MPP, and C'' the ellipse to which its sides
are tangent. If $C'' \neq C'$ (say C' is smaller than C''), consideration of the
MPP with sides tangent to C' and one vertex at m, together with the MPP
with sides tangent to C'', leads to a contradiction (figure 17.6.6.3).

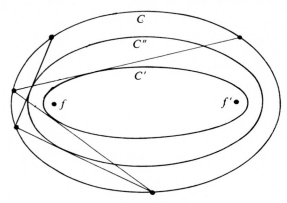

Figure 17.6.6.3

17.6.7. REMARKS. Theorem 17.6.6 implies the existence of n-sided polygons
inscribed in C and circumscribed around a conic C'_n, for every $n \geq 3$. Cayley's
formulas (cf. 16.6.12.4) give the appropriate value for λ so that C and C'_n can
be the ellipses

$$\frac{x^2}{a^2} + \frac{y^2}{b^2} - 1 = 0 \qquad \text{and} \qquad \frac{x^2}{a^2 + \lambda} + \frac{y^2}{b^2 + \lambda} - 1 = 0$$

(cf. 17.6.3.3). The reader should test the formula for $n = 3$ and 4 (see also
17.9.6).

17.7. Specific properties of ellipses

17.7.1. ELLIPSES ARE ORTHOGONAL PROJECTIONS OF CIRCLES. The
orthogonal projection of a circle from one plane in space to another is an
ellipse. On the same plane, we can obtain the ellipse

$$\frac{x^2}{a^2} + \frac{y^2}{b^2} - 1 = 0$$

from the circle $x^2 + y^2 = a^2$ by applying to it the affine transformation
$(x, y) \mapsto (x, (b/a)y)$. This gives rise to a mechanical generation for ellipses,

using a strip of paper: if two points of a line D are forced to move on two orthogonal lines Δ and Δ', every point of D describes an ellipse with axes Δ and Δ' (cf. also 17.9.14). It is also from this affine map that one deduces the construction in figure 16.7.2.

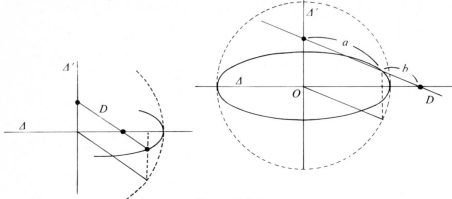

Figure 17.7.1

17.7.2. PARAMETRIZATION. A periodic parametrization for the ellipse $\dfrac{x^2}{a^2} + \dfrac{y^2}{b^2} - 1 = 0$ is suggested by the affine map in 17.7.1, namely, $t \mapsto (a\cos t, b\sin t)$. This is not a good parametrization in the sense of 16.2.9; to obtain a good parametrization, we use the change of variables $\tan(t/2) = \theta$ to obtain

$$\theta \mapsto \left(a\frac{1-\theta^2}{1+\theta^2}, \frac{2b\theta}{1+\theta^2} \right),$$

which, in homogeneous coordinates, becomes $\theta \mapsto \left(a(1-\theta^2), 2b\theta, 1+\theta^2 \right)$.

17.7.3. EXAMPLE. In the first parametrization above, four points m_i with parameters t_i ($i = 1, 2, 3, 4$) are cocyclic if and only if

$$t_1 + t_2 + t_3 + t_4 \equiv 0 \pmod{2\pi}.$$

They are the feet of four normals dropped from the same point if and only if the angles $\theta_i = \tan(t_i/2)$ of the second parametrization satisfy

$$\theta_1\theta_2 + \theta_1\theta_3 + \theta_1\theta_4 + \theta_2\theta_3 + \theta_2\theta_4 + \theta_3\theta_4 = 0 \quad \text{and} \quad \theta_1\theta_2\theta_3\theta_4 = -1.$$

In this case we also have

$$t_1 + t_2 + t_3 + t_4 \equiv \pi \pmod{2\pi}.$$

17.7.4. THE EVOLUTE OF AN ELLIPSE. We recall that the envelope of the normals to a curve, called its evolute, is the locus of the centers of curvature of the curve at each point. For our standard ellipse, the center of curvature at $m = (a\cos t, b\sin t)$ is the point

$$\left(\frac{c^2}{a}\cos^3 t, -\frac{c^2}{b}\sin^3 t \right).$$

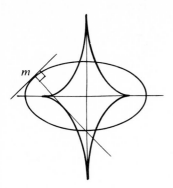
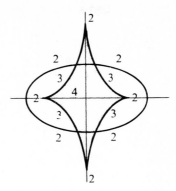

Figure 17.7.4.1

The evolute satisfies the implicit equation $(ax)^{2/3} + (by)^{2/3} = c^{4/3}$; it is the image of an astroid (cf. 9.14.34.3.F) under an affine map.

The number of normals that can be drawn from $m \in X$ to C depends on the position of m with respect to this evolute, as shown in figure 17.7.4. It is equal to 4 inside the evolute, 2 outside the evolute and at each cusp, and 3 on the rest of the curve.

Figure 17.7.4.2 indicates a construction for the center of curvature at the point m, distinct from the one given in 17.5.5.4.

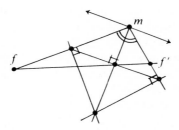

Figure 17.7.4.2

17.7.5. AREA AND LENGTH. The area of the standard ellipse $\dfrac{x^2}{a^2} + \dfrac{y^2}{b^2} - 1 = 0$ is πab (cf. 11.8.9.4). On the other hand, the length, given by

$$\int_0^{2\pi} \sqrt{a^2 \sin^2 t + b^2 \cos^2 t}\, dt,$$

cannot be expressed in simple terms. The integral above is called an *elliptic integral* exactly because it occurs in this context, but it has much wider applications.

17.7.6. See other properties of the ellipse in 17.9.6, 17.9.11 through 17.9.15, and 17.9.22.

17.8. Specific properties of hyperbolas

Hyperbolas should be thought of together with their asymptotes. There are numerous reasons for that; one is to avoid drawing a hyperbola which intersects a line in four points (cf. 14.1.3.4).

17.8.1. For example, if a line D intersects the hyperbola C at m, m' and its asymptotes at u, u', we have $(m + m')/2 = (u + u')/2$; in particular, $m = (u + u')/2$ if D is tangent to C at m. This gives a quick point-by-point construction for a hyperbola whose asymptotes and one of whose points are known. It's this property of the tangent that was used for the construction in figure 16.7.3.

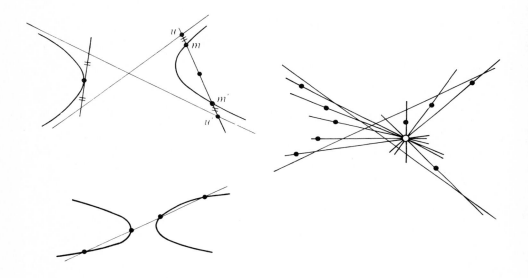

Figure 17.8.1

17.8.2. PARAMETRIZATION. The two connected components of the hyperbola $\dfrac{x^2}{a^2} - \dfrac{y^2}{b^2} - 1 = 0$ can be parametrized by $t \mapsto (a \cosh t, b \sinh t)$ and $t \mapsto (-a \cosh t, b \sinh t)$. The reader can figure out how 17.7.3 and 17.7.4 can be adapted to the case of hyperbolas.

17.8.3. EQUILATERAL HYPERBOLAS.

17.8.3.1. We have already encountered several properties of equilateral hyperbolas in 17.5.4; in particular, the equilateral hyperbolas going through three points a, b, c form a pencil, and the locus of their centers is the nine-point circle of the triangle $\{a, b, c\}$.

17.8.3.2. For the same triangle $T = \{a, b, c\}$, every asymptote of an equilateral hyperbola containing a, b, c is the Simson line of some point of the circle circumscribed around T (cf. 17.4.3.5 and 10.4.5.4). The converse also holds. From 9.14.34.3.D it follows that the envelope of the asymptotes of the equilateral hyperbolas circumscribed around T is a hypocycloid with three cusps.

17.8.3.3. Cocyclic points. If a circle Γ with center β intersects an equilateral hyperbola C with center α so that $C \sqcap \Gamma = \{(m_i, \omega_i)\}_{i=1,...,k}$ with $\sum \omega_i = 4$ (cf. 16.5.1), we have

$$\frac{1}{4} \sum_{i=1}^{k} \omega_i m_i = \frac{\beta + \alpha}{2}.$$

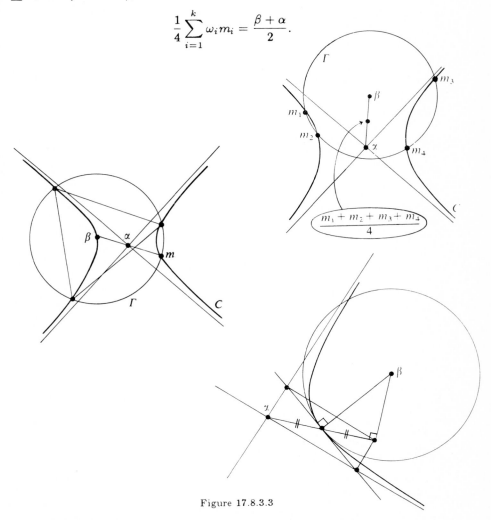

Figure 17.8.3.3

One consequence is the construction for the center of curvature shown in figure 17.8.3.3. Another is that, if the center β of Γ is on C and Γ goes through the point m such that $\alpha = (m + \beta)/2$, then Γ intersect C in three other points that form an equilateral triangle.

A necessary and sufficient condition for the points $(m_i)_{i=1,\dots,4}$ on C to be cocyclic is that

$$\prod_i d(m_i, A) = \frac{a^4}{4},$$

where A is an asymptote of A and $x^2 - y^2 = a^2$ is an equation for C.

17.8.4. NOTE. For other properties of equilateral hyperbolas, see [LM2].

17.9. Exercises

17.9.1. FOCAL CIRCLES. A *focal circle* of a proper, non-empty conic C with equation q is any circle having an equation of the form $q + k\phi^2$, where ϕ is an affine form. Study the possibilities for ϕ. Draw the different families of focal circles for the three types of conics. Are superosculating circles focal circles?

Show that, for every focal circle Γ, there exists a line D (the *associated directrix*) and a positive real number k such that

$$C = \{\, m \in X \mid p(m, \Gamma) = kd(m, D)^2 \,\},$$

where $p(m, \Gamma)$ is the power of m with respect to Γ. Prove a converse. Can you prove this property using just a three-dimensional drawing (cf. 17.3)?

If Γ, Γ' are two focal circles, show that, in some cases,

$$C = \{\, m \in X \mid \sqrt{p(m, \Gamma)} + \sqrt{p(m, \Gamma')} = \text{constant} \,\}.$$

Can you prove this property using a three-dimensional drawing?

Study the relation between focal circles and circles bitangent to C.

17.9.2. A VERTICAL BILLIARD. A little steel ball can move within the vertical disc bounded by a circle, under the action of gravity (see 17.5.5.5). Assume its collisions with the wall are perfectly elastic, and that it starts at the bottom of the circle with a certain horizontal velocity. Find examples of periodic movement.

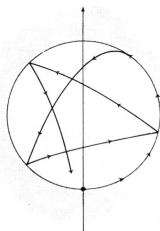

Figure 17.9.2

17.9.3. Prove 17.5.5.6 using calculus.

17.9.4. Prove 17.5.5.1 in an elementary way.

17.9.5. Prove that a necessary and condition for a circle Γ to be harmonically circumscribed around a conic C is that Γ be orthogonal to the orthoptic circle (or line) of C (if the orthoptic circle does not exist, interpret orthogonality in the sense of chapter 20). Same analysis for an equilateral hyperbola or parabola in lieu of a circle.

17.9.6. Show that there are infinitely many rectangles circumscribed around an ellipse C and inscribed in its orthoptic circle Γ. By duality with respect to C, deduce theorem 17.6.6 in the case $n = 4$ (without using 16.6.11).

17.9.7. If a convex polygon inscribed in C and having n distinct vertices is an MPP (and consequently circumscribed around C'_n), this polygon has maximal perimeter among all polygons circumscribed around C'_n.

17.9.8. Study the polygons with maximal (resp. minimal) area among the polygons inscribed in (resp. circumscribed around) a given ellipse.

17.9.9. Let C be a central conic, $m \in C$ a point and p, q the points where the normal to C at m intersects the axes of C. Show that the ratio mp/mq remains constant, and calculate it. Prove a converse. Deduce the value of the radii of curvature at the vertices of C.

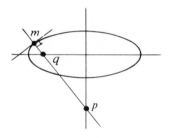

Figure 17.9.9

* **17.9.10.** JOACHIMSTAL'S THEOREM. If four points on an ellipse have concurrent normals, then the circle passing through three of them also passes through the point diametrically opposite to the fourth. See also 17.7.3.

17.9.11. THE CHASLES CIRCLES. Two points m, m' move along concentric circles with same angular velocity but in opposite directions. What happens to their midpoint $(m + m')/2$? What is the relation between this and the method of the paper slip (17.7.1)?

17.9.12. Let m, n be points on an ellipse C with center ω, and assume that $\widehat{\omega m, \omega n}$ is a right angle. Show that $\dfrac{1}{(\omega m)^2} + \dfrac{1}{(\omega n)^2}$ is a constant. What is the envelope of the line mn? See 15.7.20.

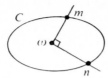 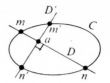

Figure 17.9.12

Let a be a point in the interior of an ellipse C with center ω, let D be a line containing a an intersecting C at m, n, and let $p \in C$ be such that ωp is parallel to D. Show that the ratio $\dfrac{am \cdot an}{(\omega p)^2}$ is constant.

Let D, D' be orthogonal lines (in the Euclidean metric), containing a and intersecting C at m, n and m', n', respectively. Show that $\dfrac{1}{am \cdot an} + \dfrac{1}{am' \cdot an'}$ is constant. What is the envelope of the lines mm'? See 15.7.9.

17.9.13. Given an ellipse C and $m \notin C$, study the variation of the function $x \mapsto xm$ as x moves around C.

17.9.14. LAHIRE'S THEOREM. Consider two points p, p' on a moving line D in such a way that $p \in \Delta$ and $p' \in \Delta'$, where Δ, Δ' are arbitrary fixed lines. Show that a third point $p'' \in D$ describes an ellipse.

17.9.15. Consider four points $(m_i)_{i=1.2.3.4}$ on an ellipse, and take the points n_i where the osculating circles at m_i intersect the ellipse (choose $n_i \neq m_i$ except in the superosculating case). Show that if the m_i are cocyclic, then so are the n_i. Does this property still hold for hyperbolas and parabolas?

17.9.16. Show that a logarithmic spiral with pole ω is self-polar with respect to an equilateral hyperbola centered at ω if it is tangent to the hyperbola.

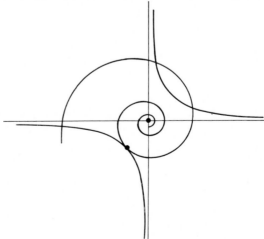

Figure 17.9.16

17.9.17. Given two concentric equilateral hyperbolas H and H', a necessary and sufficient condition for the existence of triangles inscribed in H and circumscribed around H' is that the foci of H lie on the directrices of H'.

17.9.18. PROPERTIES OF PARABOLAS

17.9.18.1. The subnormal. Let P be a parabola and $m \in P$ a point. Let the normal to P at m intersect the axis of P at s and let r be the projection of m on the axis. Show that the distance rs is constant. Deduce a geometric construction for the axis of a parabola that goes through points m, m' and has T, T' as tangents at those points.

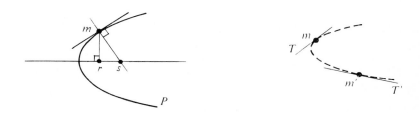

Figure 17.9.18

* 17.9.18.2. Normals from a point. Show that the normals at three points m, m', m'' of a parabola P are concurrent if and only if the barycenter $(m + m' + m'')/3$ belongs to the axis of P. Also if and only if the circle passing through m, m', m'' contains the vertex of P.

17.9.18.3. The intersection of the diagonals of a quadrilateral inscribed in a circle and circumscribed around a parabola belongs to the directrix of the parabola.

17.9.19. Consider two quadrics of revolution whose axes intersect and whose common tangent planes are tangent to a third quadric of revolution whose axis lies in the plane of the first two. Show that the six foci form a complete quadrilateral.

* 17.9.20. Given a fixed point on a conic, consider all the chords whose angle, seen from that point, is a constant. Find the envelope of these chords. Analyze the special case of a right angle.

* 17.9.21. Show that the tangent circles to two variable conjugate diameters of an ellipse, and whose center is on the ellipse, have constant radius.

17.9.22. DETERMINING THE AXES FROM TWO CONJUGATE SEMIDIAMETERS. Let Oa, Ob be two semidiameters of an ellipse C. Show that one can obtain the axes of C in the following way: Take the perpendicular to Ob

through a and mark the length of Ob on both sides of a, along that perpendicular, to obtain the points u and v. The bisectors of Ou and Ov are the axes.

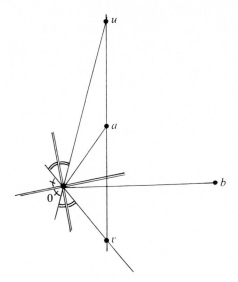

Figure 17.9.22

* **17.9.23.** Given a circle C in X and two points a and b on C, consider the ellipses E which are tangent to C, pass through a and b, and whose center is the midpoint of ab. Show that all such ellipses have the same eccentricity.

17.9.24. FINDING THE FOCI. Given the equation $f(x, y) = 0$ of a conic C in X with respect to an orthonormal affine frame, show that the coordinates (α, β) of the foci of C can be obtained by solving the equation $F(\epsilon i, -1, \beta - \epsilon i \alpha) = 0$, where $\epsilon = \pm 1$, $i = \sqrt{-1}$ and $F(u, v, w) = 0$ is the envelope equation of C in the projective base associated with the given affine frame.

Chapter 18
The sphere for its own sake

Spheres are, after subspaces, the simplest objects in Euclidean affine geometry. It is natural to expect to find them all over—not only in mathematics but also in its practical applications: navigation, astronomy, mechanics... In this chapter we have attempted to give an idea of the variety of contexts to which properties of the sphere are relevant. The properties themselves were culled from various levels: some are elementary, some are more difficult and involve outside notions that are not required for the rest of the book, such as integration theory, differential geometry, and so on. (We hope that this more advanced material will motivate the reader to delve more deeply into these theories.) The unifying criterion is that the results discussed here, as opposed to those in chapter 10, involve the sphere as an object in itself, without making reference to the ambient space. Also, we hope that they will all appear to the reader as natural and easily visualizable, with the exception of those in sections 18.8 and 18.10, which were included with the opposite purpose in view: to help the reader break free from his lower-dimensional intuition.

The first section deals with practical problems, like spherometers and charts; it includes the stereographic projection, which plays a fundamental role in the sequel. In section 18.2 we make use of notions from algebraic topology to formulate some delicate results which, in view of their simplicity and elegance, could not be omitted; proofs are not always provided.

In section 18.3 we consider spheres as differentiable manifolds. This course affords a very easy construction for the canonical measure of spheres which, in dimension two, is the measure of the earth's surface. A key result, known as Girard's formula, says that the area of a spherical triangle is equal to the sum of its angles minus π. In section 18.4 we define the intrinsic metric on S^d, which differs from the essentially fruitless distance induced from $\mathbf{R}^{d+1}$, being instead defined as the length of the shortest path between two points on the sphere—a useful concept if there ever was one. In order to calculate this intrinsic distance, or angles in astronomy, which is the same, we need formulas for spherical triangles; these are provided in a fairly complete formulary in 18.6, and we mention some of their practical applications.

At the end of the chapter we return to more mathematical concerns. In 18.7 we prove a delicate lemma, due to Cauchy, which was the key in the proof of the rigidity theorem for convex polyhedra in 12.8. Section 18.8 is a detailed study of the geometry of S^3. It is important to dwell on this study until one has a good grasp of the Hopf fibration, since this is the simplest case of more general structures ubiquitous in geometry, differential topology and algebraic topology. By stereographic projection we pass from S^3 to properties of tori of revolution in $\mathbf{R}^3$, thus returning to more visual, if non-obvious, results.

The Möbius group, introduced in 18.10, may surprise the reader and even appear somewhat artificial, but is fundamental to the larger portion of chapter 19 devoted to hyperbolic geometry, as well as to the entire chapter 20. In 18.10 we study the action of this group on S^d; this action, apart from its own geometric interest, has been an essential tool in obtaining several recent results.

In this chapter $S = S^d$ denotes the unit sphere in the Euclidean space $\mathbf{R}^{d+1}$.

18.1. Definitions, special dimensions, charts and projections

18.1.1. SPHEROMETERS. All spheres, in all Euclidean spaces of a given dimension, are similar; this is why we can limit ourselves to studying S. It is easy to modify, for a sphere of a given radius, the results and formulas stated here.

We mention, nonetheless, the practical problem of determining the radius of a "solid" sphere—a steel ball, say, or an optical lens. This is done by using an instrument called the *spherometer* (figures 18.1.1.1 and 18.1.1.2): if the equilateral triangle formed by the feet has side lengths a and the distance from the needle to the plane of the triangle is e, the desired radius is $R = \dfrac{a^2 + 3e^2}{6e}$. Figure 18.1.1.3 shows another type of spherometer.

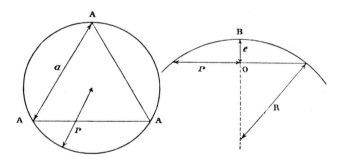

Figure 18.1.1.1

18.1.2. DEFINITIONS

18.1.2.1. A $(k\text{-})subsphere$ of S is the intersection of S with a $(k+1)$-dimensional vector subspace of $\mathbf{R}^{d+1}$. A 1-subsphere is also called a *great circle*. If $x, y \in S$ and $y \neq \pm x$, there exists a unique great circle containing x and y.

18.1.2.2. A $(k\text{-})little\ sphere$ of S is the intersection of S with a $(k+1)$-dimensional affine subspace of $\mathbf{R}^{d+1}$. A 1-little sphere is also called a *little circle*.

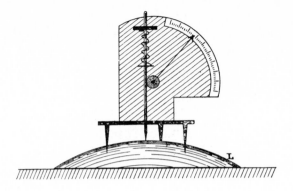

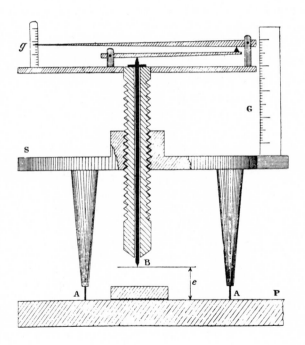

Figure 18.1.1.2 (Source: [BOU2])

182 *Le savant Cosinus.*

Cosinus mesure, à 3 ou 4 centimètres près, le diamètre de l'objet. Il trouve qu'il est égal à 0ᵐ,30. Sphéroïde n'a que du dédain pour une opération qui lui semble dénuée de tout intérêt.

Figure 18.1.1.3 (Source: [CP])

18.1.2.3. The *north pole* of S is the point $n = (0, 0, \ldots, 1)$, and the *south pole* is the point $s = (0, 0, \ldots, -1)$. The *equator* is the $(d-1)$-subsphere obtained by intersecting S^d with the hyperplane $x_{d+1} = 0$ of $\mathbf{R}^{d+1}$.

18.1.2.4. The *tangent hyperplane* to S^d at x is sometimes the vector hyperplane $x^\perp$ of $\mathbf{R}^{d+1}$ and sometimes the affine hyperplane in the direction of $x^\perp$ and containing x. The (affine) tangent hyperplane is *denoted by* $T_x S$ (cf. 10.7.4 and 18.3.2). The vectors in the (vector) tangent hyperplane are called *tangent vectors* to S at x; the *angle* between two such vectors is, by definition, the angle discussed in 8.6.3, an element of $[0, \pi]$.

18.1.2.5. The *natural topology* on S is the one induced from $\mathbf{R}^{d+1}$, and we shall consider no other.

18.1.3. SPECIAL DIMENSIONS

18.1.3.1. Dimension 1. The one-dimensional sphere is the *circle* S^1. We have seen that it can be interpreted as the multiplicative group of complex numbers with absolute value one, or as the group of rotations of $\mathbf{R}^2$ (cf. 8.3). It is also homeomorphic to the real one-dimensional projective space (cf. 4.3.6, figure 8.7.7.6 and 18.1.4.5). For its intrinsic metric, see 9.9.8 and 18.4.

18.1.3.2. Dimension 2. The two-sphere is exemplified by our planet. One of its fundamental mathematical properties is that it is homeomorphic to the Riemann sphere, that is, the field $\mathbf{C}$ of complex numbers together with one point at infinity: $S^2 \cong \mathbf{C} \cup \infty$. This extension is not merely topological (Alexandrov compactification), but also analytic, in the sense that we can extend from $\mathbf{C}$ to S^2 the notions of holomorphic and meromorphic functions, and so on (see [CH2, 90]). Since we will discuss later (18.10) the conformal group of S^d for arbitrary d, we won't insist here on the structure inherited by S^2 from $\mathbf{C}$. The identification $S^2 \cong \mathbf{C} \cup \infty$ follows from 18.1.4.5; it is also an isomorphism between S^2 and the one-dimensional complex projective space (cf. 4.3.6).

18.1.3.3. Dimension 3. The three-sphere has been interpreted as the multiplicative group of quaternions of norm one (see 8.9.1). Section 18.8 is devoted to a detailed study of S^3.

18.1.3.4. Dimension 4. The sphere S^4 can be identified with the one-dimensional quaternionic projective space (cf. 4.9.7).

18.1.3.5. Dimensions 6, 7 and 15. Spheres in dimension 6, 7 and 15 have special properties derived from the division ring of octonions (cf. [PO, 278] and [BES, chapter 3]). S^7 is almost a group, identified with the set of octonions of norm one, and S^6 is identified with the set of pure octonions of norm one. The octonionic projective plane leads to the fibration $S^{15} \to S^8$, whose fibers are equal to S^7; this generalizes the Hopf fibration $S^3 \to S^2$ with fibers equal to S^1 (cf. 18.1.3.6 and 8.8.7).

18.1.3.6. The Hopf fibrations. Hopf fibrations are maps defined from the spheres S^{2n+1} and S^{4n+3} (except for S^{15}) onto the projective spaces $P^n(\mathbf{C})$ and $P^n(\mathbf{H})$ (cf. 4.8), respectively:

$$
\begin{array}{cc}
S^{2n+1} & S^{4n+3} \\
\downarrow{\scriptstyle S^1} & \downarrow{\scriptstyle S^3} \\
P^n(\mathbf{C}) & P^n(\mathbf{H})
\end{array}
$$

As indicated next to the vertical arrow, the *fiber* of each map, that is, the inverse image of each point in the image, is homeomorphic to S^1 and S^3, respectively. There is nothing mysterious about these maps; in the case of $\mathbf{C}$ the Hopf fibration is simply the map $p : S \to P(E)$ used in 4.3.3.2. See 18.11.30.

18.1.4. The stereographic projection

18.1.4.1. The *stereographic projection* from the north pole of S^d onto $\mathbf{R}^d$ is defined as follows: identify $\mathbf{R}^{d+1}$ with $\mathbf{R}^d \times \mathbf{R}$, as usual, and associate to each $m \in S^d \setminus n$ the point $f(m)$ of $\mathbf{R}^d$ such that m, n and $f(m)$ are collinear (figure 18.1.4.1). The calculation below shows that f is bijective and its inverse f^{-1} is C^∞ as a map $\mathbf{R}^d \to S^d$.

Write $m \in \mathbf{R}^{d+1}$ as a pair (z, t) ($z \in \mathbf{R}^d$ and $t \in \mathbf{R}$); we must find $\lambda \in \mathbf{R}$ such that $f(m) = \lambda n + (1 - \lambda)m$ and $f(m) \in \mathbf{R}^d$. We get $\lambda = t/(t-1)$ and

18.1.4.2 $f : m = (z, t) \mapsto \dfrac{1}{1-t}z, \qquad f^{-1} : z \mapsto \left(\dfrac{2z}{\|z\| + 1}, \dfrac{\|z\| - 1}{\|z\| + 1} \right).$

18.1.4.3. More generally, a *stereographic projection* of S is any map $f : S \setminus m \to H$, where m is a point of S and H is a hyperplane parallel but not identical to the tangent hyperplane to S at m. By 10.8.2, such a map

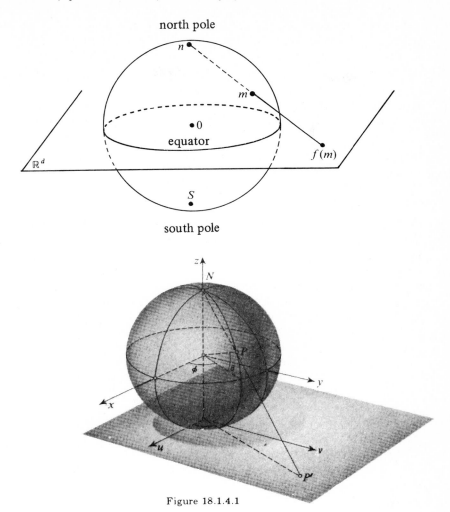

Figure 18.1.4.1

f is the restriction to S of an inversion of $\mathbf{R}^{d+1}$ with pole m; in particular, stereographic projections preserve angles (see a precise definition in 18.1.2.4 or 18.10.3 and 18.11.22), and transforms $(d-1)$-little spheres in $S \setminus m$ into spheres in H (cf. 10.8.2).

18.1.4.4. If f (resp. g) is the stereographic projection from the north pole (resp. south pole) of S^d onto $\mathbf{R}^d$, the composition $g \circ f^{-1} : \mathbf{R}^d \setminus 0 \to \mathbf{R}^d$ is exactly the inversion of $\mathbf{R}^d$ with pole 0 and power 1.

18.1.4.5. By stereographic projection we get the following homeomorphisms (cf. 4.3.8):

$$S^1 \cong \mathbf{R} \cup \infty = \tilde{\mathbf{R}} \cong P^1(\mathbf{R}),$$
$$S^2 \cong \mathbf{R}^2 \cup \infty \cong \mathbf{C} \cup \infty = \tilde{\mathbf{C}} \cong P^1(\mathbf{C}).$$

From now till the end of the section we set $d = 2$ and $S = S^2$.

18.1.5. CHARTS.

18.1.5.1. We shall understand by a *chart* of S a pair (A, f) consisting of a subset $A \subset S$ and an injective map $f : A \to \mathbf{R}^2$. In practice, f is almost always a homeomorphism onto its image. For example, stereographic projections (18.1.4.3) are charts, with $A = S \setminus m$.

First we establish that there is no chart of the whole of S, that is, no homeomorphism $f : S \to f(S) \subset \mathbf{R}^2$. To see this, take $x \in S$; by assumption and 18.2.6.5, $f(S \setminus x) = f(S) \setminus f(x)$ is an open subset of $\mathbf{R}^2$. But this contradicts the fact that $f(S)$, as a compact with non-empty interior, must have more than one frontier point (see 11.2.9 if necessary).

18.1.5.2. Note. The word "chart" has a variety of meanings, ranging from the abstract one introduced above to a printed piece of paper.

18.1.5.3. The geoid. It is clear that S is merely an approximation to the problem that really interests us, namely, the surface of the earth. More exactly, the *geoid*, a surface perpendicular at every point to the direction of gravity (and going through a reference point). This wouldn't be a problem if our charts only had to be homeomorphisms, but we are also interested in preserving metric properties of our surface, and the geoid is not isometric to S with its canonical metric defined in 18.4. A reasonable approximation to the geoid is an ellipsoid of revolution, called the geodetic reference spheroid, with major semiaxis $a =$6,378,388 m and minor semiaxis $b =$6,356,912 m (the relevant number is the flatness coefficient $\frac{a-b}{a} = 1/297$).

We will not discuss even this approximation, but work with the sphere S. The reader can have an idea of the complexity of the modifications that must be carried out in practice by thinking of the equations for parallels of latitude and meridians in the transverse Mercator projection (cf. 18.1.8.4) for the geoid. Gauss had already mastered these calculations.

18.1.5.4. References on cartography. The reader can start with [RO–SA, chapters 9 and 10], for a general and elementary treatment. Detailed discussions of all commonly used projections, as applied to the sphere, can be found in [RB] or [KY]. Finally, [BOM, 207–225] and [LS, chapter VI] are devoted to the problem of coordinates for the reference spheroid, and include a hefty dose of mathematics on the Lambert and MTU projections. See also [MI].

18.1.6. LATITUDE AND LONGITUDE

18.1.6.1. The *latitude* of a point $m \in S$ is the real number

$$\theta(m) = \pi/2 - \overrightarrow{0n}, \overrightarrow{0m}.$$

Points with latitude zero form the equator, and the north and south poles are the only points with latitude $\pi/2$ and $-\pi/2$, respectively. The *longitude* of a point $m \in S \setminus \{n, s\}$ is a measure in the interval $[-\pi, \pi]$ of the oriented angle between oriented lines $\overrightarrow{0a}, \overrightarrow{0m'}$, where $a = (1, 0, 0)$ and m' is the projection of m on the xy plane. The longitude $\phi(m)$ of m is well-determined, except when m belongs to the *international date line*, that is, the half-great-circle Γ running from n to s and containing $(-1, 0, 0)$; there the longitude can be chosen to be π or $-\pi$. For n and s the longitude is completely undefined.

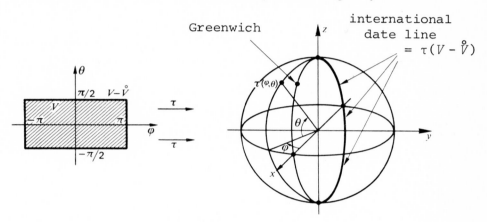

Figure 18.1.6.1 (Source: [B–G])

The *latitude-longitude chart* is the pair $(S \setminus \Gamma, f_0)$, where f_0 is given by $f_0(m) = (\phi(m), \theta(m)) \in \mathbf{R}^2$; its image is the rectangle $]-\pi, \pi[\times]-\pi/2, \pi/2[$. The map $\tau = f_0^{-1}$ is given by the formulas

18.1.6.2 $$\tau(\phi, \theta) = (\cos \phi \cos \theta, \sin \phi \cos \theta, \sin \theta) \in \mathbf{R}^3.$$

Meridians and *parallels* of S are curves of constant longitude and latitude, respectively.

18.1.6.3. Longitude and latitude can (and should) be defined for the geoid. For the longitude there is no problem; the latitude is defined as the angle the normal makes with the equator (the normal itself points in the direction of gravity).

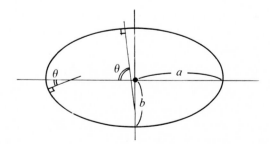

Figure 18.1.6.3

18.1.6.4. Virtually all charts f of S (or of the earth) are defined by specifying not f itself but the map

$$g = f \circ \tau = f \circ f_0^{-1}$$

(cf. 18.1.6.2), which is of the form

$$x = u(\phi, \theta), \qquad y = v(\phi, \theta).$$

18.1.6.5. In the case of the geoid, there is a natural choice for the z-coordinate because of the earth's rotation, but the origin for longitudes must be chosen arbitrarily. By convention, it is the Greenwich meridian (cf. figure 18.1.6.1); one of the reasons for this choice is that the international date line hardly touches any land, avoiding no end of jokes and tall tales...

18.1.7. WHAT YOU WANT IN A CHART

18.1.7.1. Since we're not just set theorists but also travelers, real estate brokers and land developers, we want to be able to use our charts to calculate distances (in the intrinsic metric, 18.4), surface areas (for the canonical measure, 18.3.7) and angles (cf. 18.1.2.4). Distances are apparently the more important requirement, because of fuel prices, so we want *isometric charts* (up to a scalar) between the induced metric on $A \subset S$ and the induced metric on $f(A) \subset \mathbf{R}^2$. Unfortunately, as we shall see in 18.4.4, there are no such charts, even for very small open sets A.

18.1.7.2. There are, however, *conformal charts,* those that preserve angles (see a precise definition in 18.10.3, if necessary). This means that there exist a whole slue of them, since we can compose a stereographic projection, for example (cf. 18.1.4.3), with a local conformal map of $\mathbf{R}^2$ (any local holomorphic map of $\mathbf{C} = \mathbf{R}^2$ will do, cf. 9.5.4.3).

18.1.7.3. Conformal charts—which account for most of the charts in common use—are often adequate for measuring distances. The reason is that if a chart (A, f) is conformal at m, it multiplies infinitesimal distances at m by a factor $k(m)$ that depends on m and not on the direction. If the chart is not too wild and (as is generally the case) one is interested in a very small region A of S, the factor k will not change a lot within the region of interest, and one can calculate lengths and surface areas, to the desired approximation, by introducing a correction.

18.1.7.4. There are charts, called *equivalent,* that preserve surface area; an example is given in 18.11.27.

18.1.8. A FEW TYPES OF CHARTS

18.1.8.1. Note. Charts are often called projections, because the first charts in use were really obtained by drawing rays from a point in $\mathbf{R}^3$ through S onto a plane, as is the case with the stereographic projection. An extension of this meaning includes projections on a cylinder, which is then unwrapped onto the plane (cf. the cylindric projection, 18.11.27). Starting with the Mercator "projection", charts no longer necessarily conform to this principle, especially if they are meant to depict the geoid (cf. 18.1.6.3), but the name stuck.

18.1.8.2. The classical Mercator projection. The Mercator projection is defined as a conformal chart that can be written, under the convention of 18.1.6.4, in the form $(\phi, \theta) \mapsto (\phi, V(\theta))$. Conformality uniquely determines the function V, as follows. The inverse function F of our chart is of the form

$$F : (u, v) \mapsto \tau(u, W(v)),$$

where τ is as in 18.1.6.2 and $W = V^{-1}$. The partial derivative vectors $\partial F/\partial u$ and $\partial F/\partial v$ satisfy

$$\left(\frac{\partial F}{\partial u} \,\middle|\, \frac{\partial F}{\partial v} \right) = 0, \quad \left\| \frac{\partial F}{\partial u} \right\| = |\cos W|, \quad \left\| \frac{\partial F}{\partial v} \right\| = |W'|.$$

The chart will be conformal if and only if $W' = \cos W$; solving and inverting we get

$$V(\theta) = \log\left(\tan\left(\frac{\theta}{2} + \frac{\pi}{4} \right) \right).$$

Another way of formulating these conditions is by saying that the Mercator projection is the unique conformal chart where parallels and meridians are represented by orthogonal lines and distances are respected along the equator.

As long as the factor k discussed in 18.1.7.3 does not vary too much, the Mercator projection is admirably well suited to maritime or aerial navigation. The reason is that a course that follows a constant heading, that is, makes a constant angle with the meridians, is represented on the chart as a straight line, thanks to conformality. And ships and planes do follow a fixed heading for some time: the helmsman does what is necessary to ensure that the compass shows the direction set by the captain. On the sphere, curves that make a constant angle with all meridians are called *loxodromes;* by the previous paragraph their equation is, in terms of latitudes and longitudes,

$$\phi = \alpha \log\left(\tan\left(\frac{\theta}{2} + \frac{\pi}{4} \right) \right) + \beta \qquad (\alpha, \beta \in \mathbf{R}).$$

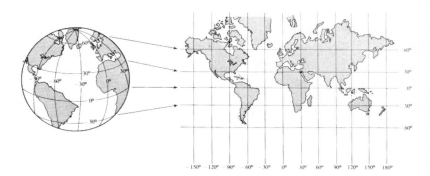

Figure 18.1.8.2

18.1.8.3. Transverse and oblique Mercator projections. Under the classical Mercator projection the factor ϕ undergoes seismic changes as the latitude gets close to $\pi/2$, since it tends to infinity at the poles. This makes distances and surface areas virtually impossible to estimate with any precision. On the other hand, near the equator the chart is excellent, since k has a minimum for $\theta = 0$ and thus tends to change very little in that region (a function changes slowly near an extremal point).

There is no difficulty in carrying the good behavior to any part of the globe. Simply apply a rotation—an isometry—to bring the interesting portion of the sphere to the equator, and apply the projection. In the composite chart meridians and parallels are no longer straight lines, but they are almost straight if the region is small. One way of visualizing this is the following: one can get the classical Mercator projection by circumscribing a cylinder around the sphere so they touch along the equator, projecting the sphere onto that cylinder (changing latitudes so that the projection is conformal) and finally unwrapping the cylinder. An *oblique Mercator projection* is obtained in the same way, except that the cylinder is tangent to the sphere along an arbitrary great circle going through the region of interest. One can also set the cylinder so that it intersects the sphere along two little circles; then the chart is isometric along the circles, but still conformal everywhere.

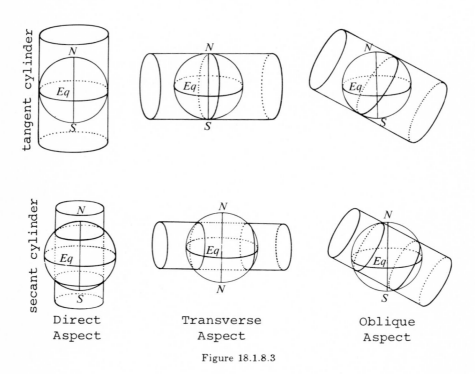

Figure 18.1.8.3

The most commonly used projection nowadays is the *transverse Mercator projection*, where the contact circle is a meridian. Formerly known as *Gauss projection*, the MTU (Mercator transverse universal) projection is used, for example, by the Swiss Federal Topographical Service. If we take the Greenwich meridian as the contact circle (cf. 18.1.6.5), formulas 18.1.6.4 become

18.1.8.4 $\qquad x = \dfrac{1}{2} \log \left(\dfrac{1 + \cos\theta \sin\phi}{1 - \cos\theta \sin\phi} \right), \qquad y = \arctan \left(\dfrac{\tan\theta}{\sin\phi} \right).$

For mathematical calculations involving the MTU projection, see [HOL, chapter 2], which deals with the case of a sphere; for the geoid, more complicated but mathematically more interesting, see [BOM, 215–220] or [LS, chapter VI].

18.1.8.5. Lambert's conformal projections. The idea behind Lambert's conformal projections is to modify the classical Mercator projection, maintaining the property that meridians and parallels are represented by simple curves. The simplest system after two families of orthogonally intersecting parallel lines is a family of concentric circles together with the lines emanating from the center—polar coordinates. Lambert's projection is given by

$$(\phi, \theta) \mapsto \big(r(\phi, \theta) = \Xi(\theta), \alpha(\phi, \theta) = \phi \sin\theta_0\big),$$

where r, α are the polar coordinates, θ_0 is the longitude of the parallel we want to concentrate on and Ξ is a function designed to make the chart conform. Along the parallel θ_0 the chart is isometric. In the spirit of 18.1.8.3, Lambert's projections can be visualized by projecting from S to a cone tangent to S

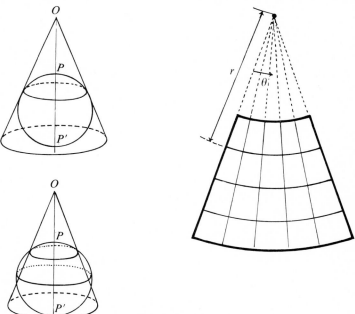

Figure 18.1.8.5

along the relevant parallel, then unwrapping the cone and adjusting the r-coordinate so that the chart is conformal. Thus meridians become straight lines and parallels of latitude become concentric circles. One can also arrange for the cone to intersect the sphere along two circles.

Presently, the official cartography of France is based on Lambert's projection; the whole country is covered by three projections, each restricted to a three-degree zone. The ratio between the largest and smallest values of k (cf. 18.1.7.3) is better than 1.001. The interested reader can verify for himself, with the help of a ruler if necessary, that parallels of latitude are not straight on the charts issued by the I.G.N. (Institut Géographique National).

18.1.8.6. Stereographic projections. Stereographic projections are not useful to mathematicians alone (see 18.10.2, for example). Cartographers use them to represent polar regions (with the source of the projection at the other pole), and astronomers for celestial charts. Stereographic projections are conformal, and depict every great circle as a circle (cf. 18.1.4.3); if the source of the projection is one pole, meridians appear as lines meeting at the other pole.

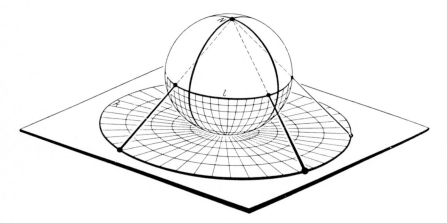

Figure 18.1.8.6 (Source: [H–C])

A single stereographic projection (with source somewhere in New Zealand!) is being debated as an alternative for the cartography of France, because the disagreement between the three Lambert projections used at present is becoming less tolerable as more precision is required.

18.1.8.7. Other systems. We have observed that any chart of the whole globe has got to be bad somewhere. The search for minimal badness has given rise to some original ideas, as shown by the figures on the next few pages.

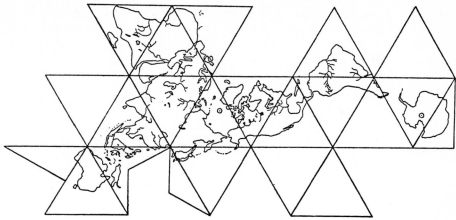

The Dymaxion Air Ocean World
©1954 R. Buckminster Fuller and Shoji Sadao

Lorgna's Transverse Projection
extended to 360°

(Hammer-Aïtoff's transformation)

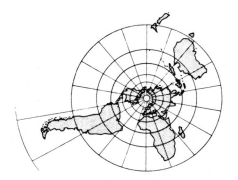

Extrapolation of the
stereographic projection

Figure 18.1.8.7 (Source: [CN])

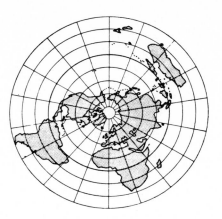

Extrapolation of G. Postel's Projection

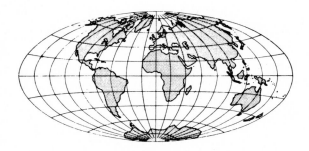

Transverse Stereographic Projection
extended to 360°

Transverse G. Postel Projection
extended to 360°
(Aïtoff's transformation)

Figure 18.1.8.7 (Source: [CN])

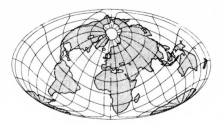

Briesemeister's Projection

Samson-Flamsteed Projection

Mollweide's Projection

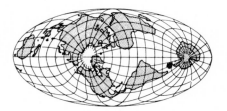

Bartholomow's "Atlantis" Projection

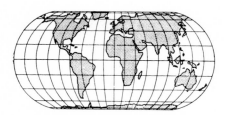

Eckert IV's Projection

Raisz's "Armadillo" Projection

Figure 18.1.8.7 (Source: [CN])

Starred projections

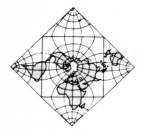

Peirce's Periodic Projection   Bertin's Compensated Projection

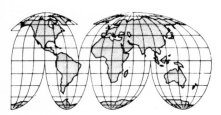

Goode's Projection   Cahill's Projection

Figure 18.1.8.7 (Source: [CN])

18.2. The topology of spheres

18.2.1. PROPOSITION. *The sphere S^d is locally homeomorphic to $\mathbf{R}^d$ at every point, and consequently locally connected. It is also compact and, for $d \geq 1$, path-connected.*

Proof. The first assertion expresses the existence of charts, for example, stereographic projections (cf. 18.1.4). Compactness derives from the definition of S^d as a closed and bounded subset of $\mathbf{R}^{d+1}$. Path-connectedness derives from the connectedness of S^1 (cf. 8.3.8), since any two points of S^d belong to at least one two-dimensional vector subspace of $\mathbf{R}^{d+1}$. $\square$

Beyond general topology, we will study the algebraic topology of the sphere, starting with the fundamental group $\pi_1(S^d)$. If you are not familiar with this notion, see [ZN, chapter 2], for example.

18.2.2. PROPOSITION. *The group $\pi_1(S^1)$ is isomorphic to $\mathbf{Z}$, and $\pi_1(S^d) = 0$ for every $d \geq 2$ (that is, S^d is simply connected for $d \geq 2$).*

18.2.3. PROOF. For the case $d = 1$, as well as applications, see [B-G, 7.6 and chapter 9] and [CH2, 62]. Now assume $d \geq 2$, and set $U_0 = S \setminus n$, $U_1 = S \setminus s$, where $S = S^d$ and n, s are the north and south poles of S (cf. 18.1.2.3). We can assume that the loop $f \in C^0([0,1]; S)$, which we must show is null-homotopic, has origin $s = f(0) = f(1)$. Since U_0 and U_1 form an open cover of S, there is, for every $t \in [0,1]$, a neighborhood $V_t \subset [0,1]$ of t in $[0,1]$ such that $f(V_t) \subset U_0$ or $f(V_t) \subset U_1$. Since $[0,1]$ is compact we can pick finitely many intervals V_{t_i} that still cover $[0,1]$, which means that there exists a sequence $0 < t_1 < t_2 < \cdots < t_m < 1$ such that $f([t_i, t_{i+1}]) \subset U_0$ or U_1, for every i. Now replace each $f_i = f|_{[t_i, t_{i+1}]}$ by a new map $f_i' : [t_i, t_{i+1}] \to S$ constructed as follows: If $n \notin f([t_i, t_{i+1}])$, set $f' = f$; if $n \in f([t_i, t_{i+1}])$, let f_s be the stereographic projection from the south pole. The composition $q : f_s \circ f : [t_i, t_{i+1}] \to \mathbf{R}^d$ exists because f avoids s in the interval $[t_i, t_{i+1}]$. Choose an arbitrary $w \in \mathbf{R}^d$ distinct from $q(t_i)$, $q(t_{i+1})$ and 0 and project $q([t_i, t_{i+1}])$ onto the union of the two half-lines with origin w and going through $q(t_i)$ and $q(t_{i+1})$. The new map $q' : [t_1, t_{i+1}] \to \mathbf{R}^d$, whose image avoids 0, is a continuous deformation of q. Now set $f_i' = f_s^{-1} \circ q'$;

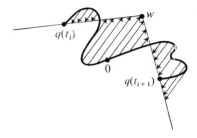

Figure 18.2.3

clearly f_i' is a continuous deformation of f_i, has the same image as f_i at the endpoints t_i and t_{i+1}, and avoids n altogether. Putting the f_i' back together we get a continuous deformation $f' : [0, 1] \to S$ of f with $f'(0) = f'(1) = s$ and $f'([0, 1]) \not\ni n$. Now use stereographic projection from n to get a loop $f_n \circ f' : [0, 1] \to \mathbf{R}^d$ in $\mathbf{R}^d$. Since $\mathbf{R}^d$ is clearly simply connected (even contractible), we can contract $f_n \circ f'$ to a point, and the same must be true of f'. This completes the proof. □

18.2.4. REMARKS.

18.2.4.1. The reader should be warned against trying to "simplify" the proof of 18.2.2. The problem is, of course, to avoid one point of the sphere in order to be able to use stereographic projection and reduce the problem to $\mathbf{R}^d$—but the image $f([0, 1])$ of a map $f \in C^0([0, 1]; S)$ can very well be the whole sphere (remember the Peano curve). If f is differentiable it cannot fill the sphere because its image has measure zero; thus another strategy to prove 18.2.2 is to approximate a continuous f by a differentiable one (cf. [GM, 17], for example).

18.2.4.2. The proof given in 18.2.3 contains the seed of the van Kampen theorem, which is much more general (cf. [MY, chapter IV] and [ZN, 45]).

The material that follows is included for the reader's general information; the interested reader should consult the references.

18.2.5. HOMOLOGY AND DEGREE

18.2.5.1. Reasonable topological spaces can be assigned *homology groups*, and this association is functorial for continuous maps. The homology groups of S^d are all zero, except for $H_0(S^d)$ and $H_d(S^d)$, which are both canonically isomorphic to $\mathbf{Z}$ (here $d \geq 1$). The fact that $H_0(S^d) = \mathbf{Z}$ is nothing extraordinary and just says that S^d is connected. On the other hand, the isomorphism $H_d(S^d) \cong \mathbf{Z}$ has as many useful consequences as $\pi_1(S^1) \cong \mathbf{Z}$ (incidentally, the two facts are equivalent for $d = 1$).

18.2.5.2. In fact, consider $f \in C^0(S^d; S^d)$. Functoriality yields a group homomorphism $f_* \in \mathrm{Hom}(H_d(S^d); H_d(S^d))$. Since $H_d(S^d)$ is canonically isomorphic to $\mathbf{Z}$, we can write $f_*(m) = km$ with $k, m \in \mathbf{Z}$. The integer k, which is an invariant of $f \in C^0(S^d; S^d)$, is called the *degree* of f and *denoted* by $\deg(f)$.

　　It is simple to show that the degree enjoys the following properties: $\deg(\mathrm{Id}_{S^d}) = 1$; $\deg(-\mathrm{Id}_{S^d}) = (-1)^{d+1}$; $\deg(f) = 0$ if f is not surjective (in particular, if f is constant); $\deg(f) = \deg(g)$ if f and g are homotopic, that is, one is a continuous deformation of the other; and $\deg(f \circ g) = \deg(f) \deg(g)$.

18.2.5.3. Corollary. *If d is even, every continuous vector field $\xi : x \mapsto \xi(x) \in T_x S^d$ on S^d is zero at one point at least.*

Proof. If $\xi(x) \neq 0$ for every $x \in S^d$, divide ξ by its norm to obtain a field

$$\eta : x \mapsto \eta(x) = \frac{\xi(x)}{\|\xi(x)\|}$$

which is still continuous; then use η to deform Id_{S^d} into $-\mathrm{Id}_{S^d}$ through the family of maps

$$f_t : x \mapsto \cos t \cdot x + \sin t \cdot \eta(x) \qquad (t \in [0, \pi]).$$

Now the degrees of Id_{S^d} and $-\mathrm{Id}_{S^d}$ should be the same, but we know one is 1 and the other is -1. $\qquad\square$

18.2.5.4. For example, on the surface of the earth there is always a point where the wind is not blowing. Or again: it is impossible to comb a hairy ball without leaving a cowlick somewhere.

18.2.5.5. For odd d there exist nowhere vanishing vector fields on S^d, for example

$$\xi(x_1, \ldots, x_{2n}) = (-x_2, x_1, -x_4, x_3, \ldots, -x_{2n}, x_{2n-1}),$$

where $d = 2n - 1$. For $n = 2$ the vector field obtained in this way points along the orbits of the action of 1.2.9 (cf. 18.8.1 also).

18.2.5.6. Corollary. *Take $f \in C^0(S^{2n}; S^{2n})$. If $\deg(f) \neq -1$, there exists at least one point $x \in S^{2n}$ fixed under f. This is certainly the case if f is homotopic to the identity* (compare with 18.2.5.3).

Proof. Assume that $f(x) \neq x$ for all $x \in S^{2n}$. Then the identity can be deformed into $f' = (-\mathrm{Id}_{S^{2n}}) \circ f$ because, for all $x \in S^{2n}$, $f'(x) \neq -x$ and x and $f'(x)$ are connected by a unique shortest piece of great circle, and such a piece depends continuously on its endpoints. Again we get a contradiction, since $\deg(f') = \deg(f)\deg(-\mathrm{Id}_{S^{2n}})$ cannot be 1. $\qquad\square$

18.2.5.7. Corollary 18.2.5.6 can be obtained from a very general result of Lefschetz on fixed points of maps. See [GG2, 224], for example.

18.2.5.8. For a simpler treatment of 18.2.5 in the framework of differentiable map, see [B–G, 7.3], for example.

18.2.6. The Jordan–Brouwer separation theorem. Here $d \geq 1$. Using the homology group $H_d(S^d) \cong \mathbf{Z}$ in a more sophisticated way, one can prove the following theorem (cf. [GG2, 81], for example):

18.2.6.1. Theorem. *Let V be a subset of S^d homeomorphic to S^{d-1}. Then $S^d \setminus V$ has exactly two connected components, each having V for frontier.* $\square$

Using stereographic projection, one deduces the following

18.2.6.2. Corollary. *Let V be a subset of $\mathbf{R}^d$ homeomorphic to S^{d-1}. Then $S^d \setminus V$ has exactly two connected components, each having V for frontier. One, called the inside of V, is relatively compact; the other, the outside, is unbounded.* $\qquad\square$

18.2.6.3. For $d = 2$ the set V is, by definition, a simple closed curve in $\mathbf{R}^2$; 18.2.6.2 is the Jordan curve theorem. See elementary proofs of this simpler case in [B–G, 9.2] and [DE4, volume 1 (*Foundations of Modern Analysis*), p. 251].

18.2.6.4. We easily deduce from 18.2.6.2 the following fundamental results:

18.2.6.5. Theorem (invariance of domain). *Let $U \subset \mathbf{R}^d$ be an open connected set, and $f : U \to \mathbf{R}^d$ an injective continuous function. The image $f(U)$ is open and f is a homeomorphism onto $f(U)$.*   □

18.2.6.6. Corollary. *Two spaces $\mathbf{R}^d$ and $\mathbf{R}^{d'}$ cannot be homeomorphic unless $d = d'$.*   □

18.2.6.7. Remarks. There are weaker but elementary versions: for example, $\mathbf{R}^d$ and $\mathbf{R}^{d'}$ cannot be locally diffeomorphic unless $d = d'$ (cf. [B–G]). Also, the cases $d = 1$ and 2 are trivial, since for an open set $U \subset \mathbf{R}^d$ and $x \in U$, the complement $U \setminus x$ is disconnected for $d = 1$, connected but not simply connected if $d = 2$, and simply connected for $d \geq 3$.

18.2.7. THEOREM OF BORSUK–ULAM. *For every continuous map $f : S^d \to \mathbf{R}^d$ there exists at least one $x \in S^d$ such that $f(x) = f(-x)$.*

For a proof, see [BN, 337].   □

18.3. The canonical measure on the sphere

18.3.1. We have already cited in 9.12.7 the difficulty in defining a measure for "k-dimensional" subsets of n-dimensional Euclidean space when $2 \leq k \leq n$. Even convex sets, the simplest kind, are somewhat involved. That's why we resort to integration theory and results on differentiable submanifolds in order to define the canonical measure on S; every rigorous theory has its price in complexity. Our exposition will be brief, however, since the canonical measure on S is not used in the rest of the book (except in 12.7.3.1 to give a proof of 12.7.3; but an elementary proof is given in 12.7.3.2).

18.3.2. The sphere is a C^∞ submanifold of $\mathbf{R}^{d+1}$; in fact, it is defined by the equation $f = \| \cdot \|^2 - 1$, whose derivative $f'(x) = 2(x \mid \cdot)$ is never zero on S. The tangent space to S at x is $\left(f'(x)\right)^{-1}(0) = x^\perp$, in accord with 18.1.2.4 (cf. [B–G, 2.1.1 and 2.5.7]).

18.3.3. Recall that $T_x S$ is the set of velocities of curves of class C^1 in $\mathbf{R}^{d+1}$ whose image is in S. Examples of curves in S are loxodromes (18.1.8.2), spherical helices (9.14.34.3.E) and so on (cf. [B–G, 8.7.12 and 9.9.6]).

18.3.4. ORIENTATION. The sphere S, as the frontier of the ball $B(0,1) \subset \mathbf{R}^{d+1}$, has a canonical orientation inherited from $\mathbf{R}^{d+1}$ (cf. [B–G, 5.3.17.2]). Another way to orient S is by using differential forms (the reader who is not familiar with them can consult [B–G]). The *canonical volume form* on S is the restriction to S of the degree-d differential form

$$\sigma = \sum_{i=1}^{d+1} (-1)^{i-1} x_i \, dx_1 \wedge \cdots \wedge \widehat{dx_i} \wedge \cdots \wedge dx_{d+1}$$

in $\mathbf{R}^{d+1}$; the form σ is just the interior product of the canonical volume form

$$\omega = dx_1 \wedge \cdots \wedge dx_{d+1}$$

on $\mathbf{R}^{d+1}$ with the unit vector field pointing away from S. For S^2 we have $\sigma = x\,dy \wedge dz + y\,dz \wedge dx + z\,dx \wedge dy$. For example, the latitude-longitude chart 18.1.6.1 is positively oriented (cf. [B–G, 5.3.22]), since we have

18.3.5 $\tau^*\sigma = \cos\theta\,d\phi \wedge d\theta$ $\left(\theta \in \,\right]-\pi/2, \pi/2\left[\right)$.

18.3.6. The form σ is canonical, since it is well-defined by S; it follows that σ is invariant under $O(d+1) = \mathrm{Is}(S)$ (cf. 18.5). On the other hand, every differential form of degree > 0 on S that is invariant under $\mathrm{Is}(S)$ is proportional to σ.

18.3.7. THE CANONICAL MEASURE.

18.3.7.1. The intuitive idea is that the tangent space to S at any point is an oriented Euclidean space, and as such possesses a canonical measure; all there remains to do is to define the appropriate concept of integration. For example, if $h : U \rightarrow \mathbf{R}^{d+1}$ is a chart of S and f is a function on S whose support lies in U, we can define the integral of f on S as

$$\int_U (f \circ h)\sqrt{\mathrm{Gram}\left(\frac{\partial h}{\partial u_1}, \ldots, \frac{\partial h}{\partial u_d}\right)}\,du_1 \cdots du_d;$$

the square root represents, by 8.11.6, the "element of volume" on the tangent space. This integral is invariant under a change of charts; since one needs several charts to cover S, a complete theory requires the introduction of partitions of unity.

18.3.7.2. Another, essentially equivalent, approach is to use integration theory on oriented manifolds. Then the canonical measure on S, still *written σ* by abuse of notation, is defined as $\int_S f\sigma$ for every function f on S, indicating the integral of the differential form $f\sigma$ on S (cf. [B–G, 6.1 and 6.6]). Since the latitude-longitude chart covers S up to a set of measure zero, we have, for every function f (cf. 18.3.5 and [B–G, 6.1]):

18.3.7.3

$$\int_S f\sigma = \int_{]0.2\pi[\times]-\pi/2,\pi/2[} f\left(\cos\theta\cos\phi, \cos\theta\sin\phi, \sin\theta\right)\cos\theta\,d\phi\,d\theta.$$

18.3.7.4. Note. The measure σ is invariant under $\mathrm{Is}(S)$, by construction. Conversely, any measure on S invariant under $\mathrm{Is}(S)$ is proportional to σ, as can be seen by applying the technique of 2.7.4.4; here the regularized measure is still continuous at x, for every $x \in S$, because the group $\mathrm{Is}(S)$ has continuous subgroups in a set of directions that spans $T_x S$.

18.3.8. VOLUMES

18.3.8.1. The *volume* of a closed set $D \subset S$ is, by definition, $\int_S \chi_D\,\sigma$, where χ_D is the characteristic function of D. If $d = 2$ we say *area* instead of volume. Look in 9.12.4.8 for the total volume of S^d.

18.3.8.2. A *spherical digon* of angle α is a subset D of S^2 bounded between two half-great-circles with same endpoints $\pm m$ and whose tangent vectors at

m form an angle α (cf. 18.1.2.4). We have

18.3.8.3 $\text{area}(D) = 2\alpha.$

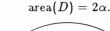

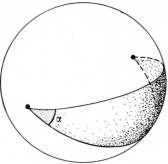

Figure 18.3.8.2

In fact, take $\pm m$ to be the north and south pole of S. In the latitude-longitude chart we have, by 18.3.7.3:

$$\text{area}(D) = \int_0^\alpha \int_{-\pi/2}^{\pi/2} \cos\theta \, d\phi \, d\theta = 2.\alpha.$$

18.3.8.4. Girard's formula (1625). *The area of a spherical triangle with angles α, β, γ is $\alpha + \beta + \gamma - \pi$.*

Proof. For definitions, see 18.6. Let x, y, z be the vertices of the triangle and α, β, γ the associated angles. Let D be the hemisphere determined by the great circle through y, z (cf. 18.1.2.1) and containing x. Up to sets of measure zero, D can be partitioned into four sets T, A, B, C, where T is our triangle, $T \cup B$ (resp. $T \cup C$) is a digon of angle β (resp. γ) and $A \cup (-T)$, which has the same area as $A \cup T$, is a digon of angle α. Thus, applying 18.3.8.3 three times, we have

$$2\pi = \text{area}(D) = \text{area}(T) + \text{area}(A) + \text{area}(B) + \text{area}(C)$$
$$= \text{area}(T) + \big(2\alpha - \text{area}(T)\big) + \big(2\beta - \text{area}(T)\big) + \big(2\gamma - \text{area}(T)\big). \quad \square$$

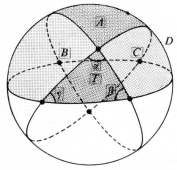

Figure 18.3.8.4

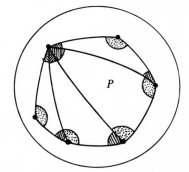

Figure 18.3.8.5

18.3.8.5. Corollary. *If P is an n-sided convex spherical polygon (cf. 18.7) having angles α_i $(i = 1, \ldots, n)$, we have*

$$\text{area}(P) = \sum_{i=1}^{n} \alpha_i - (n-2)\pi.$$

The proof is the same as that of 10.5.2.   □

18.3.8.6. Notes. Formula 18.3.8.4 is just a particular case of the Gauss–Bonnet formula, which holds for every two-dimensional Riemannian manifold and says that the angles α, β, γ of a triangle T whose sides are geodesics satisfies

$$\alpha + \beta + \gamma - \pi = \int_T K\,\sigma,$$

where K is the curvature of the manifold and σ its canonical measure. For the sphere the curvature is identically 1. We shall see when we study hyperbolic geometry (cf. 19.5.4) that the area of a triangle with angles α, β, γ is $\pi - \alpha - \beta - \gamma$; this is because the curvature of the hyperbolic plane is constant and equal to -1. For a Euclidean plane, whose curvature is zero, we recover 10.2.4. See references in 12.7.5.2.

Girard's formula shows, in particular, that if the three angles of a triangle T are in the set $\pi\mathbf{Q}$ of rational multiples of π, the area of T is also in $\pi\mathbf{Q}$. In the sphere S^d $(d \geq 3)$ the calculation of the volume of a simplex as a function of its dihedral angles (the natural generalization of Girard's formula) is a very difficult question; a recent conjecture by Cheeger and Simons is that, for $d \geq 3$, there exist simplices in S^d with all dihedral angles in $\pi\mathbf{Q}$ whose volume is not a rational multiple of π.

We have already mentioned in 12.11.5.3 the problem of extending to S^d the isoperimetric inequality 12.11.1.

18.4. The intrinsic metric on the sphere

18.4.1. The notions used here are from section 9.9. We have seen in 9.9.4.3 that the metric induced in S from $\mathbf{R}^{d+1}$ is not excellent, not even intrinsic (cf. 9.9.4.4). The discussion in 9.9.8 suggests what to do: for $x, y \in S$, set $\overline{xy} = \text{Arccos}\big((x \mid y)\big)$.

18.4.2. THEOREM. *The map $\overline{}: S \times S \rightarrow [0, \pi]$ is an excellent (hence intrinsic) metric on S. It induces on S the same topology inherited from $\mathbf{R}^{d+1}$. Let $x, y \in S$. If $y \neq \pm x$, there exists a unique shortest path from x to y, namely, the arc of great circle joining x and y (cf. 18.1.2.1 and 8.7.5.4); if $y = -x$, the shortest paths are the half-great-circles with endpoints x and $-x$. The metric $\overline{}$ is called the intrinsic metric on S, and we shall consider no other from now on.*

Proof. It is trivial that $\overline{xy} = \overline{yx}$ and that $\overline{xy} = 0$ implies $x = y$. The triangle inequality follows from 18.6.10 if x, y, z don't lie on the same plane through

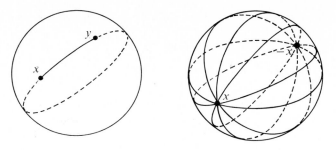

Figure 18.4.2

the origin and from 9.9.8.1 if they do. The assertions about shortest paths and the fact that $\bar{\cdot}$ is excellent can be derived from 18.6.10 and 9.9.8.1.   □

See another proof of the triangle inequality in 18.11.13.

18.4.3. NOTE. If we apply the process described in 9.9.7.1 to the metric d induced from $\mathbf{R}^{d+1}$, the metric $\bar{d}$ we obtain is exactly the intrinsic metric on S. The proof is left to the reader (see 18.11.26).

18.4.4. PROPOSITION (NON-EXISTENCE OF ISOMETRIC CHARTS). *Let U be an open subset of S. There exists no isometric map $f : U \to \mathbf{R}^d$, that is, no map satisfying $d\big(f(x), f(y)\big) = \overline{xy}$ for every x and y in U.*

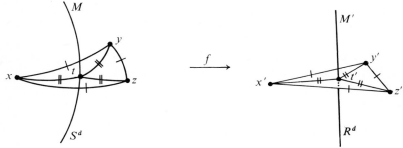

Figure 18.4.4

Proof. Notice first that, if $x, y, z \in S$ are linearly independent, the set

$$M = \big\{ m \in S \mid \overline{mx} = \overline{my} = \overline{mz} \big\}$$

is a $(d-2)$-subsphere of S orthogonal to the 2-subsphere N of S containing x, y and z. Let t be the point of $M \cap N$ which belongs to the hemisphere determined by x, y, z; then (cf. 18.6.8 and 18.11.6, for example) we have

18.4.4.1 $\overline{xt} = \inf \{ \overline{xm} \mid m \in M \}.$

Now let U be an open set in S; by 18.6.10 there exist linearly independent points $x, y, z \in S$ such that $\overline{xy} = \overline{yz} = \overline{zx}$. A bit of spherical trigonometry (cf. 18.6.13) enables us to calculate $\beta = \overline{xt} = \overline{yt} = \overline{zt}$ as a function of

$\alpha = \overline{xy}$ and to see that $\alpha < \sqrt{3}\beta$. Let x', y', z', t', m', M' be the images of x, y, z, t, m, M under f. Using the fact that f is an isometry we see that the three points x', y', z' form an equilateral triangle of side α in $\mathbf{R}^d$, that

$$M' = \{ v \in \mathbf{R}^d \mid d(v, x') = d(v, y') = d(v, z') \}$$

and that

$$d(x', t') = \inf \{ d(x', v) \mid v \in M' \},$$

this latter by 18.4.4.1. But in $\mathbf{R}^d$ it is clear that this implies that t' is the center of $\{x', y', z'\}$, and this would force $d(x', y') = \sqrt{3}d(x', t') = \sqrt{3}\beta$, a contradiction with the previously obtained inequality.

18.4.5. REMARKS. This proposition would also follow from 18.4.7. Also from an elementary study as in 18.11.7; the proof in 18.4.4 is itself based on a minimum distance problem.

18.4.6. EQUIDISTANT LOCI. By 8.3.11 and 9.7.5, the set $\{ z \in S \mid \overline{zx} = \overline{zy} \}$ for x and y distinct points in S is a $(d - 1)$-subsphere of S. For k linearly independent points, the locus of equidistant points is a $(d + 1 - k)$-subsphere.

18.4.7. CHARACTERIZATION BY THE METRIC. We can ask about $(S^d, \overline{\cdot})$ the same question answered in 9.7.4 for Euclidean spaces. The answer is yes, and a proof can be found in chapter VII of [BL]. See also exercise 18.11.14. That spheres are characterized by their metric is obviously another proof for 18.4.4.

18.5. The isometry group of the sphere

18.5.1. Now that we have a good metric on S, the most pressing problem is to find the isometry group $\mathrm{Is}(S)$. In fact, there are two groups to be considered *a priori*, $\mathrm{Is}(S, \overline{\cdot})$ and $\mathrm{Is}(S, d)$, but the two coincide by 8.3.11 and we can write $\mathrm{Is}(S)$ without ambiguity. By 9.8.2, $\mathrm{Is}(S^d)$ coincides with

$$\mathrm{Is}_{S^d}(\mathbf{R}^{d+1}) = \{ f \in \mathrm{Is}(\mathbf{R}^{d+1}) \mid f(S) = S \}.$$

On the other hand, the orthogonal group $O(d + 1)$ is included in $Is_{S^d}(\mathbf{R}^{d+1})$, so $O(d + 1) \subset \mathrm{Is}(S^d)$. There is, in fact, equality:

18.5.2. PROPOSITION. *The isometry group* $\mathrm{Is}(S^d)$ *is naturally isomorphic to* $O(d + 1)$, *the isomorphism being given by restriction to* S^d.

Proof. Given $f \in \mathrm{Is}(S^d)$, it is enough to show that $f(0) = 0$ (cf. 9.1.3). The idea is that, for $x, y \in S^d$, the condition $d(x, y) = 2$ characterizes antipodal points $y = -x$. Take an arbitrary $x \in S^d$; then $d(f(x), f(-x)) = d(x, -x) = 2$, so $f(-x) = f(x)$ and, since f is affine (cf. 3.5.1),

$$f(0) = f \left(\frac{x + (-x)}{2} \right) = \frac{f(x) + f(-x)}{2} = 0. \qquad \square$$

18.5.3. NOTATION. We *denote* by $\mathrm{Is}^+(S^d)$ the restriction of $O^+(d + 1)$ to S, and by $\mathrm{Is}^-(S^d)$ the restriction of $O^-(d + 1)$ to S (cf. 8.2.1).

18.5.4. REMARK. For any reasonable definition of orientation on S (see 18.3.4), the group $\mathrm{Is}^+(S)$ coincides with the orientation-preserving isometries of S.

The group $\mathrm{Is}(S)$, and even $\mathrm{Is}^+(S)$, are "big" in the sense that they act transitively on pairs of points:

18.5.5. PROPOSITION (TWO-TRANSITIVITY). *For any four points in S satisfying $\overline{xy} = \overline{x'y'}$ there exists $f \in \mathrm{Is}(S)$ taking x to x' and y to y'.*

(See 9.1.7 for more on this topic.) This proposition has an infinitesimal version:

18.5.6. PROPOSITION. *For any two points $x, x' \in S$ and tangent vectors $\xi \in T_x S$, $\xi' \in T_{x'}S$ such that $\|\xi\| = \|\xi'\|$ there exists $f \in \mathrm{Is}(S)$ such that $f(x) = x'$ and $f'(\xi) = \xi'$, where f' denotes the derivative of $f : \mathbf{R}^{d+1} \to \mathbf{R}^{d+1}$. In other words, $\mathrm{Is}(S)$ acts transitively on unit tangent vectors to S.*

For $d = 1$ or 2, one can be more specific:

18.5.7. PROPOSITION. *For any four points $x, y, x', y' \in S^1$ such that $0 < \overline{xy} = \overline{x'y'} < \pi$ there exists a unique $f \in \mathrm{Is}(S^1)$ taking x to x' and y to y'. For any four points $x, y, x', y' \in S^2$ such that $0 < \overline{xy} = \overline{x'y'} < \pi$ there exists a unique $f \in \mathrm{Is}^+(S^2)$ taking x to x' and y to y'.*

Propositions 18.5.5, 18.5.6 and 18.5.7 follow easily from 8.1.4 and 8.2.7.   □

The next result expresses the uniqueness of $\mathrm{Is}(S)$-invariant metrics on S:

18.5.8. PROPOSITION. *Let δ be a metric on S^d invariant under $\mathrm{Is}(S^d) = O(d+1)$. There exists an injective map $\phi : [0, \pi] \to \mathbf{R}$ such that $\delta(x, y) = \phi(\overline{xy})$ for all $x, y \in S$. If δ is also intrinsic, there exists $k \in \mathbf{R}_+^*$ such that $\delta(x, y) = k\overline{xy}$ for all $x, y \in S$.*

Proof. Fix $x \in S$ and, for every $r \in [0, \pi]$, choose $y \in S$ such that $\overline{xy} = r$. Put $\phi(r) = \delta(x, y)$. Then $\delta(u, v) = \phi(\overline{uv})$ for every $u, v \in S$; for if $\overline{uv} = r = \overline{xy}$ there exists $f \in \mathrm{Is}(S)$ such that $f(x) = u$ and $f(x) = v$ (cf. 18.5.5), whence

$$\delta(u, v) = \delta\big(f(x), f(y)\big) = \delta(x, y) = \phi(r) = \phi(\overline{uv}).$$

This takes care of the first statement; the second is harder.

18.5.8.1. For every $r, s, t \in [0, \pi/2]$ such that $r + s \in [0, \pi]$, $r > s$ and $t \in [r - s, r + s]$ we have $\phi(t) \le \phi(r) + \phi(s)$. In fact, by 18.6.10 there exist $x, y, z \in S$ with $\overline{xy} = r$, $\overline{xz} = s$ and $\overline{yz} = t$ (figure 18.5.8.1). Thus

$$\phi(t) = \phi\overline{xy} = \delta(y, z) \le \delta(x, y) + \delta(y, z) = \phi(r) + \phi(s).$$

18.5.8.2. The function ϕ is continuous. By 18.5.8.1 we just have to prove continuity at 0. Since δ is intrinsic, there exists a δ-rectifiable curve f, and, by the definition of $\mathrm{leng}(f)$ (cf. 9.9.1), there exist $x, y \in S$ with $\delta(x, y)$ as small as desired. Thus $\phi(t)$ can be taken as small as desired, which shows continuity at 0, since, by 18.5.8.1, $\phi(u) \le 2\phi(t)$ for every $u \in [0, 2t]$.

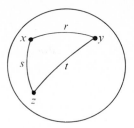

Figure 18.5.8.1

18.5.8.3. For every $x \in [0, \pi]$ there exists $\lambda(x) \in {]0, 1]}$ such that $\phi(tx) \geq t\phi(x)$ for all $t \in [0, \lambda(x)]$. Otherwise, for every $n \in \mathbf{N}^*$ there would exist $s \in {]0, 1/n]}$ such that $\phi(sx) < s\phi(x)$, so $\phi(ksx) \leq ks\phi(x)$ for every $k \in \mathbf{N}$ not too big (use 18.5.8.1 repeatedly). By the continuity of ϕ and the density of $\mathbf{Q}$ in $\mathbf{R}$ we have $\phi(tx) \leq t\phi(x)$ for all $t \in [0, 1]$. Take $u \in {]0, x[}$ with $\phi(u) < (u/x)\phi(x)$; by 18.5.8.1 we have a contradiction:

$$\phi(x) = \phi\big(u + (x - u)\big) \leq \phi(u) + \phi(x - u) < \frac{u}{x}\phi(x) + \frac{x - u}{x}\phi(x) = \phi(x).$$

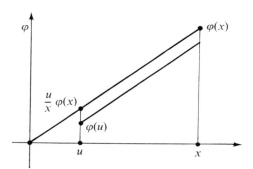

Figure 18.5.8.3

18.5.8.4. The ratio $\phi(x)/x$ tends to a finite limit k as x approaches 0. For 18.5.8.3 implies that otherwise it must tend to infinity, and this cannot happen. If it did, we'd have $\mathrm{leng}_\delta f \geq N \mathrm{leng}_{\bar{}} f$ for any δ-rectifiable curve f and any $N > 0$, which is absurd since $\mathrm{leng}_{\bar{}} f > 0$.

We finish by remarking that, by 9.9.4.4 and the technique used in 18.4.3, rectifiable curves for δ and $\bar{}$ coincide, and $\mathrm{leng}_\delta f = k \, \mathrm{leng}_{\bar{}} f$. Since f is intrinsic this implies $\delta(x, y) = k\bar{}$ for all $x, y \in S$ (cf. 9.9.4.4). □

18.5.8.5. Remarks. This result shows that $\bar{}$ is the right metric on S (up to a scalar).

The proof of uniqueness would be immediate if we restricted ourselves to Riemannian metrics invariant under S, since we could then use the transitive action of $O(d + 1)$ on unit tangent vectors (18.5.6).

18.5.9. In $\mathrm{Is}(S)$ the stabilizer of a point (cf. 1.5) is isomorphic to $O(d)$, since $\mathrm{Is}_x(S^d) = O_x(\mathbf{R}^{d+1})$.

18.5.10. For the study of finite subgroups of $\mathrm{Is}(S^1)$ and $\mathrm{Is}(S^2)$, see 1.8.2. The classification of finite subgroups of $\mathrm{Is}(S^d)$ for $d \geq 3$ is a whole new ball game; see [WF, part III].

18.6. Spherical triangles

18.6.1. The aim of this section is the study of spherical triangles in S^2, or, equivalently, of trihedra in three-dimensional Euclidean space, that is, sets of three half-lines emanating from the same point. In particular, spherical trigonometry studies the relations between six numbers: the three angles at the vertex of a trihedron and the three dihedral angles between the faces. Such relations are fundamental in astronomy, in maritime and aerial navigation, and in mechanics, and we will mention some of their applications as they come up.

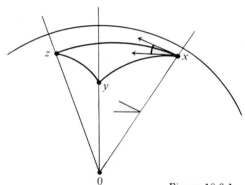

Figure 18.6.1

One possible generalization would be to study three points on S^d $(d \geq 3)$, but this case can be reduce to $d = 2$ by considering the 2-subsphere spanned by the three points. On the other hand, the figure formed by $d + 1$ points on S^d is interesting, but their study is considerably more difficult for $d \geq 3$ than for $d = 2$. See [CR3, 247] and the references therein. See also 18.3.8.6 for another example of the leap in difficulty between $d = 2$ and $d \geq 3$.

18.6.2. DEFINITION. *A spherical triangle (or just triangle) on S^2 is a triple (x, y, z) of points of S^2 that are linearly independent as vectors in $\mathbf{R}^3$. We denote it by $\langle x, y, z \rangle$.*

18.6.3. There is no harm in identifying $\langle x, y, z \rangle$ with the shaded piece of the sphere shown in figure 18.6.1, and calling that piece a *spherical triangle* as well.

18.6.4. In figure 18.6.1 we see that the dihedral angles of the triangle $\{Ox, Oy, Oz\}$ are equal to the angles between the tangent vectors at x, y and z to the arcs of great circle defined by these three points. This justifies the introduction of the following notation: if x and y are linearly independent vectors on $S^2 \subset \mathbf{R}^3$ (more generally, in any Euclidean space), we *set*

18.6.5
$$x_y = \frac{y - (x \mid y)x}{\|y - (x \mid y)x\|}.$$

Thus x_y is the second vector obtained by applying Gram-Schmidt to $\{x, y\}$ (cf. 8.1.4). In addition, x_y is the unit vector tangent at x to the arc of great circle with origin x and endpoint y. It is natural to consider $x_y \in T_x S$ as a bound vector with origin x.

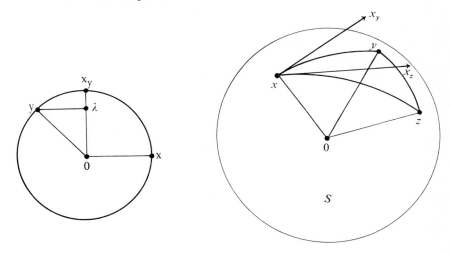

Figure 18.6.5

18.6.6. DEFINITION. *Let $\langle x, y, z \rangle$ be a spherical triangle. The points x, y and z are called the vertices of $\langle x, y, z \rangle$. By the sides of $\langle x, y, z \rangle$ we understand both the arcs of great circle determined by (x, y), (y, z) and (z, x) and their lengths, which are the real numbers $a = \overline{yz}$, $b = \overline{zx}$ and $c = \overline{xy}$. The angles of $\langle x, y, z \rangle$ are the (non-oriented) angles between vectors*

$$\alpha = \overline{x_y x_z} = \mathrm{Arccos}\big((x_y \mid x_z)\big),$$
$$\beta = \overline{y_z y_x} = \mathrm{Arccos}\big((y_z \mid y_x)\big),$$
$$\gamma = \overline{z_x z_y} = \mathrm{Arccos}\big((z_x \mid z_y)\big),$$

which are real numbers in the interval $]0, \pi[$.

18.6.7. By analogy with the case of triangles in a Euclidean plane, one expects that, generally speaking, three of the quantities $a, b, c, \alpha, \beta, \gamma$ are enough to determine the others. An intuitive reason, which will be rigorously developed in 18.6.13.10, is that, if α, b, c are given, y and z are determined up to isometry. In fact, given x, $x_y = u$ and $x_z = v$ such that $u, v \in T_x$ and $uv = \alpha$, the points y and z can be determined by marking the desired lengths $b = \overline{xz}$ and $c = \overline{xy}$ along the arcs of great circles with origin x and tangent to u and v. But if we carry out the same procedure at x', with tangent vectors $u', v' \in T_{x'}S$ and $\overline{u'v'} = \overline{uv} = \alpha$, there exits an isometry mapping x to x' and u, v to u', v', and consequently y, z to y', z'. This implies $\overline{yz} = \overline{y'z'}$. The fact that $a = \overline{yz}$ can indeed be expressed as a function of α, b, c is a consequence of the calculation that leads to 18.6.8. The whole of spherical trigonometry rests on these two keystones: formula 18.6.8 and the notion of duality arising from triangles polar to one another (18.6.12).

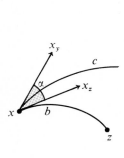
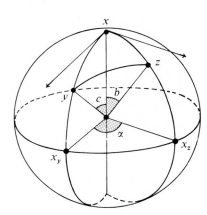

Figures 18.6.7   Figure 18.6.8

18.6.8. PROPOSITION (FUNDAMENTAL FORMULA OF SPHERICAL TRIGO-NOMETRY). *For every spherical triangle we have*

$$\cos a = \cos b \cos c + \sin b \sin c \cos \alpha.$$

Proof. Set $u = x_y$ and $v = x_z$. Then

$$y = \cos c \cdot x + \sin c \cdot u, \qquad z = \cos b \cdot x + \sin b \cdot v,$$

and since $(u \mid v) = \cos \alpha$ by the definition of α, and $(x \mid u) = (x \mid v) = 0$, we get from the definition of a that

$$\cos a = (y \mid z) = (\cos c \cdot x + \sin c \cdot u \mid \cos b \cdot x + \sin b \cdot v)$$
$$= \cos b \cos c + \sin b \sin c \cos \alpha. \qquad \square$$

18.6.9. COROLLARY. *The side a of a spherical triangle with fixed sides b and c is a strictly increasing function of α.* $\qquad \square$

Notice that the interval of variation of this function is not necessarily $\big[|b - c|,\, b + c \big]$, since there is no reason for $b + c$ to be in $]0, \pi[$. For example, if $b = c > \pi/2$, the interval of variation is $]0, 2\pi - 2b[$. More precisely:

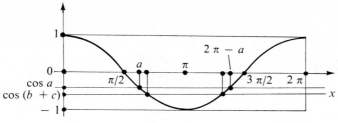

Figure 18.6.10

18.6.10. COROLLARY. *For every spherical triangle,*

$$|b - c| < a < b + c \qquad and \qquad a + b + c < 2\pi.$$

Conversely, if the three numbers $a, b, c \in \,]0, \pi[$ satisfy $|b - c| < a < b + c$ and $a + b + c < 2\pi$, there exists a spherical triangle with sides a, b, c, and such a triangle is unique up to isometries.

Proof. By 18.6.8 and since the cosine has absolute value less than 1 (remember that angles 0 and π are not allowed by 18.6.2), we have

$$\left| \frac{\cos a - \cos b \cos c}{\sin b \sin c} \right| < 1,$$

which is equivalent to

$$\cos(b + c) < \cos a < \cos(b - c).$$

For $a, b, c \in \,]0, \pi[$ this implies $a < b + c < 2\pi - a$. This says, first, that $a + b + c < 2\pi$; also that $b < c + a$ and $c < a + b$ by permutation, whence $|b - c| < a$.

Conversely, if a, b, c meet the two conditions, they satisfy

$$\left| \frac{\cos a - \cos b \cos c}{\sin b \sin c} \right| < 1,$$

so there exists $\alpha \in \,]0, \pi[$ such that $\cos a = \cos b \cos c + \sin b \sin c \cos \alpha$. Take an arbitrary $x \in S$ and $u, v \in T_x S$ with $\|u\| = \|v\| = 1$ and $\overline{uv} = \alpha$. Choosing y and z as in 18.6.7 (see also figure 18.6.7), we have $\overline{yz} = a$ by 18.6.8. Uniqueness follows from 18.6.13.10. $\square$

18.6.11. PRACTICAL APPLICATIONS. Formula 18.6.8 has countless practical applications: in astronomy, where most of the formulas given below are useful (cf. [NU, appendix A]); in navigation, where, in the form known as the cosine-haversine formula, it is used to determine lines of position (cf. [AE, 379-386] and [B–W, 336 and 301–306]); in explaining the behavior of a car when turning (cf. 18.11.16); and even in Riemannian geometry, where spherical trigonometry has been recently used in the solution of several problems (see [G–K–M, §6], for example). Corollary 18.6.9 will be essential in the proof of Cauchy's lemma (section 18.7).

18.6.12. POLAR TRIANGLES. If $\langle x, y, z \rangle$ is a spherical triangle, we define the point $x' \in S$ as the pole of the great circle passing through y and z that lies in the same side—or hemisphere—as x. We can write the conditions as follows:

$$(x' \mid y) = (x' \mid z) = 0 \qquad \text{and} \qquad (x' \mid x) > 0$$

(the case $(x' \mid x) = 0$ is excluded by definition 18.6.2). This leads to the following

18.6.12.1. Proposition and definition. *Let $\langle x, y, z \rangle$ be a spherical triangle. The triple (x', y', z') defined by the conditions*

$$(x' \mid y) = (x' \mid z) = 0, \qquad (x' \mid x) > 0,$$
$$(y' \mid z) = (y' \mid x) = 0, \qquad (y' \mid y) > 0,$$
$$(z' \mid x) = (z' \mid y) = 0, \qquad (z' \mid z) > 0,$$

is a spheric triangle $\langle x', y', z' \rangle$, denoted by $\langle x, y, z \rangle'$ and called the polar triangle of $\langle x, y, z \rangle$.

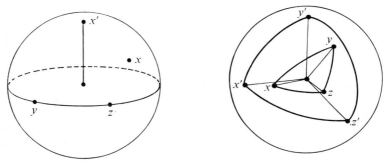

Figure 18.6.12.1

Proof. If x', y', z' were linearly dependent we would have, say, $z' \in \mathbf{R}x' + \mathbf{R}y'$, contradicting $(z' \mid z) > 0$. $\qquad \square$

18.6.12.2. Proposition. *For an arbitrary spheric triangle $\langle x, y, z \rangle$ we have $\left(\langle x, y, z \rangle' \right)' = \langle x, y, z \rangle$. If the six elements of $\langle x, y, z \rangle'$ are $a', b', c', \alpha', \beta', \gamma'$, we have*

$$a + \alpha' = b + \beta' = c + \gamma' = a' + \alpha = b' + \beta = c' + \gamma = \pi.$$

Proof. Duality is a direct consequence of the conditions in 18.6.12.1. Among the six relations, three are derived from the other three by duality, so we just have to show that $a' + \alpha = \pi$, for example.

We work in the plane orthogonal to x (see figure 18.6.12.2, where the circled letters represent projections of the respective vector on that plane). We have the conditions

$$(z' \mid u) = 0, \qquad (y' \mid v) = 0, \qquad (z' \mid v) > 0, \qquad (y' \mid u) > 0,$$

from which it easily follows (cf. 8.7.5.3, for example) that $y'z' + uv = \pi$. $\qquad \square$

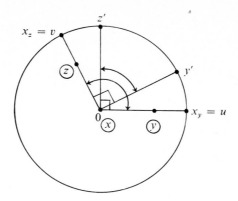

Figure 18.6.12.2

18.6.12.3. Corollary. *For every spherical triangle we have*

$$\alpha + \beta + \gamma > \pi, \quad \alpha + \pi > \beta + \gamma, \quad \beta + \pi > \gamma + \alpha, \quad \gamma + \pi > \alpha + \beta. \quad \square$$

18.6.12.4. Remarks. We see that spherical geometry differs profoundly from plane Euclidean geometry, in which the sum of the angles of a triangle is always equal to π! We have seen in 18.3.8.4 that the excess $\sigma = \alpha + \beta + \gamma - \pi$ is no other than the area of $\langle x, y, z \rangle$. For a geometry where the sum of the angles is less than π, see 19.5.4.

18.6.13. FORMULARY. In the formulary below we have generally omitted formulas that can be derived by just permuting (a, b, c) and (α, β, γ). We have also *set* (cf. 10.3 and 18.11.9 for analogies):

18.6.13.1
$$\begin{cases} p = \dfrac{a+b+c}{2}, \quad P = \dfrac{\alpha+\beta+\gamma}{2}, \quad \delta = \dfrac{1}{2}\left|\det(x, y, z)\right|, \\[2mm] \Delta = \dfrac{1}{2}\left|\det(x', y', z')\right|, \quad \sigma = \alpha + \beta + \gamma - \pi = 2P - \pi. \end{cases}$$

We then have, for every spherical triangle:

18.6.13.2
$$\begin{aligned} \delta &= \sqrt{\sin p \sin(p - a) \sin(p - b) \sin(p - c)} \\ \Delta &= \sqrt{\sin P \sin(P - \alpha) \sin(P - \beta) \sin(P - \gamma)} \end{aligned}$$

18.6.13.3
$$\tan \frac{\sigma}{4} = \sqrt{\tan \frac{p}{2} \tan \frac{p-a}{2} \tan \frac{p-b}{2} \tan \frac{p-c}{2}}$$

18.6.13.4
$$\frac{\sin a}{\sin \alpha} = \frac{\sin b}{\sin \beta} = \frac{\sin c}{\sin \gamma} = \frac{\delta}{\Delta}$$

18.6.13.5
$$\delta = \frac{1}{2}\sin b \sin c \sin \alpha, \qquad \Delta = \frac{1}{2}\sin \beta \sin \gamma \sin a$$

18.6.13.6

$$\tan \frac{\beta - \gamma}{2} = \left(\cot \frac{\alpha}{2}\right) \frac{\sin \frac{b-c}{2}}{\sin \frac{b+c}{2}},$$

$$\tan \frac{\beta + \gamma}{2} = \left(\cot \frac{\alpha}{2}\right) \frac{\cos \frac{b-c}{2}}{\cos \frac{b+c}{2}}$$

18.6.13.7

$$\cos \alpha = \frac{\cos a - \cos b \cos c}{\sin b \sin c},$$

$$\tan \frac{\alpha}{2} = \sqrt{\frac{\sin(p - b) \sin(p - c)}{\sin p \sin(p - a)}}$$

Proof. The first formula in 18.6.13.7 is just 18.6.8; the second is deduced from the first and the identities

$$\cos^2 \frac{\alpha}{2} = \frac{\cos \alpha + 1}{2}, \qquad \sin^2 \frac{\alpha}{2} = \frac{1 - \cos \alpha}{2}$$

by means of the usual algebraic manipulations involving the formulas for the sum of angles. To prove 18.6.13.5 (whose second formula uses the first and 18.6.12.2) we consider the orthonormal frame $\{e_1, e_2, e_3\}$ in $\mathbf{R}^3$ given by $e_1 = x$, $e_2 = x_y$ and $e_3 = e_1 \times e_2$. In this frame the coordinates of x, y and z are easy to write down (see the proof of 18.6.8) and we have

$$\det(x, y, z) = \begin{vmatrix} 1 & \cos c & \cos b \\ 0 & \sin c & \cos \alpha \sin b \\ 0 & 0 & \sin \alpha \sin b \end{vmatrix} = \sin \alpha \sin b \sin c.$$

The first formula in 18.6.13.2 requires a trick involving the Gram determinant (cf. 8.11.5):

$$4\delta^2 = \begin{vmatrix} (x|x) & (x|y) & (x|z) \\ (y|x) & (y|y) & (y|z) \\ (z|x) & (z|y) & (z|z) \end{vmatrix} = \begin{vmatrix} 1 & \cos c & \cos b \\ \cos c & 1 & \cos a \\ \cos b & \cos a & 1 \end{vmatrix};$$

we conclude by expanding the determinant and using the formulas for the sum of angles. The second formula in 18.6.13.2 is derived from the first by considering the polar triangle (cf. 18.6.12.2).

Formula 18.6.13.4 is trivial once we have 18.6.13.5. Formulas 18.6.13.6 and 18.6.13.3 are obtained by manipulating 18.6.13.7; the manipulations are long in the case of 18.6.13.3, but involve nothing beyond usual trigonometry.
$\square$

18.6.13.8. Notes. Formulas 18.6.13.2 through 18.6.13.7 are used all the time in positional astronomy (see details in [NU, appendix A] or [AE, 367 ff.]). Notice that $\sin \alpha$ does not determine α, since $\alpha \in {]}0, \pi[$. On the other hand, $\cos \alpha$, $\tan \alpha/2$, $\cos \alpha/2$ or $\sin \alpha/2$ do, and that is one of the main reasons for the usefulness of formulas involving half-arcs (known among astronomers as "second-order" formulas, whereas those involving quarter-arcs, like 18.6.13.3, are "third-order" formulas). Formula 18.6.13.3, and also 18.6.13.2, are reminiscent of the formula for the area of a Euclidean triangle as a function of its sides (10.3.3). Exercise 18.11.9 is very instructive in this respect.

The first formula in 18.6.13.7 is due to Gauss, those in 18.6.13.6 are known as "Napier's analogies," and formula 18.6.13.3 is due to Simon Lhuillier.

18.6.13.9. Corollary. *In a spherical triangle, $b = c$ if and only if $\beta = \gamma$* (compare with 10.2.2).

Proof. Use 18.6.13.7, not 18.6.13.4!   □

In particular, a spherical triangle is *equilateral* (that is, $a = b = c$) if and only if its three angles are equal ($\alpha = \beta = \gamma$). But, contrary to Euclidean triangles, the three angles can take any value in $]\pi/3, \pi[$—see 18.11.11 for details, and 18.4.4 for an application.

18.6.13.10. Corollary (equality between spherical triangles). *Let $\langle x, y, z \rangle$ and $\langle \tilde{x}, \tilde{y}, \tilde{z} \rangle$ be spherical triangles, with respective elements $a, b, c, \alpha, \beta, \gamma$ and $\tilde{a}, \tilde{b}, \tilde{c}, \tilde{\alpha}, \tilde{\beta}, \tilde{\gamma}$. The five conditions below are equivalent:*
  i) *There exists $f \in \mathrm{Is}(S)$ such that $f(x) = \tilde{x}$, $f(y) = \tilde{y}$, $f(z) = \tilde{z}$;*
  ii) *$a = \tilde{a}$, $b = \tilde{b}$, $c = \tilde{c}$;*
  iii) *$\alpha = \tilde{\alpha}$, $b = \tilde{b}$, $c = \tilde{c}$;*
  iv) *$a = \tilde{a}$, $\beta = \tilde{\beta}$, $\gamma = \tilde{\gamma}$;*
  v) *$\alpha = \tilde{\alpha}$, $\beta = \tilde{\beta}$, $\gamma = \tilde{\gamma}$.*

Proof. The first condition obviously implies the others. By introducing polar triangles (cf. 18.6.12.1 and 18.6.12.2) we see that (ii) and (v) are equivalent, as are (iii) and (iv). By 18.6.8, (ii) implies (iii). To show that (iii), say, implies (i), we develop the idea introduced in 18.6.7: setting

$$u = x_y, \qquad v = x_z, \qquad \tilde{u} = \tilde{x}_{\tilde{y}}, \qquad \tilde{v} = \tilde{x}_{\tilde{z}},$$

we have $\overline{uv} = a = \tilde{a} = \overline{\tilde{u}\tilde{v}}$ and 18.6.7 ensures the existence of $f \in \mathrm{Is}(S)$ such that

$$f(x) = \tilde{x}, \qquad f'(u) = \tilde{u}, \qquad f'(v) = \tilde{v}.$$

Since $f \in \mathrm{Is}(S)$, we get $\overline{xy} = \overline{\tilde{x}\tilde{y}}$ and $\overline{xz} = \overline{\tilde{x}\tilde{z}}$, whence $f(y) = \tilde{y}$ and $f(z) = \tilde{z}$.   □

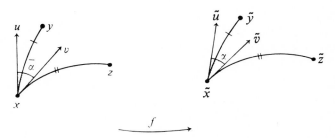

Figure 18.6.13.10

18.6.13.11. Remarks.

1. Case (v) is radically different from plane Euclidean geometry, where three triangles with the same angles only have to be similar.

2. As an exercise, the reader can study the existence and uniqueness of spherical triangles for which the data (a, b, γ), (a, b, α) or (α, β, a) are given (cf. 10.2.6).

3. See 18.11.11 for a study of right spherical triangles, and of those having one side equal to $\pi/2$.

4. See in 18.11.10 a generalization of the useful theorem 10.9.3.

5. See in [DX, 203 ff.] a study of spherical trigonometry in the complexification $\mathbf{C}^3$ of $\mathbf{R}^3$.

18.6.14. IDEAS FOR EXERCISES. The general plan is to take any subject from chapter 10 and decide whether an analogous problem makes sense in S^2, then solve it and compare the results. We have already encountered this kind of question in 18.4.4. We encourage the reader to do the same thing for 10.3.10, 10.4.1, 10.4.3 and 10.4.5. See also 18.11.5.

18.6.15. NOTE. See the hyperbolic counterpart of 18.6.13 in 19.3.4.

18.6.16. REMARK. Problems of a completely different nature involving S^2 have also been studied; see, for example, [FT1, chapter VI], [BL, chapter VIII]. The problem of distributing a certain number of points on a sphere so that they are as far apart as possible is still open (see [FT1, 171]). See also [RS].

18.7. Convex spherical polygons and Cauchy's lemma

This section is geared toward Cauchy's lemma (18.7.16), which is one of the two key ideas in the proof of Cauchy's theorem on the rigidity of convex polyhedra in $\mathbf{R}^3$ (cf. 12.8.1). Its proof rests on proposition 18.7.7, but, being a bit tricky, it requires a certain amount of preparation.

We still write $S = S^2 \subset \mathbf{R}^3$.

18.7.1. DEFINITION. *A hemisphere is a subset of S of the form $S \cap H$, where H is a half-space bounded by a plane through the origin. A convex spherical polygon (or just a polygon) is a subset P of S obtained as the intersection of finitely many hemispheres and such that $\overset{\circ}{P} \neq \emptyset$ or $P \cap (-P) = \emptyset$ (that is, P does not contain antipodal points).*

For example, a spherical triangle has a naturally associated polygon (18.6.3).

In order to talk about sides and vertices of P, we will use some results on convex polyhedra.

18.7.2. If H is a hemisphere of S, we *let $\hat{H}$ be the half-space of $\mathbf{R}^3$ spanned* by H. If $P = \bigcap_i H_i$ we set $\hat{P} = \bigcap_i \hat{H}_i$. Thus $\hat{P}$ is a convex polyhedron, in

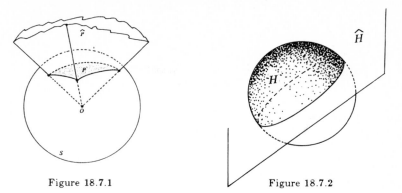

Figure 18.7.1   Figure 18.7.2

fact, a convex cone with vertex O. The condition $\overset{\circ}{P} \neq \emptyset$ implies $\overset{\circ}{\hat{P}} \neq \emptyset$, so we can apply 12.1.5 and 12.1.8 and consider the faces and edges of $\hat{P}$. Since $P \cap (-P) = \emptyset$, the edges are half-lines originating at O and the faces are sectors with origin O in planes passing through O.

18.7.3. DEFINITION. *The vertices of P are the intersections of S with the edges of $\hat{P}$. The sides of P are the intersections of S with the faces of $\hat{P}$. A polygon is called a triangle if it has three vertices.*

By 12.1.12, each side of P contains exactly two vertices and each vertex belongs to exactly two edges. Let the sequence of vertices of P be $(x_i)_{i=1,\ldots,n}$, where two successive vertices lie on the same edge; P is well-determined by this sequence. The number of vertices is equal to the number of sides. The angles and side lengths of a spherical polygon are defined in the same way as for a spherical triangle:

18.7.4. DEFINITION AND NOTATION. *The sides of a convex spherical polygon $P = (x_i)_{i=1,\ldots,n}$ are the real numbers $\overline{a_i} = \overline{x_i x_{i+1}}$, with $i = 1, \ldots, n$ (and the convention $x_{n+1} = x_1$). The angles (cf. 8.6.3) are given by $\alpha_i = \widehat{(x_i)_{x_{i-1}} (x_i)_{x_{i+1}}}$, with $i = 1, \ldots, n$ and $x_{n+1} = x_1$.*

Figure 18.7.4

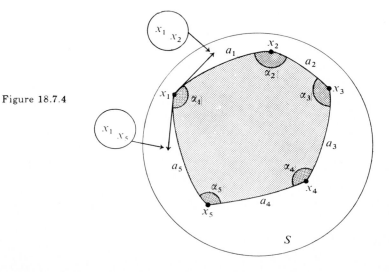

18.7.5. PROPOSITION. *Let x, y be two vertices of a polygon P that do not belong to the same side. Let γ be the arc of great circle joining x and y (by definition, $y \neq -x$, so γ is well-defined). Then γ is contained in P and $P = P' \cup P''$, where P' and P'' are two polygons having γ as a side and as their intersection.*

Proof. If x, y both lie in a hemisphere H, so does γ (since $y \neq -x$); this shows $\gamma \subset P$. The polygons P' and P'' are defined as $P \cap H'$ and $P \cap H''$, where H' and H'' are the two hemispheres determined by γ. ☐

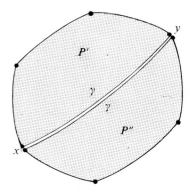

Figure 18.7.5

18.7.6. Let P and $\overline{P}$ be polygons with vertices $(x_i)_{i=1.....n}$ and $(\overline{x}_i)_{i=1.....n}$, respectively (here the labeling of the vertices is supposed to be simultaneous, that is, we're also fixing a bijection between the vertices of the two polygons). The sides and angles of P and $\overline{P}$ are denoted by $a_i, \overline{a}_i, \alpha_i, \overline{\alpha}_i$ $(i = 1, \ldots, n)$. Cauchy's lemma will be an easy consequence of the following

18.7.7. PROPOSITION. *If $a_i = \overline{a}_i$ for all $i = 1, \ldots, n-1$ and $\alpha_i \leq \overline{\alpha}_i$ for all $i = 2, \ldots, n-1$, we have $a_n \leq \overline{a}_n$. If, in addition, there is $i \in \{2, \ldots, n-1\}$ with $\alpha_i < \overline{\alpha}_i$, then $a_n < \overline{a}_n$.*

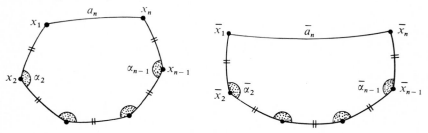

Figure 18.7.7

18.7.8. This result is quite intuitive: if we increase the angles of a polygon all of whose sides except one are fixed, the last side can only increase. Even for a quadrilateral in a Euclidean plane, however, the result is not evident. For a triangle it has already been proven (18.6.9):

18.7.9. LEMMA. *Proposition 18.7.7 holds for $n = 3$.* □

The induction step is easy in the following particular case:

18.7.10. LEMMA. *If 18.7.7 holds for $(n-1)$-sided polygons, it also holds for any pair $P, \overline{P}$ of n-sided polygons for which there exists $i \in \{2, \dots, n-1\}$ with $\alpha_i = \overline{\alpha}_i$.*

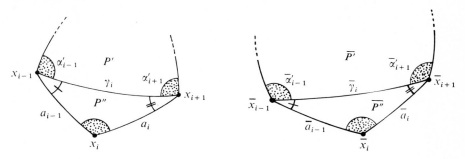

Figure 18.7.10

Proof. We apply 18.7.5 to the arcs of great circle γ_i and $\overline{\gamma}_i$ with endpoints x_{i-1}, x_{i+1} and $\overline{x}_{i-1}, \overline{x}_{i+1}$, respectively. We obtain two $(n-1)$-sided polygons $P', \overline{P}'$ and two triangles $P'', \overline{P}''$. By 18.6.13.10 these two triangles are equal; in particular, $\overline{x_{i-1}x_{i+1}} = \overline{\overline{x}_{i-1}\overline{x}_{i+1}}$ and the angles at x_{i-1} and $\overline{x}_{i-1}$ (resp. at x_{i+1} and $\overline{x}_{i+1}$) are equal. This implies that the angles $\alpha'_{i-1}, \alpha'_{i+1}, \overline{\alpha}'_{i-1}, \overline{\alpha}'_{i+1}$ of P' and $\overline{P}'$ at $x_{i-1}, x_{x+1}, \overline{x}_{i-1}, \overline{x}_{i+1}$ satisfy the inequalities $\alpha'_{i-1} \leq \overline{\alpha}'_{i-1}$ and $\alpha'_{i+1} \leq \overline{\alpha}'_{i+1}$. Thus P' and $\overline{P}'$ satisfy the assumptions of 18.7.7, and the conclusion follows. □

Now proposition 18.7.7 is proven by induction, but there is one pitfall.

18.7.11. PROOF OF 18.7.7. The idea is to increase the angle α_{n-1} of P at x_{n-1} until it gets equal to $\overline{\alpha}_{n-1}$, while maintaining the sides constant; one obtains a new polygon P', with vertices $x_1, \dots, x_{n-1}, x'_n$, sides $a_1, \dots, a_{n-1}, a'_n = \overline{x'_n x_1}$ and angles at $x_2, \dots, x_{n-2}, x_{n-1}$ equal to $\alpha_2, \dots, \alpha_{n-2}, \overline{\alpha}_{n-1}$. Unfortunately there is no reason to assume that P' is convex, as shown in figure 18.7.11. We shall therefore distinguish two cases.

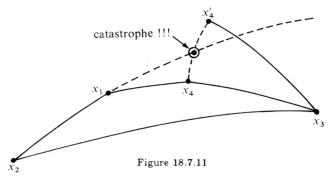

Figure 18.7.11

18.7.12. FIRST CASE: P' IS CONVEX. Thanks to 18.7.5, we can apply 18.7.9 to the two triangles x_1, x_{n-1}, x_n and x_1, x_{n-1}, x'_n, whence

$$a_n = \overline{x_1 x_n} \leq \overline{x_1 x'_n} \quad \text{and} \quad \overline{x_1 x_n} < \overline{x_1 x'_n} \quad \text{if} \quad \alpha_{n-1} < \overline{\alpha}_{n-1}.$$

We then apply the induction assumption and lemma 18.7.10 to the polygons P' and $\overline{P}$ which have the same angle at x_{n-1} and $\overline{x}_{n-1}$ by construction, and obtain

$$\overline{x_1 x'_n} \leq \overline{\overline{x}_1 \overline{x}_n} = \overline{a}_n.$$

This concludes the proof in the first case.

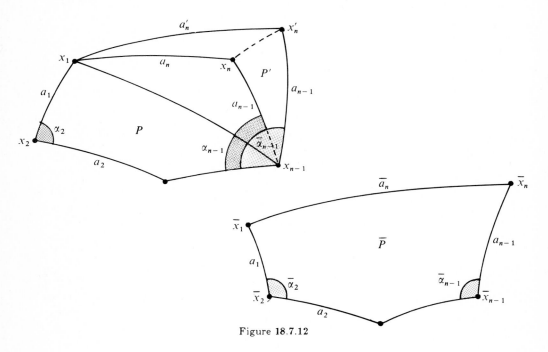

Figure 18.7.12

18.7.13. SECOND CASE: P' IS NOT CONVEX. In this case, as we increase α_{n-1} by rotating $x_{n-1}x_n$ around x_{n-1}, there exists a smallest value α'_{n-1} of the angle for which P' ceases to be convex. This value lies between α_{n-1} and $\overline{\alpha}_{n-1}$.

Let x'_n be the point thus obtained; by construction x'_n belongs to the great circle determined by x_2 and x_1. We have

$$\overline{x_1 x'_n} = \overline{x_2 x'_n} - \overline{x_1 x_2}.$$

Applying 18.6.10 and 18.7.5 we find

$$\overline{a}_n = \overline{\overline{x}_1 \overline{x}_n} \geq \overline{\overline{x}_2 \overline{x}_n} - \overline{\overline{x}_1 \overline{x}_2} = \overline{x_2 \overline{x}_n} - \overline{x_1 x_2}.$$

Now we apply the induction assumption and 18.7.5 again to the convex polygons $(x_2, x_3, \ldots, x_{n-1}, x_n)$ and $(\overline{x}_2, \overline{x}_3, \ldots, \overline{x}_{n-1}, \overline{x}_n)$ to get

$$\overline{x_2 \overline{x}_n} \geq \overline{x_2 x'_n}.$$

Finally, again thanks to 18.7.5, we apply 18.7.9 to the triangles (x_1, x_n, x_{n-1}) and (x_1, x'_n, x_{n-1}) to get

$$\overline{x_1 x'_n} > \overline{x_1 x_n}.$$

Comparing all the inequalities we conclude that

$$\overline{a}_n \geq \overline{\overline{x}_2 \overline{x}_n} - \overline{x_1 x_2} \geq \overline{x_2 x'_n} - \overline{x_1 x_2} \geq \overline{x_1 x'_n} > \overline{x_1 x_n} = a_n. \qquad \square$$

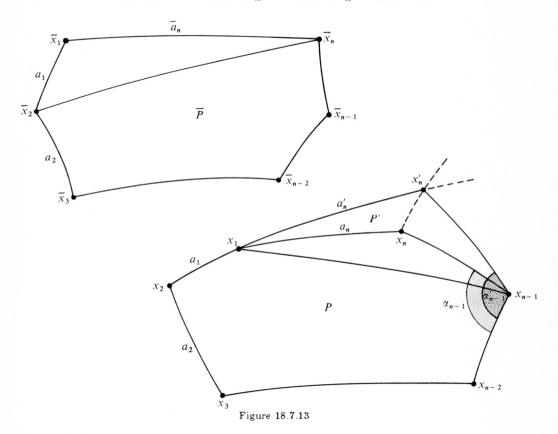

Figure 18.7.13

18.7.14. Consider again two polygons P and $\overline{P}$ as in 18.7.6, that is, having the same number of (simultaneously labeled) sides. The notation remains the same.

18.7.15. DEFINITION AND NOTATION. *For $i = 1, \ldots, n$, set*

$$\mathrm{sgn}(i) = \begin{cases} +1 & \text{if } \alpha_1 > \overline{\alpha}_i, \\ 0 & \text{if } \alpha_1 = \overline{\alpha}_i, \\ -1 & \text{if } \alpha_1 < \overline{\alpha}_i. \end{cases}$$

The number of sign flips of the pair $(P, \overline{P})$ *is the number of indices* i *such that* $\operatorname{sgn}(i) \operatorname{sgn}(i+1) = -1$ *(with the convention* $\operatorname{sgn}(n+1) = \operatorname{sgn}(1).)$

18.7.16. THEOREM (CAUCHY'S LEMMA). *If* $P, \overline{P}$ *are such that* $a_i = \overline{a}_i$ *for* $i = 1, \ldots, n$, *either* $\alpha_i = \overline{\alpha}_i$ *for all* i *or the number of strict sign flips is* ≥ 4.

Proof. Since there must be an even number of sign flips, we just have to prove this number is not two. If it were, there would exist non-consecutive indices $i, j \in \{1, \ldots n\}$ such that (say) $\alpha_k \geq \overline{\alpha}_k$ for $k \in \{i+1, \ldots, j-1\}$ and $\alpha_h \leq \overline{\alpha}_h$ for $h \in \{j+1, \ldots, n\} \cup \{1, \ldots, i-1\}$, the inequality being strict for at least one k and one h. Introduce the arcs of great circle $\gamma, \overline{\gamma}$ that connect x_i to x_j and $\overline{x}_i$ to $\overline{x}_j$. By 18.7.5 they divide P and $\overline{P}$ into polygons P', P'' and $\overline{P}', \overline{P}''$, respectively. Applying proposition 18.7.7 to $(P', \overline{P}')$ and $(P'', \overline{P}'')$ we reach the contradiction $\overline{x_i x_j} > \overline{\overline{x}_i \overline{x}_j}$ and $\overline{x_i x_j} < \overline{\overline{x}_i \overline{x}_j}$. □

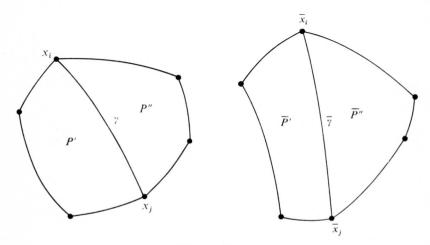

Figure 18.7.16

18.8. The three-sphere and Clifford parallelism

> In this section $d = 3$, that is, $S = S^3$.

18.8.1. INTRODUCTION. Identify $\mathbf{R}^4$ with $\mathbf{C}^2$ and recall the action of $\mathbf{R}$ on S defined in 1.2.9 by

$$\mathbf{R} \ni t \mapsto \left\{ (z, z') \mapsto (e^{it} z, e^{it} z') \right\} \in S_{S^3}$$

(see also 4.3.3.2 and 4.3.6.2). Orbits under this action are great circles, since $(e^{it} z, e^{it} z')$ belongs to the one-dimensional vector subspace of $\mathbf{C}^2$ spanned by (z, z'), which is also a real vector subspace. The action is also isometric and (by definition) transitive on each orbit. Thus, for two orbits C, C', we have

18.8.1.1 $\qquad d(m, C') = d(n, C')$ for any $m, n \in C$.

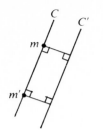 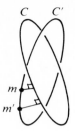

Figure 18.8.1

This property is reminiscent of two parallel lines D, D' in a Euclidean plane. It implies that the common distance in 18.8.1.1 is $d(C, C')$ and that if $m \in C$ and $m' \in C'$ satisfy $\overline{mm'} = d(C, C')$ any great circle containing m and m' intersects C and C' orthogonally (cf. 18.11.6). This leads to the following

18.8.1.2. Definition. *Two great circles C, C' in S are said to be Clifford parallel if $d(m, C')$ does not depend on $m \in C$. If C, C' are Clifford parallel we write $C \parallel C'$.*

The object of this section is the study of the relation $C \parallel C'$. We will see (18.8.2.5) that $\parallel$ is not an equivalence relation as for parallel lines in the plane, but it can be decomposed into two equivalence relations, called Clifford parallelism of first and second kind. The proof will combine a metrical study in 18.8.2 with a study of the orbits under certain **R**-actions on S^3 which are an intrinsic version of the action mentioned above. See also 19.1.4.

18.8.2. METRICAL STUDY.

18.8.2.1. For every great circle C, we denote by $\overline{C}$ the plane in $\mathbf{R}^4$ such that $C = \overline{C} \cap S$, and by $C^\perp$ the great circle such that $\overline{C^\perp} = (\overline{C})^\perp$ (observe that this definition only works for $d = 3$). Further, for every great circle C and every $\alpha \in [0, \pi]$ we set

18.8.2.2 $$C_\alpha = \{ n \in S \mid d(n, C) = \alpha \}.$$

We have $C_0 = C$, $C_{\pi/2} = C^\perp$, $C_{\pi - \alpha} = C_\alpha$ (because C contains antipodal points and $C_{\pi/2 - \alpha} = C_\alpha^\perp$); thus we can restrict ourselves to $\alpha \in]0, \pi/2[$. Now introduce orthonormal coordinates (x, y, z, t) whose first two basis vectors are in C and the other two in $C^\perp$, and set $u = x + iy$, $v = z + it$. Then we have, for $m = (x, y, z, t)$,

18.8.2.3 $$d(m, C) = \alpha \iff x^2 + y^2 = |u|^2 = \cos^2 \alpha.$$

In fact, the distance $\overline{(u, v)(w, 0)}$ is minimal when the scalar product $((u, v) \mid (w, 0))$ is maximal, and this happens for $w = u/|u|$, in which case

$$\cos \alpha = \left((u, v) \,\middle|\, \left(\frac{u}{|u|}, 0 \right) \right) = \frac{|u|^2}{|u|} = |u|.$$

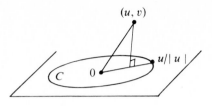

Figure 18.8.2.3

Notice that then $|v|^2 = z^2 + t^2 = \sin^2 \alpha$ because $|u|^2 + |v|^2 = 1$. In fact,

18.8.2.4. Lemma. *We have* $C_\alpha = S \cap Q_\alpha$, *where* Q_α *is the cone in* $\mathbf{R}^4$ *defined by the equation*

$$\sin^2 \alpha (x^2 + y^2) - \cos^2 \alpha (z^2 + t^2) = 0.$$

Proof. One direction has just been shown; conversely, if

$$\sin^2 \alpha \cdot |u|^2 = \cos^2 \alpha \cdot |v|^2$$

and $|u|^2 + |v|^2 = 1$ we have $|u|^2 = \cos^2 \alpha$, and 18.8.2.3 leads to the desired conclusion. $\square$

The key point now is that the form in 18.8.2.4 is neutral (cf. 13.1.4.3). Applying 13.7.11 we obtain:

18.8.2.5. Proposition. *For every* $\alpha \in \,]0, \pi/2[$, *every great circle* C *and every* $m \in C_\alpha$ *there exist exactly two Clifford parallels to* C *containing* m. $\square$

This shows that $\|$ is not an equivalence relation: the two Clifford parallels C', C'' to C cannot themselves be Clifford parallel because $d(C', C'') = 0$ and $C' \neq C''$.

18.8.3. GROUP ACTIONS ASSOCIATED WITH A GREAT CIRCLE.

18.8.3.1. Our aim is to define, for each great circle C, a group action on S^3 that generalizes the one mentioned in 18.8.1. Fix an orientation for $\overline{C}$, and orient $\overline{C}^\perp$ in such a way that the composite orientation in $\overline{C} \oplus \overline{C}^\perp$ is the canonical orientation of $\mathbf{R}^4$. Let $f \in \mathrm{Isom}^+(\overline{C}; \overline{C}^\perp)$ be an orientation-preserving isometry between the Euclidean spaces $\overline{C}$ and $\overline{C}^\perp$. For any $\rho \in O^+(\overline{C})$ and any $f \in \mathrm{Isom}^+(\overline{C}; \overline{C}^\perp)$ we have $f\rho f^{-1} \in O^+(\overline{C}^\perp)$, and $f\rho f^{-1}$ depends on ρ but not on f, since $O^+(\overline{C})$ is abelian. Finally, observe that, if we switch the orientation of $\overline{C}$, the orientation of $\overline{C}^\perp$ is also reversed.

All this shows that the following definition is intrinsic:

18.8.3.2. Definition. *For each great circle* C *in* S *we define the subgroups* G_C^+ *and* G_C^- *of* $O^+(\overline{C}) \times O^+(\overline{C}^\perp) \subset O^+(\mathbf{R}^4) = \mathrm{Is}^+(S)$ *as follows:*

$$G_C^+ = \big\{ (\rho, f\rho f^{-1}) \in \mathrm{Is}^+(S) \mid \rho \in O^+(\overline{C}) \text{ and } f \in \mathrm{Isom}^+(\overline{C}, \overline{C}^\perp) \big\},$$

$$G_C^- = \big\{ (\rho, f\rho f^{-1}) \in \mathrm{Is}^+(S) \mid \rho \in O^+(\overline{C}) \text{ and } f \in \mathrm{Isom}^-(\overline{C}, \overline{C}^\perp) \big\}.$$

If C and C' are great circles in S, we say that C' is a Clifford parallel of the first kind (resp. second kind) to C, and write $C \parallel^+ C'$ (resp. $C \parallel^- C'$) if C' is an orbit under G_C^+ (resp. G_C^-).

18.8.4. THEOREM.
i) *The relations $\parallel^+$ and $\parallel^-$ are equivalence relations.*
ii) *The condition $C \parallel C'$ is equivalent to $C \parallel^+ C'$ or $C \parallel^- C'$. In particular, for each $m \in S$, there exist two Clifford parallels to C through m, one of the first kind and one of the second (and $C' \neq C''$ if $m \notin C \cup C^\perp$).*
iii) *The two Clifford parallels C', C'' to C through $m \in S \setminus (C \cup C^\perp)$ are characterized by the following geometric conditions: let $m_0 \in C$ be such that $d(m_0, m) = d(m, C) = \alpha$, let $P = \langle m \cup C \rangle$ be the two-subsphere containing m and C, and D the great circle determined by m_0 and m. Then C' and C'' are orthogonal to D and make an angle α with P; in particular, the angle between C' and C'' is 2α. (See figure 18.8.4.1.)*

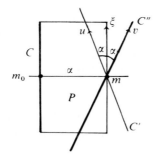

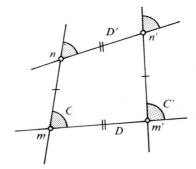

Figure 18.8.4.1   Figure 18.8.4.2

iv) *Let C, C', C'' be great circles such that $C' \parallel^+ C$, $C'' \parallel^- C$ and $d(C', C) = d(C'', C)$. Then $C' \cap C'' \neq \emptyset$.*
v) *(Clifford parallelograms.) Let C, C' be great circles such that $C' \parallel^+ C$, and $m, n \in C$, $m' \in C'$ points. Let D be a great circle containing m and m' and D' a great circle such that $n \in D'$ and $D' \parallel^- D$. Then $D' \cap C'$ consists of a single point n' and $\{m, n, m', n'\}$ is a Clifford parallelogram, in the sense that $\overline{mm'} = \overline{nn'}$, $\overline{mn} = \overline{m'n'}$, and the four angles at m, n, m', n' are equal. (See figure 18.8.4.2.)*
vi) *(Uniqueness.) Let the four great circles C, C', D, D' be such that $C \parallel C'$ and $D \parallel D'$. There exists $f \in \mathrm{Is}(S)$ such that $f(C) = D$ and $f(C') = D'$. If, in addition, $C \parallel^+ C'$ and $D \parallel^+ D'$, the map f can be chosen in $\mathrm{Is}^+(S)$.*

Proof. (i) The condition $C \parallel^+ C'$ is equivalent to $G_C^+ = G_{C'}^+$, as can be seen by observing that G_C^+ can be written in the form 18.8.1 in the appropriate coordinates. This shows that $\parallel^+$ is an equivalence relation, and the reasoning for $\parallel^-$ is analogous.

(ii) If $C \parallel^+ C'$ or $C \parallel^- C'$, we have $C \parallel C'$ by definition 18.8.1.2 and because C and C' are two orbits under a subgroup of $\mathrm{Is}(S)$. Conversely, given $m \in S \setminus (C \cup C^\perp)$, there exists one orbit of G_C^+ and one of G_C^- going through m, and they are both Clifford parallel to C, so the conclusion follows from 18.8.2.5.

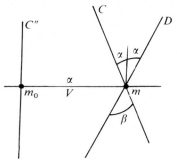

Figure 18.8.4.3

(iii) Orthogonality to D derives from the fact that $d(m_0, m) = d(C, C') = d(C, C'')$. As to the angles, we will explicitly calculate tangent vectors u, v to C', C'' at m and a vector ξ orthogonal to D at m and tangent to P (see figure 18.8.4.3). In the appropriate coordinate system, the orbits C' and C'' of a point $m = (z, z')$ are given by $t \mapsto (e^{it}z, e^{it}z')$ and $t \mapsto (e^{it}z, e^{-it}z')$; u and v can be taken as the derivatives of these maps with respect to t, namely, $u = (iz, iz')$ and $v = (iz, -iz')$. The vector $\xi = (u+v)/2 = (iz, 0)$ is certainly tangent to P and orthogonal to D. Then

$$(u \mid \xi) = (v \mid \xi) = (iz \mid iz) = |z|^2 = \cos^2 \alpha,$$

and (cf. 18.8.2.3)

$$\frac{(u \mid \xi)}{\|\xi\|} = \frac{\cos^2 \alpha}{\cos \alpha} = \cos \alpha.$$

(iv) Follows from 18.8.2.4 and 13.7.11.

(v) The idea is to find a great circle C'' such that the union $C \cup C' \cup D$ sits inside C''_α (cf. 18.8.2.2) for an appropriate α. Part (iii) dictates our choice. Take $\alpha = \beta/2$, where β is the acute angle between C and D at m, and draw the great circle V perpendicular at m to the two-sphere Q spanned by C and D. Mark on V a point m_0 such that $\overline{m_0 m} = \alpha = \beta/2$, and call C'' the great circle through m such that $C'' = C$ (cf. part (ii)). By construction and parts (iii), (i) and (ii) we have $C \subset C''_\alpha$, $C' \subset C''_\alpha$ and $D \subset C''_\alpha$. Then $D' \subset C''_\alpha$ as well, and C' intersects D' in a single point n' by 13.7.11. By the definition of $\parallel^-$, there exists $g \in G_{C''}^-$ such that $g(m) = m'$; thus $g(D) = D$ and $g(D') = D'$, and also $g(C) = C'$ because $g(C)$ is a great circle in C''_α containing m'. In particular, $g(n) = n'$, and $\overline{mn} = \overline{m'n'}$ because $g \in \mathrm{Is}(S)$. Similarly, we see that $\overline{mm'} = \overline{nn'}$. To see that the four angles are equal, use the same g to show that the angles at m, m' and n, n', respectively, are equal, or use part (iii) to show that all the angles are equal to $2\alpha = \beta$.

(vi) We can assume $C = D$, by 8.2.7, then apply 18.8.2.5. If $C' = D'$ we're done; otherwise part (iii) shows that the reflection through the two-subsphere containing $C = D$ and $C \cap D$ takes C' into D'. □

18.8.5. NOTES. We know that the quotient of $S = S^3$ under the action 18.8.1 can be identified with $P^1(\mathbf{C}) \cong S^2$; see 18.1.3.6 and 4.3.6.2.

Part of (iii) can be rephrased by saying that, given C and C' with $C \parallel C'$, any two-subsphere P containing C intersects C' under the same angle $\big($equal to $2d(C, C')\big)$.

Part (v) is analogous to the properties of a plane parallelogram. For further analogies, see 18.11.17.

18.8.6. MORE ON C_α. By 18.8.2.2, the subset C_α, for any C and α, is invariant not only under G_C^+ and G_C^- but even under $O^+(\overline{C}) \times O^+(\overline{C}^\perp)$. Invariance under $O^+(\overline{C}) \times \mathrm{Id}_{\overline{C}^\perp}$ means that C_α is a surface of revolution with axis $C^\perp$; similarly, C_α is a surface of revolution with axis C. Thus C_α is a surface of revolution in two ways.

Topologically speaking, $C_\alpha \cong \big\{ (u, v) \in \mathbf{C}^2 \mid |u| = \cos \alpha \text{ and } |v| = \sin \alpha \big\}$ (cf. 18.8.2.3), so it is homeomorphic to the torus $S^1 \times S^1$.

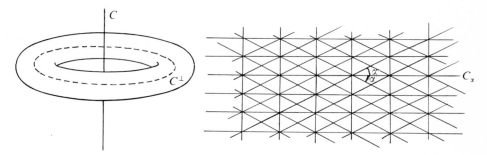

Figure 18.8.6

The orbits of C_α under $O^+(\overline{C}) \times \mathrm{Id}_{\overline{C}^\perp}$ are small circles in S, the intersections of C_α and two-subspheres containing $C^\perp$; the same holds for orbits under $\mathrm{Id}_{\overline{C}} \times O^+(\overline{C}^\perp)$. Each of these orbits is a *meridian* of C_α (though they could also be thought of as parallels of latitude), and the two families of meridians intersect at right angles. The Clifford parallels to C (or $C^\perp$) contained in C_α also form two families which, according to 18.8.4 (iii), intersect the meridians along their bisectors, forming a constant angle α with both families. Notice, though, that two Clifford parallels (of same kind) on C_α are linked, whereas two meridians of same kind are not. And Clifford parallels are not at all parallels of latitude of C_α.

18.8.7. DECOMPOSITION INTO SOLID TORI. By fixing some α, say $\pi/4$, and some great circle C, we obtain a decomposition of the sphere into two solid tori, namely, $\Xi = \bigcup_{\alpha \in [0,\pi/4]} C_\alpha$ and $\Xi' = \bigcup_{\alpha \in [\pi/4,\pi/2]} C_\alpha$. The two tori are glued to one another along their common frontier $C_{\pi/4}$. This way of looking at S^3 is often useful, especially in algebraic topology; it's in this way that one gets the famous Reeb foliation, for instance ([RE, 19 and 25]).

18.8.8. USE OF QUATERNIONS. Theorem 18.8.4 can be partly proved by using quaternions, and one can even obtain some sharper results (see, for example, [DE2, 172–173]). But this approach seems less intrinsic and more artificial than the one we've followed.

18.9. Villarceau circles and parataxy

18.9.1. Here we make good the promise made in 10.12, to prove geometrically the phenomena described in 10.12.1, 10.2.2 and 10.2.3. All we have to do is consider the stereographic projection $f : S^3 \to \mathbf{R}^3$ from the north pole $n \in S^3$, and fix a great circle C going through n. For a fixed coordinate system in $\mathbf{R}^3$ we can assume that $f(C \setminus n)$ is the z-axis and $f(C^\perp)$ is the unit circle on the xy-plane. We will make essential use of the properties of inversion, cf. 10.8 and 18.1.4.3.

18.9.2. What are the images $f(C_\alpha)$, for $\alpha \in]0, \pi/2[$? Notice that $f(C_\alpha)$ is a surface of revolution around the z-axis, so one meridian of $f(C_\alpha)$ is enough to determine the whole curve. But such a meridian is the image of a meridian of C_α in the sense of 18.8.6, which is a small circle in S; thus a meridian of $f(C_\alpha)$ is a circle in $\mathbf{R}^3$ (cf. 18.1.4.3). We conclude that $f(C_\alpha)$ is a torus of revolution in $\mathbf{R}^3$. In fact, up to a similarity, every torus in $\mathbf{R}^3$ is of the form $f(C_\alpha)$, for the appropriate value of α.

18.9.3. Now 18.8.6 shows that the torus $\Theta = f(C_\alpha)$ contains four families of circles: the usual meridians and parallels of latitude (no need to use 18.8.6 to know that they intersect at right angles!), plus the images under f of the two kinds of Clifford parallels to C contained in C_α. These two last families are the Villarceau circles mentioned in 10.12.1; they satisfy the intersection properties stated in 10.12.2 as a consequence of 18.8.4 (iv). Theorem 18.8.6 also shows that they are helices, making an angle α with each meridian, since the stereographic projection preserves angles. The parataxy property stated at the end of 10.12.3 is deduced from 18.8.5.

18.9.4. NOTES

18.9.4.1. In 20.5.4 and 20.7 we will round off the subject with an algebraic proof of the same results, and we will even encounter connected surfaces, analogous to tori and of degree four, but having six families of circles.

18.9.4.2. As α ranges from 0 to $\pi/2$ we get a nice picture of S^3 which we can elaborate as follows: Consider on each C_α parallels of the first kind only. Starting from $C = C_0$ and going all the way to $C_{\pi/2} = C^\perp$, we fill S^3 with

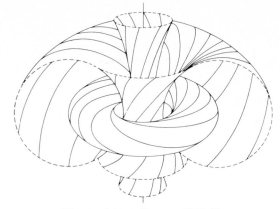

Figure 18.9.4.2 (Source: [PEN])

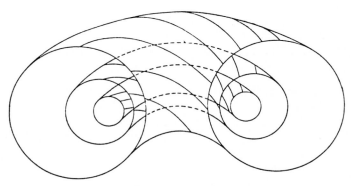

Figure 18.9.4.3

pairwise disjoint circles (the parallels of first kind to C). Needless to say, this partition of S^3 into circles is exactly the Hopf fibration introduced in 4.3.7.

The two solid tori Ξ and Ξ' (cf. 18.8.7) are glued along their Villarceau circles (in $\mathbf{R}^3$), and as we pass from one to the other the meridians of Ξ become parallels of latitude of Ξ'.

18.9.4.3. Notice that, in a fixed plane containing the z-axis, the meridians of $f(C_\alpha)$ form a pencil of circles whose limit points are the intersections of the plane with $f(C^\perp)$.

18.9.4.4. The reader may be wondering if there exist generalizations of Clifford parallelism. He should consult [WO].

18.10. The Möbius group

18.10.1. THE SPHERE AS A PROJECTIVE QUADRIC. We have seen in 14.3.3 that the image $C(1, n)$ of the quadric α in $P''(\mathbf{R})$ having $q = -\sum_{i=1}^{n} x_i^2 +$

x_{n+1}^2 for equation is a topological space homeomorphic to the sphere S^{n-1}. Here we use a slight modification of the identification given in the proof of 14.3.3 and fix from now on a bijection between the sphere S^d and the image of a projective quadric in $P^{d+1}(\mathbf{R})$.

18.10.1.1. Notation. Let $q = -\sum_{i=1}^{d+1} x_i^2 + x_{d+2}^2$ be a quadratic form in $\mathbf{R}^{d+2}$. We identify $\mathbf{R}^{d+2}$ with $\mathbf{R}^{d+1} \times \mathbf{R}$ by writing $\xi = (z, t)$, where $\xi \in \mathbf{R}^{d+2}$, $z \in \mathbf{R}^{d+1}$ and $t \in \mathbf{R}$. In particular, $q(\xi) = -\|z\|^2 + t^2$. We define a bijection Σ from the image $\mathrm{im}(\alpha)$ of the quadric α in $P^{d+1}(\mathbf{R})$ having q for equation into S^d as follows:

18.10.1.2 $\Sigma : \mathrm{im}(\alpha) \ni p(z, t) \mapsto z/t \in S^d.$

18.10.1.3. This identification is excellent, as shown by 18.10.1.4, which is a first geometric realization of the group $\mathrm{PO}(\alpha)$ of α (cf. 14.7); for other realizations see chapters 19 and 20. For a geometric picture of Σ, it is often more convenient to identify S^d with the unit sphere in the hyperplane $H = \{(z, t) \mid t = 1\}$ of $\mathbf{R}^{d+2}$ by the map $z \mapsto (z, 1)$; then Σ associates to a line in $\mathbf{R}^{d+2}$ contained in the isotropic cone of q the point where this line pierces H (we know this trick from 5.0, and shall encounter it again in 19.2).

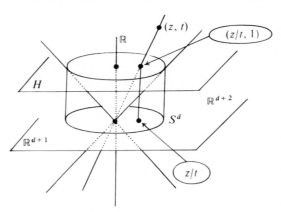

Figure 18.10.1.3

18.10.1.4. Definition. *An inversion of S^d is the restriction to S^d of an inversion of $\mathbf{R}^{d+1}$ that leaves S^d globally invariant, or of a reflection through a vector hyperplane of $\mathbf{R}^{d+1}$. The Möbius group of S^d, denoted by $\mathrm{M\ddot{o}b}(d)$, is the group generated by all inversions of S^d.*

18.10.1.5. Proposition. *Under the identification Σ, the Möbius group of S^d is the restriction of $\mathrm{PO}(\alpha)$ to $\mathrm{im}(\alpha)$. In symbols,*

$$\mathrm{M\ddot{o}b}(d) = \{ \Sigma \circ (f|_{\mathrm{im}(\alpha)}) \circ \Sigma^{-1} \mid f \in \mathrm{PO}(\alpha) \}.$$

Proof. By 13.7.12, it suffices to show that, if $f \in \mathrm{PO}(\alpha)$ comes from a reflection through a hyperplane associated with q, the composition $\Sigma \circ (f|_{\mathrm{im}(\alpha)}) \circ \Sigma^{-1}$ is an inversion of S^d. But this is just 14.7.4, since an inversion g of S^d

is characterized by the fact that the line $\langle s, g(s) \rangle$ goes through a fixed point of $\mathbf{R}^{d+1}$ or is parallel to a fixed direction. $\square$

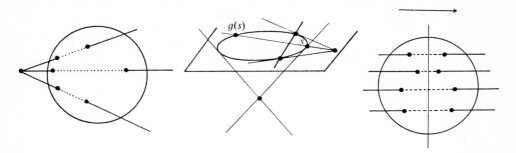

Figure 18.10.1.5

18.10.1.6. Notes. Restriction to $\operatorname{im}(\alpha)$ defines a map $\mathrm{PO}(\alpha) \to \mathrm{M\ddot{o}b}(\alpha)$ which is a group isomorphism. To see this, observe that elements of $\mathrm{PO}(\alpha)$ come from linear maps of $\mathbf{R}^{d+2}$, and that the cone $q^{-1}(0)$ spans $\mathbf{R}^{d+2}$.

18.10.2. THE MÖBIUS GROUP AND STEREOGRAPHIC PROJECTION.

18.10.2.1. Let n be the north pole of S^d and $g : S^d \setminus n \to \mathbf{R}^d$ the stereographic projection. Let f be an inversion of S^d and $n' = f(n)$ the image of n under f. If $n' = n$, the map $g \circ f \circ g^{-1}$ (where f is, in fact, restricted to $S^d \setminus n$) is a hyperplane reflection of $\mathbf{R}^d$; if $n' \neq n$, the map $g \circ f \circ g^{-1}$ (where f is restricted to $S^d \setminus \{n, n'\}$) is an inversion of $\mathbf{R}^d$ with pole $g(n')$. The proof is left to the reader—it's a good way to get familiar with inversions.

18.10.2.2. Consequence. For every similarity $h \in \mathrm{Sim}(\mathbf{R}^d)$ the map $f : S^d \to S^d$ defined by $f(n) = n$ and $f = g^{-1} \circ h \circ g$ on $S^d \setminus n$ belongs to $\mathrm{M\ddot{o}b}(d)$. In fact, the remarks above show that the extension of $g^{-1} \circ h \circ g$ to S^d is in $\mathrm{M\ddot{o}b}(d)$ if h is an inversion or a hyperplane reflection; but, by 9.3.3, 9.5.2 and 10.8.1.2, such maps generate $\mathrm{Sim}(\mathbf{R}^d)$. To show that $g^{-1} \circ h \circ g$ can be extended to S^d, the easiest thing is to add to $\mathbf{R}^d$ a point at infinity and set $g(n) = \infty$, etc. We will return to this point in section 20.6.

18.10.2.3. Example. In particular, we find in $\mathrm{M\ddot{o}b}(d)$ maps $f_\lambda = g^{-1} \circ H_{0,\lambda} \circ g$ associated with vector homotheties of $\mathbf{R}^d$. If $\lambda > 1$, the north pole acts as a *sink* and the south pole as a *source* for f_λ, that is, the iterates f_λ^n ($n \in \mathbf{N}$) make every point in $S^d \setminus s$ converge to n—only the south pole escapes the attraction of n.

18.10.2.4. The two-dimensional case. For $d = 2$ the group $\mathrm{M\ddot{o}b}(2)$ is made up of exactly those maps which can be written in one of the forms below (where the identification given in 4.3.8 and 18.1.4.5 is in effect):

$$z \longmapsto \frac{az + b}{cz + d} \qquad \text{or} \qquad z \longmapsto \frac{a\bar{z} + b}{c\bar{z} + d},$$

for $ad - bc \neq 0$. Maps of the second form are sometimes called *antihomographies*.

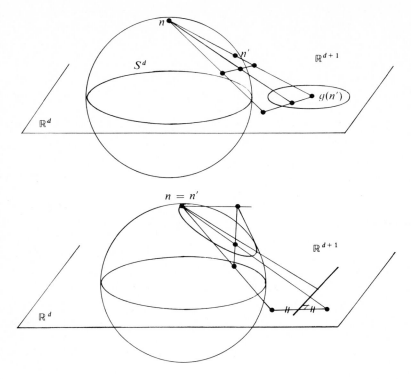

Figure 18.10.2

18.10.3. CONFORMAL MAPS. We have studied in 9.5.4 the notion of conformal maps for open subsets of Euclidean spaces. In the more general definition below we resort to the language of differential geometry (cf. [B–G, chapter 2]), but a more elementary exposition for the case of spheres, the only one we will study, can be found in 18.11.22.

18.10.3.1. Definition. *Let E and F be Euclidean affine spaces and $M \subset E$, $N \subset F$ differentiable submanifolds of class C^1 of E, F having the same dimension. A map $f : M \to N$ of class C^1 is called conformal if, for every $x \in M$, the derivative $f'(x) : T_x M \to T_{f(x)} N$ is a similarity between the tangent spaces $T_x M \subset E$ and $T_{f(x)} N \subset F$ with the Euclidean structures induced by E and F. The set of conformal maps from M to N is denoted by $\mathrm{Conf}(M; N)$; we also write $\mathrm{Conf}(M) = \mathrm{Conf}(M; M)$. If M and N are oriented, we denote by $\mathrm{Conf}^+(M; N)$ the subset of $\mathrm{Conf}(M; N)$ consisting of maps f such that $f'(x)$ preserves orientation for every $x \in M$.*

18.10.3.2. Examples. The composition of two conformal maps is conformal. In particular, $\mathrm{Conf}(M)$ and $\mathrm{Conf}^+(M)$ have a natural group structure.

The Mercator projection (18.1.8.2) is conformal.

The restriction of a conformal map (in particular, an inversion or composition of inversions) to a differentiable submanifold is conformal. For example, stereographic projections are conformal, and so are inversions of a sphere (cf. 18.10.1.4).

18.10.3.3. Notes. The imaginative reader will have realized that one can define the notion of a conformal map $f : M \to N$ whenever the tangent spaces to M and N have a Euclidean structure, that is, whenever M and N are Riemannian manifolds. For general theorems on conformal maps between Riemannian manifolds, see 18.10.9.

18.10.4. THEOREM. *For every $d \geq 2$ we have $\mathrm{Möb}(d) = \mathrm{Conf}(S^d)$. Set $\mathrm{Möb}^{\pm}(d) = \mathrm{Conf}^{\pm}(S^d)$. For any little sphere σ in S^d and any $f \in \mathrm{Möb}(d)$ the image $f(\sigma)$ is a little sphere in S^d; conversely, any bijection of S^d that takes spheres into spheres is an element of $\mathrm{Möb}(d)$. Every $f \in \mathrm{Möb}(d)$ is the product of at most $d + 2$ inversions of S^d. The group $\mathrm{Möb}(d)$ is naturally isomorphic to the group $\mathrm{PO}(\alpha)$ of the quadric having $q = -\sum_{i=1}^{d+1} x_i^2 + x_{d+2}^2$ for equation; in particular, $\mathrm{Möb}(d)$ is a non-compact Lie group of dimension $(d + 1)(d + 2)/2$. The follow isomorphisms hold: $\mathrm{Möb}(1) \cong \mathrm{GP}(1; \mathbf{R})$; $\mathrm{Möb}^{+}(2) \cong \mathrm{GP}(1; \mathbf{C}) \cong$ the group of automorphisms of the Riemann sphere.*

Proof. By 10.8.5.2 and 18.10.3.2, we have $\mathrm{Möb}(d) \subset \mathrm{Conf}(S^d)$. To prove the converse, we may assume that $f \in \mathrm{Conf}(S^d)$ leaves the north pole n fixed, since $\mathrm{Möb}(d) \subset \mathrm{Conf}(S^d)$ and $\mathrm{Möb}(d)$ acts transitively on S^d. Since $f(n) = n$, we can apply the stereographic projection g from n to deduce that

$$g \circ f \circ g^{-1} \in \mathrm{Conf}(\mathbf{R}^d),$$

since g and f are conformal (cf. 18.10.3.2); by the Liouville theorem (9.5.4.6), we have $g \circ f \circ g^{-1} \in \mathrm{Sim}(\mathbf{R}^d)$, so $f \in \mathrm{Möb}(d)$ by 18.10.2.2 (observe that we are taking for granted a delicate result: if f is conformal, f is of class C^4; cf. 9.5.3.4).

The properties of inversions show that f takes little spheres into little spheres (cf. 10.8.2). To prove the converse we assume again that $f(n) = n$; then $g \circ f \circ g^{-1}$ transforms lines in $\mathbf{R}^d$ into lines, since lines in $\mathbf{R}^d$ are the images under g of little circles containing n. The result follows from theorem 2.6.5, the fundamental theorem of affine geometry (compare with 9.5.3.4).

The remaining statements follow from 18.10.1.5, 13.7.12 and 16.3.9. $\square$

18.10.5. NOTES. The difference between the dimensions of $\mathrm{Möb}(d) = \mathrm{Conf}(S^d)$ and $\mathrm{Is}(S^d)$ is equal to

$$\frac{(d + 1)(d + 2)}{2} - \frac{d(d + 1)}{2} = d + 1;$$

it corresponds to the maps f_λ associated in 18.10.2.3 with points of S^d and constants $\lambda \in \mathbf{R}^{*}$ (d dimensions for the points, 1 for λ) which must be added to $\mathrm{Is}(S^d)$ to generate $\mathrm{Möb}(d)$.

For $d = 1$ the dimension of $\mathrm{Möb}(d)$ is three, and for $d = 2$ it is six; this is not surprising in view of the next result:

18.10.6. PROPOSITION. *The group* Möb(1) (*resp.* Möb(2)) *acts simply transitively on triples of distinct points of S^1 (resp. S^2). For every $d \geq 2$ the group* Möb(d) *acts transitively on triples of distinct points of S^d.*

Proof. The cases $d = 1$ and $d = 2$ follow from 4.5.10 via 16.3.9. For $d \geq 3$ it suffices to observe that three points span a two-subsphere and that Möb(d) contains Is$^+(S^d)$, which acts transitively on two-subspheres, by 8.2.7.   □

18.10.7. THE MÖBIUS INVARIANT. If we compare 18.10.6 with 6.1, a question springs to mind: Does the same result hold for four points (say for $d \geq 3$) and, if not, is there an invariant of quadruples of points that classifies their orbits under the Möbius group? The answer is easy: for $a, b, c, d \in S^d$, set

18.10.8
$$\mu(a, b, c, d) = \frac{ca}{cb} \bigg/ \frac{da}{db},$$

where the distances are Euclidean distances in $\mathbf{R}^{d+1}$. Then μ is invariant under Möb(d), that is, $\mu(f(a), f(b), f(c), f(d)) = \mu(a, b, c, d)$ for any four distinct points $a, b, c, d \in S^d$ and $f \in$ Möb(d). By 18.10.1.5, it is enough to prove that inversions of S^d leave μ invariant. For reflections this is trivial; for inversions it follows from a direct calculation using 10.8.1.3. But μ, called the *Möbius invariant*, is not enough to classify the orbits, as can be seen from figure 18.10.8. See also 19.6.10.

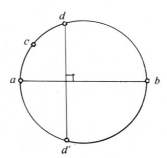

Figure 18.10.8

18.10.9. NOTES. The conformal group of the sphere is strictly larger than its isometry group. This property, although perhaps not surprising *a priori*, is not shared by any other compact Riemannian manifold, as has been recently proved ([LF2] or [OA]). See also 19.8.6.

   The Möbius group of S^d has been fundamental in the proof of several recent results, including Mostow's theorem on the rigidity of certain negatively curves spaces ([MW1] and [MW2]) and Hersch's theorem on the smallest eigenvalue of the laplacian on a Riemannian sphere S^2 ([HS]). See also [E-S, 130–131].

18.11. Exercises

* **18.11.1.** Explain the function of the lever system at the top of the spherometer shown in figure 18.1.1.2.

18.11.2. Show that a chart of the sphere that is both conformal and area-preserving is isometric.

* **18.11.3.** Show that the image of a loxodrome under the stereographic projection centered at the north pole is a logarithmic spiral.

18.11.4. Study the construction of a little circle $C \subset S^2$ satisfying each of the following conditions: C contains three given points; C contains two points and is tangent to a given circle; C contains one point and is tangent to two circles; C is tangent to three circles.

18.11.5. Study the notions of perpendicular bisectors, altitudes, medians and bisectors for spherical triangles. Are each of these concurrent?

18.11.6. Study 9.2.2 on the sphere.

18.11.7. Find an analog for 9.7.6 for a family $(x_i)_{i=1.....k}$ of points in S^d.

18.11.8. Study the "dubious" cases of equality of spherical triangles, for example, when two sides and the angle opposite one of them are equal.

* **18.11.9.** Generalize formulas 18.6.13 for the intrinsic metric of a sphere of radius R. Find out what the formulas become when R approaches infinity.

18.11.10. LEXELL'S THEOREM. Given two points $x, y \in S^2$, study the set of points $z \in S^2$ such that the angles α, β, γ of $T = \langle x, y, z \rangle$ satisfy $\alpha + \beta - \gamma = $ constant (compare with 10.9.4). Deduce the set of points z such that area of T is a constant. For generalizations, see [DX, 227].

18.11.11. Compile a complete formulary for special kinds of spherical triangles (right, isosceles, equilateral, rectilateral).

18.11.12. Study quadrilaterals in S^2 having three right angles. Find the remaining angle as a function of the two sides that sit between right angles. Compare with 19.8.7.

* **18.11.13.** THE TRIANGLE INEQUALITY FOR THE SPHERE. Prove the triangle inequality on the sphere using the Gram determinants of three points (cf. 8.11.5).

* **18.11.14.** RELATIONS BETWEEN THE DISTANCES OF $d+2$ POINTS. Show that if $(x_i)_{i=1.....d+2}$ are $d+2$ points in S^d, their distances $\overline{x_i x_j}$ always satisfy the relation

$$\det\big(\cos(\overline{x_i x_j})\big) = 0.$$

18.11.15. SPHERICAL CONICS. Let $q \in Q(\mathbf{R}^3)$ be a non-degenerate quadric and $C = q^{-1}(0)$ its isotropic cone; assume $C \neq \emptyset$. Show that there exist one or two directions of affine planes intersecting C in circles.

Show that there exists a non-degenerate quadric $q^* \in Q(\mathbf{R}^3)$ such that $x^\perp$ is a tangent plane to C for any $x \in C^* \setminus 0$, where $C^* = (q^*)^{-1}(0)$ is the *polar cone* of C.

Suppose that there are two distinct directions of affine planes intersecting C in circles (cf. 15.7.14). Show that the same holds for C^*. Let D, D' be the lines of $\mathbf{R}^3$ perpendicular to the circular sections of C^*, and set $D \cap S = \{f, g\}$, $D' \cap S = \{f', g'\}$, where $S = S^2$. Show that $C \cap S$ is equal to either $\{m \in S \mid \overline{mf} + \overline{mf'} = \text{constant}\}$ or $\{m \in S \mid \overline{mf} + \overline{mg'} = \text{constant}\}$.

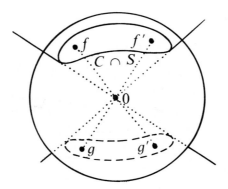

Figure 18.11.15

Conversely, show that, given two distinct points f, f' in S, the set $\{m \in S \mid \overline{mf} + \overline{mf'} = a\}$ is of the form $C \cap S$ for some $a \in \mathbf{R}_+$. The intersection $C \cap S$ is called a *spherical conic*, and f, f' (as well as g, g') are called its *foci*. For more results on spherical conics, see [DX, 230–231].

* **18.11.16.** HOOKE JOINTS, HOMOKINETIC JOINTS. Consider a Hooke joint (figure 18.11.16.2) whose axes make an angle θ. The ratio between the instant angular velocities of the two shafts is a function of the angle between the plane of either fork and the plane containing the axes of the shaft; find the worst possible value for this ratio. To do this, take two great circles C, D in S^2 making an angle θ, and two moving points $m(t), n(t)$ on the circles, so that $\overline{m(t)n(t)} = \pi/2$. Find the value of the worst possible ratio when $\theta = \pi/3, \pi/4, \pi/6$.

Show that if two shafts A and A' are joined by Hooke joints to a third shaft B whose forks are in the same plane, in such a way that A, B, A' are in the same plane and the angles of B with A and A' are the same (*homokinetic joint*), then A and A' always have the same angular velocity.

18.11.17. FLAT TORI AND CLIFFORD PARALLELS. Let Λ be a lattice in $\mathbf{R}^2$ (cf. 9.14.29) and $\Theta_\Lambda = \mathbf{R}^2/\Lambda$ the quotient of $\mathbf{R}^2$ by the subgroup Λ, that is,

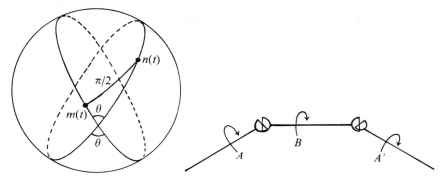

Figure 18.11.16.1

Figure 18.11.16.2

HOOKE JOINT

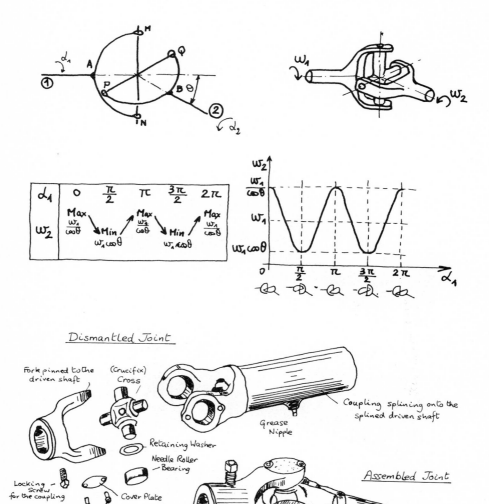

Figure 18.11.16.3

SPICER-GLAENZER JOINT

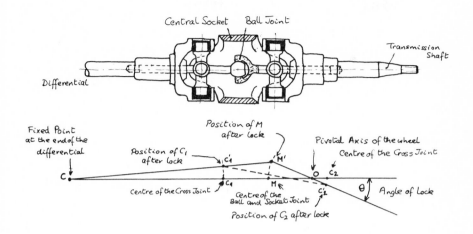

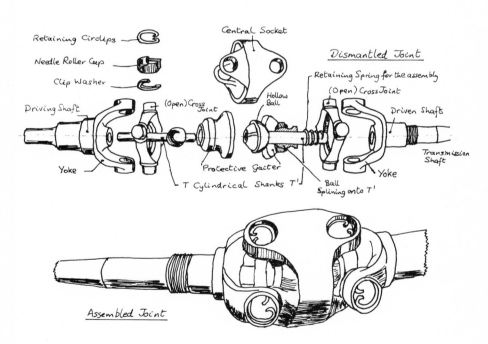

Figure 18.11.16.4

by the equivalence relation $x \sim y$ if and only if $m - n \in \Lambda$. Let $p : \mathbf{R}^2 \to \Theta_\Lambda$ be the canonical projection. For $u, v \in \Theta_\Lambda$, set

$$d(u, v) = d\big(p^{-1}(u), p^{-1}(v)\big) = \inf \big\{ \, d(x, y) \mid p(x) = u, p(y) = v \, \big\}.$$

Show that d is a metric on Θ_Λ, that it defines the same topology as the quotient topology and that it makes Θ_Λ locally isometric to $\mathbf{R}^2$. Study shortest paths in Θ_Λ according to the nature of Λ (cf. 9.14.29). The spaces Θ_Λ are called *flat tori*.

Show that the surfaces C_α ($\alpha \in \,]0, \pi/2[$) of 18.8.6, endowed with the metric induced from S^3, are isometric to flat tori. For a given value of α, find an explicit lattice Λ such that C_α is isometric to Θ_Λ.

18.11.18. Given a great circle C of S^3, left Σ be the set of parallels of the first kind to C. For $C', C'' \in \Sigma$, set $\delta(C', C'') = 2d(C', C'')$. Show that δ makes Σ into a metric space isometric to S^2 (cf. 4.3.6.2).

18.11.19. DUPIN CYCLIDS. Let $\Sigma, \Sigma', \Sigma''$ be spheres in $\mathbf{R}^3$; show that, for certain configurations of these spheres, the set of spheres tangent to $\Sigma, \Sigma', \Sigma''$ has for envelope the surface obtained from a torus of revolution by inversion relative to an appropriate point. Deduce several properties of such surfaces, which we will encounter again in 20.7 and which are called *Dupin cyclids*.

18.11.20. Show that every sphere bitangent to a torus intersects it in two Villarceau circles.

18.11.21. What are the orbits of the action of $\text{Möb}^+(S^1)$ on triples of distinct points of S^1?

18.11.22. For $S = S^d$, let $f : S \to S$ be a map of class C^1, and $\phi : S^n \setminus \mathbf{R}^d$ (resp. $\psi : S^s \setminus \mathbf{R}^d$) the stereographic projection from the north (resp. south) pole. Show that f is conformal if and only if the four maps

$$\phi \circ f \circ \phi^{-1}, \quad \psi \circ f \circ \phi^{-1}, \quad \phi \circ f \circ \psi^{-1}, \quad \psi \circ f \circ \psi^{-1}$$

are conformal in the sense of 9.5.4.2 (on their domain of definition).

18.11.23. Study the spherical analogues of the questions discussed in 9.8.1 and 9.8.5.

18.11.24. Make a critical study of Cosinus's spherometer (figure 18.1.1.3).

18.11.25. Calculate, as a function of a, the angle α and the side b of the spherical quadrilaterals shown in figures 18.11.25.1 and 18.11.25.2.

18.11.26. Show that, if γ is a curve of $\mathbf{R}^{n+1}$ with image in S^n, the length of γ in the Euclidean metric is equal to its length in the intrinsic spherical metric.

18.11.27. Show that the chart obtained by projecting from the sphere onto a cylinder circumscribed around the equator, along lines orthogonal to the north-south axis, is area-preserving. (This projection is sometimes wrongly referred to as Lambert projection.)

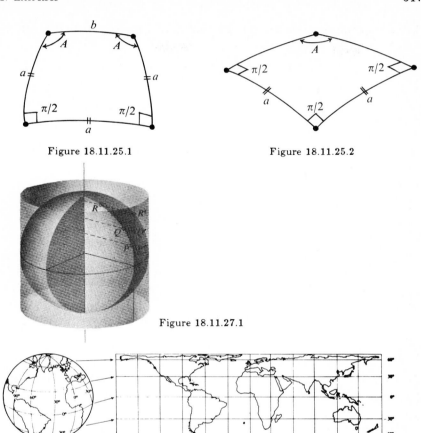

Figure 18.11.25.1   Figure 18.11.25.2

Figure 18.11.27.1

Figure 18.11.27.2

18.11.28. Prove the formulas in 18.1.8.4.

18.11.29. Determine the function Ξ in 18.1.8.5.

18.11.30. Show that every great circle of S^3 projects to a (little or great) circle of S^2 under the Hopf fibration (4.3.6.3).

Chapter 19
Elliptic and hyperbolic geometry

Elliptic and hyperbolic geometry are important from the historical and contemporary points of view. Historically, they provided counterexamples for Euclidean geometry: in elliptic geometry, there are no parallels to a line through a point outside the line, and the sum of the angles of a triangle is greater than π; in hyperbolic geometry, there are infinitely many parallels, and the sum of the angles is always less than π. But these geometries share with the traditional, Euclidean one its rich structure, with lots of symmetries and excellent lines. Thus the three form a system having a certain unity. For a history of non-Euclidean geometries, see [CR3, chapter I].

From the contemporary point of view, hyperbolic geometry has been the basis of a great number of important results in mathematics, including many recent ones. We list some references in 19.7.3.

In section 19.1 we study elliptic geometry. This study is much simplified by the introduction of quotient spaces; but nineteenth-century mathematicians did not yet possess that tool, and this explains why elliptic geometry trailed hyperbolic geometry for some fifty years. Clifford parallelism can be easily expressed in the context of elliptic geometry.

All the rest of the chapter is devoted to hyperbolic geometry, which, even in two dimensions, exhibits a certain difficulty. As we mention in section 19.7, it is possible to give an elementary exposition of hyperbolic geometry within the framework of the half-plane model; but a student who has mastered this model alone will have trouble with certain calculations, like those involving triangles. Besides, this approach masks the fundamental ties between hyperbolic geometry and the conformal group of the sphere.

For this reason we have opted for a comprehensive, if perhaps demanding, presentation, which should enable the reader to solve most questions from hyperbolic geometry: toward the end of a book, a reader should have more opportunities to display his maturity. The mainstays of our presentation are the projective model, the Klein model and the Poincaré model. Each model is best suited to a particular problem, but it is the projective model that best shows what's really going on and can be counted on for the more difficult analyses. We have made use of results from chapter 13, but we don't need them in full generality; they can be proved more simply, if one wants to restrict oneself to the framework of hyperbolic geometry.

$$\boxed{\text{PART I: ELLIPTIC GEOMETRY}}$$

19.1. Elliptic geometry

In this section E denotes a $(d + 1)$-dimensional Euclidean vector space, $P = P(E)$ its associated projective space, $p : E \setminus 0 \to P$ the canonical projection, and $S = S(E)$ the unit sphere in D. We identify P with the set $G_{E,1}$ of vector lines of E, also denoted by $\mathcal{D}(E)$. We denote by $\overline{DD'} \in [0, \pi/2]$ the angle between $D, D' \in P$ (cf. 8.6.3). The projective group of the quadric with equation $\| \cdot \|^2$ is denoted by $\mathrm{PO}(E)$ (cf. 14.7).

19.1.1. ELLIPTIC SPACES.

19.1.1.1. Proposition and definition. *The map $\overline{\cdot\cdot} : P \times P \to \mathbf{R}$ is a metric on P. With this structure, P is called the elliptic space associated with E. The topology associated with this metric coincides with the one given in 4.3.1, and will be the only one considered on P.*

Proof. Use 4.3.3.2: the restriction $p : S \to P$ is surjective, and we compare $\overline{xy}$, the distance between x and y in S (cf. 18.4) with $\overline{p(x)p(y)}$, the distance in P. For $m, n \in P$ and $p^{-1}(m), p^{-1}(n) \in S$, we have

19.1.1.2
$$\overline{mn} = d\big(p^{-1}(m), p^{-1}(n)\big) = \inf \big\{ \overline{xy}, \overline{(-x)y} \big\}$$

for any x, y such that $p(x) = m$ and $p(y) = n$.

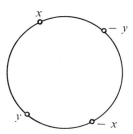

Figure 19.1.1.2

This is a consequence of 8.6.3, because either $(x \mid y) \geq 0$ or $(-x \mid y) \geq 0$. Now let $m, n, s \in P$ be points, and take $x \in S$ such that $p(x) = m$; we can find $y \in p^{-1}(n)$ and $z \in p^{-1}(s)$ such that $\overline{xy} = \overline{mn}$ and $\overline{xz} = \overline{ms}$. By 18.4.2 this implies $\overline{yz} \leq \overline{yx} + \overline{xz}$, hence

$$\overline{ns} = \inf \big\{ \overline{xz}, \overline{(-x)z} \big\} \leq \overline{yz} = \overline{mn} + \overline{ns}. \qquad \square$$

19.1.1.3. Remarks. For $x \in S$ and $m = p(x)$, we have

$$B_S(x, \pi/4) = \{\, y \in S \mid \overline{xy} \leq \pi/4 \,\}, \qquad B_P(m, \pi/4) = \{\, n \in S \mid \overline{mn} \leq \pi/4 \,\}.$$

The restriction of p to $B_S(x, \pi/4)$ is an isometry between $B_S(x, \pi/4)$ and $B_P(m, \pi/4)$. This is no longer true for $B_S(x, r)$ if $r > \pi/4$.

All elliptic spaces of same dimension are isometric (cf. 8.1.6); their diameter is $\pi/2$.

We have explained in 4.3.9.1 why it is difficult to make drawings of $P^2(\mathbf{R})$; to get an idea of things, the reader can ruminate 19.1.2.1.

19.1.1.4. Tangent vectors and tangent bundle. The *tangent bundle* to P, denoted by TP, is the quotient set $\Theta/\sim$, where Θ is the set of pairs (x, u) with $x \in S$ and $u \in x^{\perp} \in E$, and $\sim$ is the equivalence relation given by $(x, u) \sim (x', u')$ if and only if $x' = \epsilon x$ and $u' = \epsilon u$, with $\epsilon = \pm 1$.

The *tangent space* $T_m P$ to P at $m \in P$ is the subset of $\Theta/\sim$ corresponding to pairs (x, u) such that $p(x) = m$. Since $-\mathrm{Id}_E$ is an isometry, $T_m P$ has, for every $m \in P$, a natural Euclidean vector space structure; the angle between two non-zero vectors in $T_m P$ is as defined in 8.6.3. Our tangent space $T_m P$ can be identified with the tangent space to the differentiable manifold P (cf. 4.2.6).

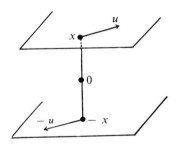

Figure 19.1.1.4

19.1.1.5. Lines and half-lines. Defining lines (and more generally subspaces) in P presents no difficulty: just consider P as a projective space (cf. 4.6). A *half-line* with origin m and direction y, where $m \in P$ is a point and $y \in T_m P$ is a unit tangent vector to P at m, is the subset of P given by

$$\{\, p(\cos t \cdot x + \sin t \cdot u) \mid t \in [0, \pi/2] \,\},$$

where $(x, u) \in \Theta$ is a representative of the tangent vector y. The *endpoint* of this half-line is $n = p(u)$. (We also say that m and n are the endpoints of the half-line.) Conversely, given two points $m, n \in P$, there exists a unique half-line with origin m and containing n if $\overline{mn} < \pi/2$; otherwise there exist two, with opposite directions (figure 19.1.1.5).

19.1.1.6. Notes. In an elliptic space, there is a unique line going through any two distinct points, and two distinct lines have at most one common

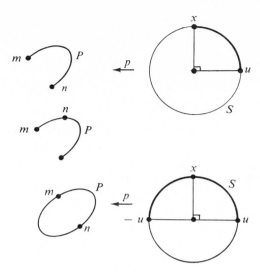

Figure 19.1.1.5

point (exactly one in the case of the elliptic plane, $d = 2$). This is not true for the sphere, which is why we introduce P as a quotient space of S. In elliptic geometry there are no parallels (in the sense of non-intersecting lines, cf. 2.4.9.5 and 19.3.2); on the other hand, all properties from Euclidean geometry that do not involve the fifth postulate remain true here. The historical importance of elliptic and hyperbolic geometry was exactly to have provided a counterexample to Euclid's fifth postulate, proving its independence from the remaining ones. For more history, see [GG1, 367 and *passim*].

From 19.1.1.4 we see that P has a natural Riemannian structure.

19.1.1.7. The canonical measure. Using the fact that $p^{-1}(m) = 2$ for every $m \in P$, we define the canonical measure on P, *denoted by* τ, as follows:

$$\int_P f\tau = \frac{1}{2} \int_S (f \circ P)\sigma,$$

where σ is the canonical measure on S (cf. 18.3.7). The volume of P, for example, is half the volume of S, that is, $\alpha(d+1)/2$.

19.1.2. METRICAL PROPERTIES

19.1.2.1. The strict triangle inequality and shortest paths. The proof of 19.1.1.1 shows that $\overline{ns} = \overline{nm} + \overline{ms}$ if and only if m belongs to a half-line with endpoints n and s (cf. 19.1.1.5). It follows that, given $m, n \in P$, there exists a unique shortest path from m to n, contained in a half-line with origin m, as long as $\overline{mn} < \pi/2$. If $\overline{mn} = \pi/2$, there exist two shortest paths, the half-lines with endpoints m and n.

19.1.2.2. The isometry group. By the definition of the metric on P, the isometry group $\mathrm{Is}(P)$ contains $\mathrm{PO}(E)$ (cf. 14.7). We show the two are equal.

For $f \in \mathrm{Is}(P)$, we can assume there exists m such that $f(m) = m$, since $\mathrm{PO}(E) \subset \mathrm{Is}(P)$ acts transitively on P. Fix x with $p(x) = m$; by 19.1.1.3, $f \in \mathrm{Is}(P)$ induces an isometry on $B_S(x, \pi/4)$, which, by 9.8.2, comes from a unique isometry $\overline{f} \in O(E)$. We now show that $\overline{f}$ induces an isometry on the whole of P, and not only on $B_P(m, \pi/4)$. Since $\mathrm{PO}(E) = \mathrm{Is}(P)$, we can assume $\overline{f} = \mathrm{Id}_E$, and we have to show that if $f \in \mathrm{Is}(P)$ is the identity on $B_P(m, \pi/4)$, it is the identity on P. Take $n \in P$ and let D be a shortest path from m to n; the image $f(D)$ is still a shortest path from m to $f(n)$, so it is contained in a half-line, which shows that $f(D) = D$ since $D \cap B_P(m, \pi/4) = f(D) \subset B_P(m, \pi/4)$.

The group of isometries is two-transitive on P, as can be seen from 18.5.5; see also 9.1.7.

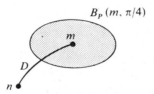

$$B_P(m, \pi/4)$$

Figure 19.1.2.2.

19.1.2.3. Equidistant locus. In contrast with the Euclidean and hyperbolic cases (cf. 9.7.5 and 19.4.2), the *equidistant locus* of two distinct points $m, n \in P$, given by $M = \{\, s \in P \mid \overline{ms} = \overline{ns}\,\}$ is not a hyperplane. In fact, take x and y such that $p(x) = m$ and $p(y) = n$; by 19.1.1.2 we have

$$M = p\big(\{\, z \in S \mid \overline{zx} = \overline{zy} \text{ or } \overline{z(-x)} = \overline{zy}\,\}\big).$$

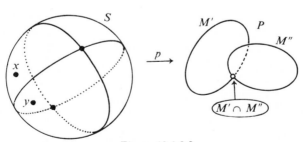

Figure 19.1.2.3

Thus M is the union of two orthogonal hyperplanes M', M'' in P, and the intersection $M' \cap M''$ is the set $\{\, s \in P \mid \overline{sm} = \overline{sn} = \pi/2\,\}$. When $d = 2$, for example, the equidistant locus consists of two lines that intersect at right angles.

19.1.2.4. Reflections. We can define reflections through subspaces of P. Hyperplane reflections generate $\mathrm{Is}(P)$, by 8.2.12.

19.1.2.5. In analogy with 9.7.2 and 9.7.4 for Euclidean spaces and 18.4.7 for spheres, there are universal relations for distances between $d + 2$ points in P, and purely metrical characterizations of elliptic spaces; see [BL, chapters IX, X and XI], [B-K, 117] and [PV2].

19.1.3. TRIANGLES. To study triangles in P we will assume that $d = 2$, since three points in P span an elliptic plane.

19.1.3.1. A *triangle* $T = \{m, n, s\}$ in P is a set of three projectively independent points in P (cf. 4.6.6) satisfying $\overline{mn} < \pi/2$, $\overline{ns} < \pi/2$ and $\overline{sm} < \pi/2$. By 19.1.1.5 triangles have well-determined *sides*, contained in half-lines; let M, N and S be the sides. The *angle* α of T at m is defined as the angle between the unit vectors at m in the directions of the unique half-lines with origin m and containing N and S; the angles β at n and γ at s are defined similarly. Finally, we *denote* by $a = \overline{ns}$, $b = \overline{sm}$ and $c = \overline{mn}$ the lengths of the sides of T.

19.1.3.2. The reader should notice that, given a triangle T of P, there isn't always a (spherical) triangle in S having the same angles and side lengths as T. There are many explanations for this fact. First consider the following example: for $\theta \in \,]\pi/3, \pi/2[$, there exist by 18.6.10 two spherical triangles $\langle x, y, z \rangle$ and $\langle x', y', z' \rangle$ with sides $\pi/3$, $\pi/3$, θ and $\pi/3$, $\pi/3$ and $\pi - \theta$, respectively. The two triangles

$$\{m = p(x), n = p(y), s = p(z)\} \quad \text{and} \quad \{m' = p(x'), n' = p(y'), s' = p(z')\}$$

of P have same side lengths, namely, $\pi/3$, $\pi/3$ and θ. If the angles of $\langle x, y, z \rangle$ are (α, β, β) those of $\langle x', y', z' \rangle$ are $(\alpha', \beta', \beta')$, we see that the angles of $\{m, n, p\}$ are (α, β, β), but the angles of $\{m', n', p'\}$ are $(\alpha', \pi - \beta', \pi - \beta')$. Thus case (ii) of equality of spherical triangles is false in P (cf. 18.6.13.10).

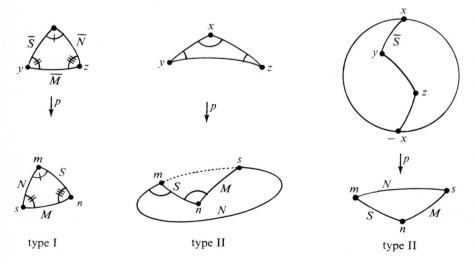

type I   type II   type II

Figure 19.1.3.2

19.1.3.3. Working the other way around, let $T = \{m, n, p\}$ be a triangle in P, and let's try to construct a triangle in S that projects to T. Take $x \in S$ such that $p(x) = m$. There exists a unique arc of great circle $\overline{S}$ with origin y and such that $p(\overline{S}) = S$. Let $\overline{M}$ be the arc of great circle whose origin y is the endpoint of $\overline{S}$ and which projects to M, and let $\overline{N}$ the arc of great circle whose origin z is the endpoint of $\overline{M}$ and which projects to N. The endpoint x' of $\overline{N}$ satisfies $p(x') = m$, so there are two possibilities: if $x' = x$, we say that T is of *type I*, and if $x' = -x$ we say T is of *type II*. It is easy to see that the type of T does not depend on the vertex where we start the lifting; in fact, the two types of triangle correspond to the two elements of the fundamental group of $P^d(\mathbf{R})$, for $d \geq 2$ (cf. 4.3.9.3). The triangle T is of type I if and only if the loop $M \cup N \cup S$ is null-homotopic. In practice, one can calculate the number

$$\cos b \cos c + \sin b \sin c \cos \alpha,$$

which is positive (resp. negative) when T is of type I (resp. II).

If T is of type I, we can define the *interior* of T as the projection of the interior of a spherical triangle associated with T; the area of the interior (in the sense of 19.1.1.7) is given by $\alpha + \beta + \gamma - \pi$ (cf. 18.3.8.4).

For more on triangles in P see 19.8.4 or [CR3, 232–237].

19.1.4. Clifford parallelism. Assume $d = 3$. The discussion in 18.8.4 can be carried over to $P = P(\mathbf{R}^4) = p(S^3)$, and the statements become more elegant in projective terms.

Let $\mathcal{D} = \mathcal{D}(P)$ the set of lines of P. For $D, D' \in \mathcal{D}$, we write $D \parallel D'$ if $d(m, D') = d(D, D')$ for all $m \in D$. The relation $\parallel$ is not an equivalence relation, but there exist two equivalence relations $\parallel^+$ and $\parallel^-$ such that $D \parallel D'$ implies $D \parallel^+ D'$ or $D \parallel^- D'$. For every $m \in P$ and $D \in \mathcal{D}$ there exist unique lines $D' \ni m$ and $D'' \ni m$ such that $D' \parallel^+ D$ and $D'' \parallel^- D$. If $d(D, D') = d(D, D'')$ and $D \parallel^+ D'$, $D \parallel^- D''$ the lines D' and D'' intersect in one point. A line $D' \ni m$ such that $D' \parallel D$ is characterized as follows: if $m_0 \in D$ with $d(m, D) = \overline{m_0 m}$, then D' is orthogonal to $\langle m_0, m \rangle$ and makes an angle $\overline{m_0 m}$ with the plane $\langle D, m \rangle$.

The reader should translate 18.8.4 (v) and (vi) into this language.

19.1.5. Note. For an analogous metric on complex projective spaces, see 19.8.22.

PART II: HYPERBOLIC GEOMETRY

19.2. The projective model and the ball model

19.2.1. NOTATION. From now till the end of the chapter, we identify $\mathbf{R}^{n+1}$ with the product $\mathbf{R}^n \times \mathbf{R}$ and write $\xi = (z,t)$ for $\xi \in \mathbf{R}^{n+1}$, $z \in \mathbf{R}^n$ and $t \in \mathbf{R}$. We let H stand for the hyperplane $\{(z,t) \mid t = 1\}$. By q we denote the quadratic form $-\sum_{i=1}^n x_i^2 + x_{n+1}^2$ with signature $(n,1)$, by P its polar form and by $Q = q^{-1}(0)$ its isotropic cone. We thus have

$$q(z,t) = -\|z\|^2 + t^2, \qquad P\big((z,t),(z',t')\big) = -(z \mid z') + tt'.$$

Orthogonality, expressed by the symbol $\perp$, will always refer to q (cf. 13.3) and not to the Euclidean structure in $\mathbf{R}^{n+1}$. Further, $\mathcal{B} = U(0,1) \subset \mathbf{R}^n$ will be the open unit ball in $\mathbf{R}^n$. The projective group $\mathrm{PO}(n)$ of the quadric α with equation q will be *denoted* by $G(n)$. We recall that $P^n(\mathbf{R}) = P(\mathbf{R}^{n+1})$ can be identified with the set $G_{n+1,1} = \mathcal{D}(\mathbf{R}^{n+1})$ of vector lines in $\mathbf{R}^{n+1}$.

19.2.2. THE METRIC. The action of $G(n)$ on $\mathcal{D}(\mathbf{R}^{n+1})$ has three orbits: the set of isotropic lines, the set $\mathcal{P}$ of lines where q is positive definite, and the set of lines where q is negative definite. The action of $G(n)$ on the first of these orbits corresponds to the Möbius group $\mathrm{M\ddot{o}b}(d)$ studied in 18.10 (cf. 18.10.1.5). The second orbit will be studied here, and the third in chapter 20.

19.2.3. We *denote* by $\mathcal{P}$ the set of $D \in \mathcal{D}(\mathbf{R}^{n+1})$ such that $q(\xi) > 0$ for all $\xi \in D \setminus 0$. The equation $q(\xi) = -\|z\|^2 + t^2 > 0$ implies that $t \neq 0$; in particular, $D = \mathbf{R}\xi$ intersects H in a unique point $(z/t, 1)$, and this point satisfies $\|z/t\| < 1$. This leads to a map

19.2.4 $\Phi : \mathcal{D} \ni p(z,t) \mapsto z/t \in \mathcal{B}.$

This map is injective and its inverse will be *denoted* by $\Lambda = \Phi^{-1}$.

It is sometimes convenient to identify $\mathcal{B}$ with the subset $\{(z,t) \in H \mid \|z\| < 1\}$, and we have done so in some of the figures.

Figure 19.2.3

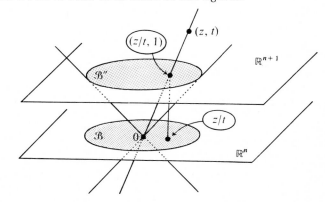

19.2.5. PROPOSITION AND DEFINITION. *Let D and D' be lines in $\mathcal{P}$. Every plane containing D and D' (there is exactly one if $D \neq D'$) intersects Q in two distinct lines U, U'. For every $\xi \in D \setminus 0$ and $\xi' \in D' \setminus 0$ we have $P(\xi, \xi') \geq \sqrt{q(\xi)q(\xi')}$ and*

$$\text{Arccosh } \frac{P(\xi, \xi')}{\sqrt{q(\xi)q(\xi')}} = \frac{1}{2}\left|\log([D, D', U, U'])\right|.$$

This number is denoted by $d(D, D')$ and $d : \mathcal{P} \times \mathcal{P} \to \mathbf{R}$ makes $\mathcal{P}$ into a metric space. The metric space $(\mathcal{P}, d)$ is called the (projective model of) the n-dimensional hyperbolic space. The metric space $(\mathcal{B}, d)$, where d is the metric defined by $d(z, z') = d(\Lambda^{-1}(z), \Lambda^{-1}(z'))$ is called the (Klein model of) hyperbolic space.

Proof. The cross-ratio in the equation is taken on the projective line coming from the vector plane P spanned by D and D' (if the lines are distinct). Showing that d is a metric is trivial, except for the triangle inequality, which will be proven in 19.3.2. The other properties follow from 13.8.6 and 13.8.9 if we show that P is an Artinian plane; but this follows from 13.4.7, since the restriction of q to P is non-degenerate, and its signature can be neither $(2, 0)$ nor $(0, 2)$.

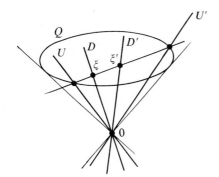

Figure 19.2.5

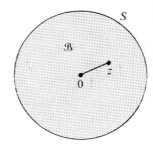

Figure 19.2.6

19.2.5.1. Note. Compare with 11.9.4.

19.2.6. EXAMPLE. For $z \in \mathcal{B}$, set $r = d(0, z)$. Then

$$r = \text{Arccosh } \frac{1}{\sqrt{1 - \|z\|^2}},$$

or $\|z\| = \tanh r$. Thus $d(0, z)$ approaches infinity as $\|z\|$ approaches unity. It is reasonable to ask whether the hyperbolic space is complete.

19.2.7. THE TANGENT SPACE. We mimic 19.1.1.4. Notice first that, by 2.4.8.2, $\xi^\perp$ is a hyperplane in $\mathbf{R}^{n+1}$ if $q(\xi) > 0$. The *tangent bundle* to $\mathcal{P}$, denoted by $T\mathcal{P}$, is the quotient of the set of pairs (ξ, u) for which $q(\xi) = 1$ and $u \in \xi^\perp$ by the equivalence relation $(\xi, u) \sim (\xi', u')$ if $\xi' = \epsilon\xi$ and $u' = \epsilon u$, with

$\epsilon = \pm 1$. The *tangent space* to P at D, denoted by $T_D P$, is the image in TP of the pairs (ξ, u) such that $q(\xi) = 1$ and $\xi \in D$. The tangent space has an *anti-Euclidean* structure, that is, it is endowed with a negative definite quadratic form. In fact, for $q(\xi) > 0$, the restriction of q to $\xi^\perp$ is negative definite by 13.4.7, and $- \operatorname{Id}_{\xi^\perp}$ is an isometry of $(\xi^\perp, q|_{x^\perp})$. In particular we can define the *angle* between two vectors of $T_D P$: if the vectors are represented by pairs (ξ, u) and (ξ, v), their angle $\alpha \in [0, \pi]$ is defined by

19.2.8
$$\cos \alpha = - \frac{P(u, v)}{\sqrt{q(u)q(v)}}.$$

19.2.9. In the Klein model B, the tangent space to $z \in B$ is defined by carrying over the tangent space to $\Lambda(z)$, but it can also be identified with $\mathbf{R}^n$. Observe that the angle between two vectors in $T_z B$ differs from their angle as vectors in $\mathbf{R}^n$, except for $z = 0$ (check this). This is one of the reasons why we later introduce the models C and $\mathcal{H}$, where hyperbolic and Euclidean angles coincide; on the other hand the new models do not share with P and B the property that hyperbolic straight lines coincide with the usual lines (cf. 19.6.2).

19.2.10. LINES, HALF-LINES AND SUBSPACES. In the projective model, a *line* is the non-empty intersection of P with a projective line in $P^n(\mathbf{R})$. In the Klein model, lines are the images under Φ of the lines in P, that is, open segments whose endpoints are points in S^{n-1} (the frontier of B).

In P a *half-line* with origin D is a set consisting of the lines spanned by the vectors $\cosh t \cdot \xi + \sinh t \cdot u$ $(t \in \mathbf{R}_+)$, where $q(\xi) = 1$, $q(u) = -1$ and $u \in \xi^\perp$. The *direction* of this line is the tangent vector associated with (ξ, u).

Applying Φ we deduce the notion of a half-line in the Klein model: half-lines are just half-open segments, with the closed endpoint in B and the open endpoint in S^{n-1}.

The angle between two half-lines with the same origin (in P or B) is the angle between their directions as tangent vectors in the sense of 19.2.7 (cf. 19.2.9 for B).

There is a unique line going through two distinct points in P or B, and a unique half-line originating at one and containing the other.

The definition of a k-dimensional *subspace* of B is analogous; $(n-1)$-dimensional subspaces are called *hyperplanes*.

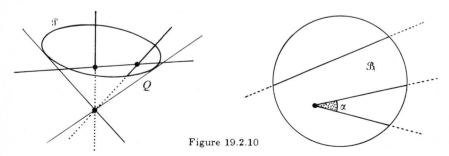

Figure 19.2.10

19.2.11. NOTE. For questions analogous to 9.7.4, 18.4.7 and 19.1.2.5, see 19.8.16 and [BL, chapter XII].

19.3. The fundamental formula

Take $z, z', z'' \in \mathcal{B}$ such that $z \neq z'$, $z \neq z''$. Set $d(z', z'') = a$, $d(z'', z) = b$, $d(z, z') = c$, and let α be the angle at z between the half-lines originating at z and going through z' and z''. Then

19.3.1 $\cosh a = \cosh b \cosh c - \sinh b \sinh c \cos \alpha.$

We can mimic the proof of 18.6.8. Set $D = \Lambda(z)$, $D' = \Lambda(z')$, $D'' = \Lambda(z'')$ and choose $\xi \in D$, $u, v \in \xi^{\perp}$ and $k, h \in \mathbf{R}_{+}$ such that $q(\xi) = 1$, $q(u) = q(v) = -1$ and

$$\xi' = \xi + ku \in D', \qquad \xi'' = \xi + hv \in D''.$$

By 19.2.5, 19.2.6 and the fact that k and h are positive we obtain:

$$q(\xi') = 1 - k^2, \qquad q(\xi'') = 1 - h^2,$$

$$\cosh c = \frac{1}{\sqrt{1 - k^2}}, \qquad \sinh c = \frac{k}{\sqrt{1 - k^2}},$$

$$\cosh b = \frac{1}{\sqrt{1 - h^2}}, \qquad \sinh b = \frac{k}{\sqrt{1 - h^2}},$$

$$P(u, v) = -\cos \alpha, \qquad P(\xi', \xi'') = 1 - \cos \alpha \cdot kh.$$

Putting everything together we get

$$\cosh a = \frac{1 - \cos \alpha \cdot kh}{\sqrt{1 - k^2}\sqrt{1 - h^2}} = \cosh b \cosh c - \sinh b \sinh c \cos \alpha. \qquad \square$$

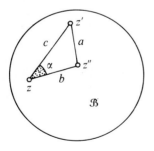
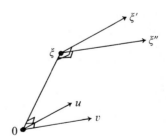

Figure 19.3.1

19.3.2. THE STRICT TRIANGLE INEQUALITY. Since $\cos \alpha \geq -1$, we have

$$\cosh a \leq \cosh b \cosh c + \sinh b \sinh c = \cosh(b + c),$$

and $a \neq b + c$, with $a = b + c$ if and only if $\cos \alpha = -1$, since $\cosh$ is an strictly increasing function in $\mathbf{R}_{+}$. The condition $\cos \alpha = -1$, or $\alpha = \pi$, is equivalent to saying that, in $\mathcal{B}$, the point z belongs to the segment with endpoints z', z''.

It follows that, given two points $z', z'' \in \mathcal{B}$, there exists a unique shortest path joining z' and z'', namely, the segment $[z, z']$ (with uniform parametrization). Its length is $d(z', z'')$; in particular, the metric in $\mathcal{B}$ is excellent, and consequently intrinsic (cf. 9.9.4.5). Thus $\mathcal{B}$ shares with Euclidean affine spaces the property that the shortest path between two points is a line. On the other hand (cf. also 19.1.1.6), given a point z outside a line Δ, there exist infinitely many lines (an "interval" of them, as shown in figure 19.3.2) through z that do not intersect Δ: parallels are no longer unique.

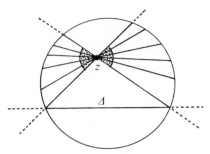

Figure 19.3.2

19.3.3. EXAMPLES

19.3.3.1. Distance to a line. Assume $\alpha = \pi/2$. (Observe that, by 19.2.9, the geometric condition for two lines through z to be orthogonal is that they be conjugate, in the sense of 10.7.11, with respect to the circle obtained by intersecting the plane that they determine with the frontier S^{n-1} of $\mathcal{B}$; cf. figure 19.3.3.1.2.) Formula 19.3.1 becomes $\cosh a = \cosh b \cosh c$, whence $a \geq b$, and equality takes place only when $z' = z$. This proves an analogue for 9.2.2, since, given a line Δ a point z'', there exists, by compactness, a point $z \in \Delta$ such that $d(z, z'') = d(z'', \Delta)$.

Notice that, if Δ and Δ' are lines and $z \in \Delta$, $z \in \Delta'$ are points such that $d(z, z') = d(\Delta, \Delta')$, the angle that Δ (resp. Δ') makes with $\langle z, z' \rangle$ at z (resp. z') is equal to $\pi/2$, but the distance $d(w, \Delta')$ increases as w moves away from z on Δ, in contrast with the Euclidean case. See also 19.8.8 and 19.8.11.

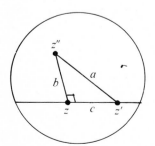 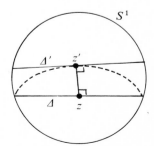

Figure 19.3.3.1.1

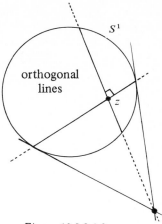

Figure 19.3.3.1.2

19.3.3.2. Circles. The length of a curve in $\mathcal{B}$ is easily calculated in polar coordinates (cf. 19.8.18). We define a *circle* of *radius* r in the hyperbolic plane to be a set of the form

$$\{\, z \in \mathcal{B} \mid d(z, z_0) = r \,\};$$

the length of a circle of radius r is $2\pi \sinh r$, whereas in the Euclidean plane it is 2π, and in the sphere S^2 it is $2\pi \sin r$. In $\mathcal{B}$ circles are represented by ellipses (cf. 19.8.12), but in the conformal models they look like Euclidean circles (19.6.8.2).

19.3.4. TRIANGLES. Calculating from 19.3.1 we see that, for a hyperbolic triangle with sides a, b, c and angles α, β, γ, we have

19.3.5 $\qquad \dfrac{\sin \alpha}{\sinh a} = \dfrac{\sin \beta}{\sinh b} = \dfrac{\sin \gamma}{\sinh c}, \qquad \cos \alpha = \sin \beta \sin \gamma \cosh a - \cos \beta \cos \gamma.$

The reader can find the hyperbolic analogues of all our results concerning Euclidean and spherical triangles: formulas 10.3, 10.13.2 and 18.6.13, the construction of a triangle with given elements, cases of equality (under isometry), the concurrence of medians, bisectors and altitudes. He can seek the link between the concurrence of bisectors and Brianchon's theorem (cf. 16.2.13 and the frontispiece of [CR3]). Given the three angles of a triangle, formulas 19.3.5 allow the calculation of the sides; but an interesting additional formula will be given in 19.5.4. The reader can also study the analytic method developed in [CR3, chapter XII].

19.4. The isometry group

19.4.1. The *isometry group* $\mathrm{Is}(\mathcal{B})$ of $\mathcal{B}$ clearly contains the group derived from $G(n)$ (cf. 19.2.1) via Φ, since the metric on $\mathcal{B}$ is defined using q, and

$G(n)$ preserves q by definition (cf. 19.2.3 and 19.2.5). This subgroup will still be *denoted* by $G(n)$, by abuse of notation. After a short detour, needed also for other purposes, we will see that, in fact, $\text{Is}(\mathcal{B}) = G(n)$.

19.4.2. HYPERPLANE REFLECTIONS AND EQUIDISTANT HYPERPLANES. If η satisfies $q(\eta) < 0$, then $\eta^\perp$ is a hyperplane in $\mathbf{R}^{n+1}$ and $\eta^\perp \cap P$ is a hyperplane in P, by 13.4.7 and 19.2. This corresponds in $\mathcal{B}$ to a hyperplane $\Phi(\eta^\perp \cap P)$. The associated orthogonal reflection in $P^n(\mathbf{R})$ (cf. 13.6.6 and 14.7.4) lies in $\text{PO}(\alpha) \setminus G(n)$; it is an involutive isometry of P whose sole fixed points are the points in $\eta^\perp \cap P$. In $\mathcal{B}$ (or P), such a reflection is called a *hyperplane reflection* through the hyperplane $W = \Phi(\eta^\perp \cup P)$ (or $\eta^\perp \cap P$). In particular, $d(a, z) = d\big(a, f(z)\big)$ for all $a \in W$ and $z \in \mathcal{B}$.

Conversely, for any distinct points $a, b \in \mathcal{B}$, there exists a unique hyperplane W in $\mathcal{B}$ such that the hyperplane reflection through W switches a and b, and W is characterized by

$$W = \big\{ z \in \mathcal{B} \mid d(a, z) = d(b, z) \big\}$$

(that is, W is the hyperplane *equidistant* from a and b). To see this, take $\xi \in \Lambda(a)$ and $\xi' \in \Lambda(b)$ such that $q(\xi) = q(\xi')$ and $\eta = \xi - \xi'$; then $q(\xi - \xi') < 0$ by 13.4.7. The reflection through W in $\mathcal{B}$ takes a and b into each other by 13.6.6.2; further, the condition $d(a, z) = d(b, z)$, for $z = \Phi(\mathbf{R}_\varsigma)$, is equivalent, by 19.2.5, to $P(\xi, \varsigma) = P(\xi', \varsigma)$, that is, $P(\xi - \xi', \varsigma) = 0$, which is the same as saying that $\varsigma \in (\xi - \xi')^\perp$.

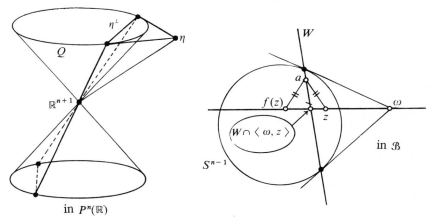

Figure 19.4.2

A geometric procedure to find where the reflection f through a hyperplane W takes a point $z \in \mathcal{B}$ can be derived from 14.7.4 and 15.5. Let ω be the pole of W with respect to S^{n-1} (cf. 10.7.11; if W contains the center 0 of $\mathcal{B}$, the point ω is at infinity, in the direction of $W^\perp$). The point $f(z)$ sits on the line $\langle \omega, z \rangle$, and is characterized by the condition that $\omega, \langle \omega, z \rangle \cap W, z, f(z)$ are in harmonic division.

19.4.3. EXAMPLE: THE MIDPOINT. From 19.4.2 and 19.3.3.1 we see that, given two points $a, b \in \mathcal{B}$, there exists a unique point m such that

$$d(m, a) = d(m, b) = \frac{1}{2}d(a, b);$$

this point is the intersection of the segment $[a, b]$ with the plane equidistant from a and b. The geometric construction for m is the same as that for the double points of the involution of $\langle a, b \rangle$ swapping the pairs (a, b) and (u, v), where u and v are the intersections of $\langle a, b \rangle$ with the sphere S^{n-1}.

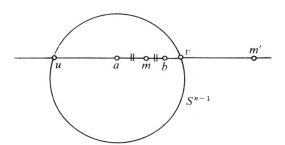

Figure 19.4.3

19.4.4. PROPOSITION. *Let* $(z_i)_{i=1,\ldots,n+1}$ *be affinely independent points in* $\mathcal{B}$. *If* $d(z, z_i) = d(z', z_i)$ *for all* $i = 1, \ldots, n+1$, *we have* $z = z'$. *Let* $(z_i)_{i=1,\ldots,n+1}$ *and* $(z'_i)_{i=1,\ldots,n+1}$ *be* k-*tuples of points in* $\mathcal{B}$ *such that* $d(z_i, z_j) = d(z'_i, z'_j)$ *for all* $i, j = 1, \ldots, k$; *there exists* $f \in G(n)$ *taking* z_i *to* z'_i *for all* $i = 1, \ldots, k$.

Proof. Using our knowledge about equidistant planes, we proceed exactly as in 9.7.1. □

19.4.5. COROLLARY. $\mathrm{Is}(\mathcal{B}) = G(n)$. □

19.4.5.1. **Note.** In particular, $\mathrm{Is}(\mathcal{B})$ acts transitively on pairs of points (cf. 9.1.7).

19.4.6. CONSEQUENCES.

19.4.6.1. From the proof of 19.4.4 it follows that every element of $\mathrm{Is}(\mathcal{B})$ is the product of at most $n+1$ hyperplane reflections. Notice that using 13.7.12 would also give us at most $n + 1$ reflections, but not necessarily through hyperplanes in $\mathcal{B}$: a hyperplane in $\mathbf{R}^{n+1}$ of the form $\eta^{\perp}$ with $q(\eta) > 0$ corresponds to a reflection through a *point* of $\mathcal{B}$, namely, $\Phi(\mathbf{R}\eta)$.

19.4.6.2. From 13.7.1 it follows that two k-dimensional subspaces of $\mathcal{B}$ can be taken into one another by some $f \in \mathrm{Is}(\mathcal{B})$. Thus the metric induced by d on any k-subspace is the same as the metric of the standard k-hyperbolic space (which is what we obtain by taking a k-dimensional space through the origin of $\mathcal{B}$).

19.4.6.3. As in 18.5.8, one could show that d is, up to a scalar, the only intrinsic metric on $\mathcal{B}$ invariant under $\mathrm{Is}(\mathcal{B})$.

19.4.6.4. The group $\mathrm{Is}(\mathcal{B})$ is isomorphic to $\mathrm{M\ddot{o}b}(n-1)$ (cf. 18.10), but its action on $\mathcal{B}$ (or $\mathcal{P}$, or the other models below) looks very different than on S^{n-1}.

19.4.6.5. The stabilizer $\mathrm{Is}_z(\mathcal{B})$ of a point z is naturally isomorphic to the orthogonal group $O(T_z\mathcal{B})$ of the tangent space to $\mathcal{B}$ at z (cf. 19.2.7). At the origin, $\mathrm{Is}_0(\mathcal{B})$ can be identified with $O(n)$, since for $\xi = (0,1)$ we have $\xi^\perp = \mathbf{R}^n$.

19.4.6.6. The group $\mathrm{Is}(\mathcal{B})$ is conformal ($G(n)$ preserves angles by definition).

19.4.7. COMPACT SUBGROUPS OF THE ISOMETRY GROUP. A compact subgroup of $\mathrm{Is}(\mathcal{B})$ leaves at least one point invariant under all its elements. In fact, it is easy to see using 19.3.1 that the midpoint of two points (cf. 19.4.3) satisfies the condition in 9.8.6.5.

19.5. The canonical measure of hyperbolic space

19.5.1. PROPOSITION. *Every measure on $\mathcal{B}$ that is invariant under $\mathrm{Is}(\mathcal{B})$ is of the form $k\lambda\omega$, where $k \in \mathbf{R}^*_+$, ω is the restriction to $\mathcal{B}$ of the Lebesgue measure on $\mathbf{R}^n$, and λ is the function*

$$z \mapsto \left(1 - \|z\|^2\right)^{-(d+1)/2}.$$

The measure $\lambda\omega$ is called the canonical measure of $\mathcal{B}$; its ratio to the Lebesgue measure at the origin is unity.

Proof. The proof that every measure on $\mathcal{B}$ invariant under $\mathrm{Is}(\mathcal{B})$ is of the form $g\omega$, where $g : \mathcal{B} \to \mathbf{R}^*_+$ is continuous, was outlined in 2.7.4.4 and 18.3.7.4. Here we also have that $\mathrm{Is}(\mathcal{B})$ acts continuously in each direction and at each point of $\mathcal{B}$, showing that two invariant measures differ by a constant, since $\mathrm{Is}(\mathcal{B})$ is transitive.

There remains to see that $\lambda\omega$ is invariant under $\mathrm{Is}(\mathcal{B})$. Since $\| \cdot \|^2$ is invariant under $\mathrm{Is}_0(\mathcal{B}) = O(n)$, so is $\lambda\omega$; thus it is enough to show that, for every $z \in \mathcal{B}$, there exists $f \in \mathrm{Is}(\mathcal{B})$ such that $f^*(\lambda\omega) = \lambda\omega$ and $f(0) = z$. In fact, we can restrict ourselves to points $z = (x,0,\ldots,0)$. The function f defined by

$$f(x_1,\ldots,x_n) = \frac{1}{\sinh t \cdot x_1 + \cosh t}(\cosh t \cdot x_1 + \sinh t, x_2, \ldots, x_n)$$

takes the origin into $(x,0,\ldots,0)$ and comes from a map $F : \mathbf{R}^{n+1} \to \mathbf{R}^{n+1}$ given by

$$F(x_1,\ldots,x_n,x_{n+1}) = (\cosh t \cdot x_1 + \sinh t \cdot x_{n+1},$$
$$x_2,\ldots,x_n,\sinh t \cdot x_1 + \cosh t \cdot x_{n+1}),$$

which clearly leaves $q = -\sum_{i=1}^n x_i^2 + x_{n+1}^2$ invariant. Calculating the jacobian of f at the origin we see that $\det f'(0) = (\cosh t)^{-(d+1)}$, so indeed $f^*(\lambda\omega) = \lambda\omega$. $\qquad\square$

19.5.2. EXAMPLE. Let T be a triangle in the hyperbolic plane B having angles $\alpha, \beta, \pi/2$. Then

19.5.3 $$\text{area}(T) = \pi/2 - \alpha - \beta.$$

By the transitivity of $\text{Is}(B)$, we can prove this formula for the particular triangle $T = ((0,0), (u,0), (u, u/\sin\alpha))$. For T the angle at the origin is α because at 0 angles in B and $\mathbf{R}^2$ coincide, and the angle at $(u,0)$ is $\pi/2$ by symmetry. Thus, by 19.2.6, we have $u = \tanh b$, where b is the side of T opposite to β, and $\cosh b = \cos\beta/\sin\alpha$, by 19.3.5. In polar coordinates (ρ, θ), the canonical measure is $(1 - \rho^2)^{-3/2}\rho\,d\rho\,d\theta$, whence

$$\text{area}(\theta) = \int_0^\alpha \int_0^{u/\cos\theta} \rho(1 - \rho^2)^{-3/2}\,d\rho\,d\theta = \int_0^\alpha (1 - \rho^2)^{-1/2}\big]_0^{u/\cos\theta}\,d\theta$$

$$= \int_0^\alpha \left(\left(1 - \frac{u^2}{\cos^2\theta}\right)^{-1/2} - 1\right)d\theta = \int_0^\alpha \frac{\cos\theta}{\sqrt{\cos^2\theta - u^2}}\,d\theta - \alpha$$

$$= \text{Arcsin}\left(\frac{\sin\theta}{\sqrt{1 - u^2}}\right)\Big]_0^\alpha - \alpha = \text{Arcsin}(\sin\alpha\cosh b) - \alpha$$

$$= \text{Arcsin}(\cos\beta) - \alpha = \frac{\pi}{2} - \beta - \alpha.$$

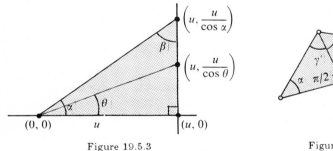

Figure 19.5.3   Figure 19.5.4

19.5.4. COROLLARY. *The area of any hyperbolic triangle T with angles α, β and γ is given by*

$$\text{area}(T) = \pi - \alpha - \beta - \gamma.$$

Proof. Cut the given triangle into two right triangles, using 19.3.3.1, and apply 19.5.3 to each piece. ☐

19.5.5. NOTE. See 18.3.8.6 for a general formula of which 19.5.4 is a very particular case. Notice also that, for every triangle, $\alpha + \beta + \gamma < \pi$.

19.6. The Poincaré model

19.6.1. Consider the stereographic projection $f : S^n \setminus n \to \mathbf{R}^n$ from the north pole n, and call Σ the open southern hemisphere of S^n,

$$\Sigma = \big\{\, (z, t) \in S^n \subset \mathbf{R}^{n+1} \mid t < 0 \,\big\}.$$

Denote by π the projection $(z, t) \to z$ from $\mathbf{R}^{n+1}$ to $\mathbf{R}^n$. The restriction of π to Σ and B is bijective; *let g be its inverse*. Finally, *set*

$$\Xi = f \circ g : B \to B, \qquad \Omega = \Xi^{-1} : B \to B;$$

the maps Ξ and Ω are bijections. The *Poincaré* (or *conformal*) *model of hyperbolic space, denoted by* $\mathcal{C}$, *is the set* B together with the metric δ defined by

$$\delta(x, y) = d\big(\Omega(x), \Omega(y)\big),$$

where d is the hyperbolic distance in B.

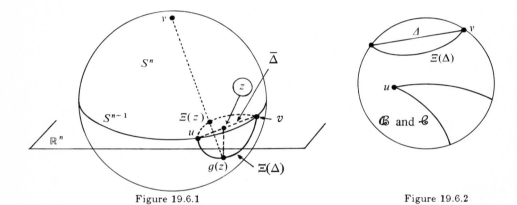

Figure 19.6.1   Figure 19.6.2

19.6.2. Lines, half-lines, tangent vectors and angles in $\mathcal{C}$ are defined by applying Ξ and Ω to the corresponding objects in B. Lines, for example, no longer look like affine lines as they did in B: if Δ is a line in B, contained in an affine line $\overline{\Delta}$ that intersects S^{n-1} in u and v, the image $g(\Delta)$ is a circle in S^n that intersects S^{n-1} at right angles in u and v, so that $\Xi(\Delta) = f\big(g(\overline{\Delta})\big)$ is a circle in $\mathbf{R}^n$, still intersecting S^{n-1} at right angles in u and v (cf. 18.1.4.3). Thus the lines of $\mathcal{C}$ are the open arcs obtained by intersecting B with circles orthogonal to S^{n-1}. Half-lines are the half-open subarcs of these arcs.

19.6.3. Our next step is to show that angles in $\mathcal{C}$ are the same as in $\mathbf{R}^n$, which justifies the name "conformal". We have the following

19.6.4. LEMMA. *Let h be a reflection of B through the hyperplane W. If W contains the center 0 of B (and $\mathcal{C}$) the composition $\Xi \circ h \circ \Omega = h$ is the Euclidean hyperplane reflection through W. Otherwise $\Xi \circ h \circ \Omega$ is the*

restriction to C of an inversion of $\mathbf{R}^n$ whose pole is the pole of W with respect to S^{n-1} and which leaves S^{n-1} globally invariant.

Proof. The first case is trivial. If W does not contain 0, let w be the pole of W with respect to S^{n-1}. Let $\overline{W}$ be the hyperplane of $\mathbf{R}^{n+1}$ containing W and orthogonal to $\mathbf{R}^n$; this is the polar hyperplane of w with respect to S^n. For $z \in B$ and $\theta = \langle w, z \rangle \cap W$, we have $[w, \theta, z, h(z)] = -1$ by 19.4.2, so $g(z)$, $g(h(z))$ and z are collinear by 14.5.2.6. Thus passing from $g(z)$ to $g(h(z))$ is achieved by an inversion of S^n with pole w (cf. 18.10.1.4); but $f \circ k \circ f^{-1}$ is exactly the inversion of $\mathbf{R}^n$ with pole w and leaving S^{n-1} invariant. □

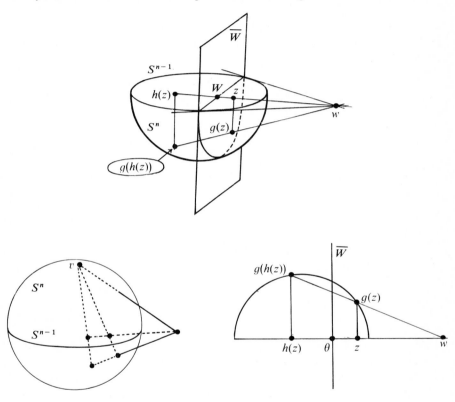

Figure 19.6.4

19.6.5. THEOREM. *The group $\mathrm{Is}(C)$ is generated by the restrictions to C of inversions of $\mathbf{R}^n$ leaving C invariant and having a pole outside C, and by reflections through hyperplanes containing the origin.*

Proof. This follows from 19.6.4 and 19.4.6.1. □

19.6.6. Notice that $\mathrm{Is}(C)$ has been expressed as the extension to C (the interior of S^{n-1}) of the group $\mathrm{M\ddot{o}b}(n-1)$. Conversely, $\mathrm{M\ddot{o}b}(n-1)$ is obtained by extending the isometries of C to the frontier S^{n-1} of C.

19.6.7. COROLLARY. *The Poincaré model is conformal, that is, hyperbolic and Euclidean angles coincide in* C.

Proof. We have seen that the two kinds of angles agree at the origin in B, so they agree at the origin in C. But Is(C) preserves (Euclidean) angles by 19.6.5, 8.6.6 and 10.8.5.2.   □

19.6.8. PLANE EXAMPLES. By the discussion above, the Poincaré model is very practical since it allows us to read angles off right away.

19.6.8.1. The area of a triangle in C can be calculated by using 19.5.4. It is easy to construct circles $\Gamma, \Gamma', \Gamma''$ orthogonal to S^1 and forming a "triangle" with vertices on S^1; thus there exist triangles in C (or B) with arbitrarily small angles, and area arbitrarily close to π, but the limit π is not achieved. Notice that in figure 19.6.8.1 the shaded region has sides with infinite length, but its area is finite and equal to π.

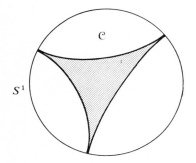
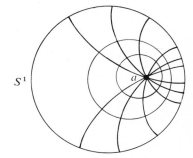

Figure 19.6.8.1   Figure 19.6.8.2

19.6.8.2. Circles. By 19.3.3.1, any circle with center $a \in C$ (cf. 19.3.3.2) is orthogonal to all lines containing a. By 10.10 such a hyperbolic circle is a circle in the pencil determined by S^1 and the limit point a. Conversely, every circle that belongs to this pencil and lies inside C is a hyperbolic circle with center a.

19.6.8.3. Horocycles. The previous paragraph takes care of Euclidean circles strictly interior to S^1; how about circles interior to S^1, but tangent to S^1 at w? Such a circle Γ is orthogonal to every line in C having w as a point at infinity; we can call it a circle with center at infinity, but it is not a straight line. One can obtain Γ as the limit of the circles passing through a fixed point $a \in C$ and with center on the line $\langle a, w \rangle$, as their center moves off toward w (figure 16.6.8.3). Such curves are called *horocycles;* they play an important role in several applications of hyperbolic geometry (see for example [A–A, 52 and Appendix 20]).

The reader can prove that in B horror cycles are exactly the superosculating ellipses to S^1.

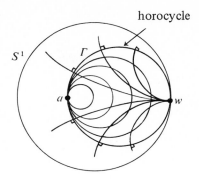

Figure 19.6.8.3

19.6.9. FORMULAS. The map Ω (cf. 19.6.1) is given by

$$\Omega(x) = \frac{2x}{1 + \|x\|^2}.$$

In two dimensions, if we identify $\mathbf{R}^2$ with $\mathbf{C}$, it is natural to expect that the elements of $\mathrm{Is}(\mathcal{C})$ have a simple expression in terms of the complex variable z. In fact, the elements of $\mathrm{Is}^+(\mathcal{C})$ are the restrictions to $\mathcal{C} = \{\, z \mid |z| < 1 \,\}$ of maps $\mathbf{C} \to \mathbf{C}$ of the form

$$z \mapsto e^{i\theta}\, \frac{z + z_0}{1 + \bar{z}_0 z},$$

where θ is real and $|z_0| = 1$ (see 6.8.16, and compare [CH2, 187]).

19.6.10. THE POINCARÉ METRIC. It is reasonable to ask what is the distance δ between two points in $\mathcal{C}$ whose Euclidean coordinates are given. Consider $a, b \in \mathcal{B}$ and let u, v be the points where $\langle a, b \rangle$ intersects S^{n-1}; by 19.2.5 we have

$$d(a, b) = \frac{1}{2}\big|\log([u, v, a, b])\big|.$$

Thus $[u, v, a, b] = [u, v, g(a), g(b)]^2$, by the definitions of g (cf. 19.6.1) and of the cross-ratio (cf. 6.2.4), and because, from figure 19.6.10,

$$\frac{au}{av} = \left(\frac{g(a)u}{g(a)v}\right)^2 \qquad \text{and} \qquad \frac{bu}{bv} = \left(\frac{g(b)u}{g(b)v}\right)^2.$$

By the proof of 18.10.7 we have

19.6.11
$$\delta(x, y) = \left|\log\left(\frac{xu}{xv} \Big/ \frac{yu}{yv}\right)\right|$$

for all $x, y \in \mathcal{C}$, where u, v denotes the points in S^{n+1} where the circle orthogonal to S^{n-1} and passing through x and y intersects S^{n-1}. The remarks

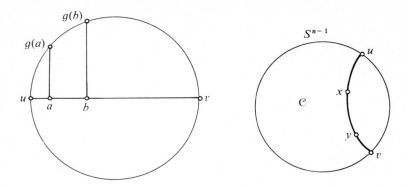

Figure 19.6.10

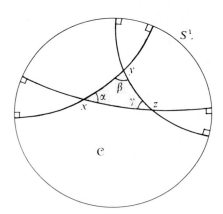

Figure 19.6.12.1

in 18.10.7 show that $\mu(a,b,c,d)$ is an invariant of $\mathbf{R}^n$ under any inversion, and in particular under stereographic projection. See also 16.8.6.

As an application, the reader can give a rigorous proof of the description of circles given in 19.6.8.2.

19.6.12. HYPERBOLIC TILINGS. Let p, q and r be integers ≥ 3 satisfying $\frac{1}{p} + \frac{1}{q} + \frac{1}{r} < 1$. By 19.3.4, there exists a triangle T in the hyperbolic plane C with angles $\alpha = \pi/p$, $\beta = \pi/q$ and $\gamma = \pi/r$. It can be proved that reflections through the sides of T generate a discrete subgroup of $\mathrm{Is}(C)$ that gives rise to a tiling of C (cf. [RM] and the references therein). Thus the situation is completely different from the Euclidean and elliptic cases (cf. 1.8.6, 1.7.4, 1.8.2), where there exists only a finite number of tilings.

It is possible to tile C with polygons other than triangles (see figure 19.6.12.3). Such tilings are interesting not only from the esthetic point of view, but also because they play a fundamental role in analysis (Fuchsian

Figure 19.6.12.2   Figure 19.6.12.3

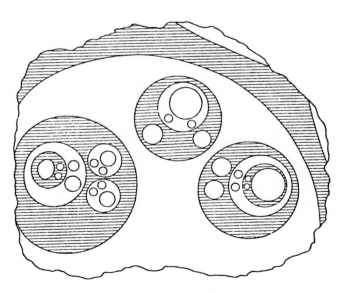

Figure 19.6.12.4

Source: [CR1]

M. C. Escher, *Circle Limit IV (Heaven and Hell)*, 1960. Woodcut in two colors, diameter 42 cm.
Escher Foundation–Haags Gemeentemuseum (The Hague)

Figure 19.6.12.5 (See [ER])

groups, cf. [RM], for instance) and in differential geometry, where they give rise to compact surfaces of constant negative curvature (if the polygon has $4k$ sides, the quotient under the appropriate identifications is a torus with k holes, 12.7.5.4): see [WF, 69], for example.

For the study of necessarily non-periodic hyperbolic tilings (cf. 1.7.2), see [RN2].

19.7. Other models

19.7.1. THE HALF-SPACE MODEL. The half-space model $\mathcal{H}$ (or half-plane, if $n = 2$), also due to Poincaré, is derived from the Poincaré model $\mathcal{C}$ by applying to it an inversion of $\mathbf{R}^n$ whose pole is on S^{n-1}; the frontier of $\mathcal{H}$ is a hyperplane H, the image of S^{n-1} under this inversion. The metric on $\mathcal{H}$ (and consequently lines, half-lines and angles) are transported from $\mathcal{C}$ by the same inversion. The half-space model is conformal.

The model $\mathcal{H}$ is less elegant than $\mathcal{C}$, in that statements generally have to be broken up into two cases. For example, lines in $\mathcal{H}$ are of two types: half-circles and Euclidean lines orthogonal to H. One solution is, of course, to add a point at infinity to H (cf. 20.1 and 20.6 for $H \cup \infty$); then lines can be characterized as Euclidean circles orthogonal to $H \cup \infty$. Similarly, Is($\mathcal{H}$) cannot be extended to H (cf. 19.6.6), only to $H \cup \infty$.

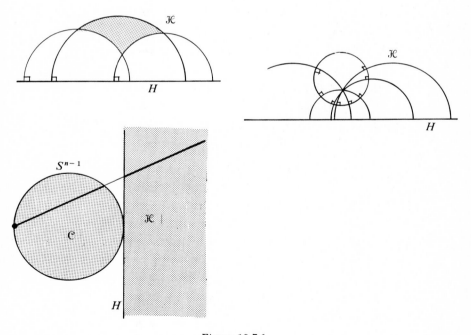

Figure 19.7.1

To make up for it, the half-space model is often the one that allows the simplest calculations; the case $n = 2$ is especially elementary, and can be used as the basis for an elementary exposition of hyperbolic geometry. The best thing is to take

$$\mathcal{H} = \{\, z \in \mathbf{C} \mid \operatorname{Re}(z) > 0 \,\};$$

then the elements of $\operatorname{Is}(\mathcal{H})$ are of the form $z \mapsto (az + b)/(cz + d)$, with $a, b, c, d \in \mathbf{R}$ and $ad - bc \neq 0$ (this highlights the isomorphism between $\operatorname{Is}(\mathcal{H})$ and $GP(1; \mathbf{R})$—cf. 18.10.4 and 5.2.4). The invariant measure is given simply by $\dfrac{1}{y}\, dx\, dy$.

19.7.2. OTHER MODELS. By analogy with chapter 18, we could consider the model consisting of a connected component E of the hypersurface in $\mathbf{R}^{n+1}$ with equation $q(\xi) = 1$ (cf. 15.4.3), as long as we consider on E not the intrinsic metric associated with the metric induced by $\mathbf{R}^{n+1}$ (cf. 9.9.7 and 18.4.3), but the Riemannian metric obtained by giving each $T_x E$ ($x \in E$) the Euclidean structure induced by the restriction of $-q$ to $\xi^\perp$.

Obtaining surfaces in $\mathbf{R}^3$, for example, whose intrinsic metric is hyperbolic is trickier. One can take surfaces of revolution, like those in figure 19.7.2, but they only yield local models since, by a theorem of Hilbert (cf. [KM, 476]) no surface in $\mathbf{R}^3$ can be isometric to the whole of $\mathcal{H}$. The models in figure 19.7.2 are due to Beltrami; the historical models of Lobatchevsky and Bolyai were axiomatic.

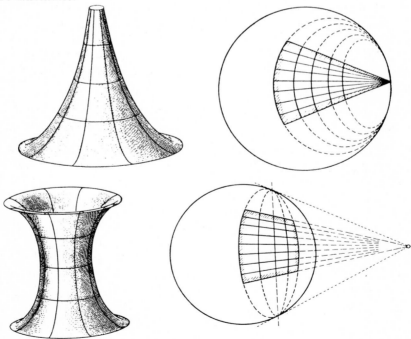

Figure 19.7.2 (Source: [KN2])

19.7.3. NOTE. Hyperbolic geometry is used in analysis ([G–G–V, chapter IV]), in arithmetic ([SA]), in differential geometry ([G–K–M, 195]), in ergodic theory ([A–A]), and in relativity (the Lorentz group is no other than the group $O(q)$ for $n = 3$).

For more on hyperbolic geometry, see [B–K], an elementary but very pedagogic reference; [CR3], which contains a very complete formulary; and [GR].

We have already remarked in 12.11.5.3 that the isoperimetric inequality 12.11.1 can be generalized to hyperbolic spaces.

19.8. Exercises

19.8.1. Study the subset

$$\{\, m \mid \overline{mm_1} = \cdots = \overline{mm_k} \,\}$$

of elliptic space, where the k points m_i are given.

19.8.2. Let D, D' be lines in elliptic space. Study the set

$$\{\, m \mid d(m, D) = d(m, D') \,\}.$$

(Distinguish between the cases $d = 2$ and $d > 2$.)

19.8.3. Study the notion of polar triangles in the elliptic plane.

19.8.4. Study the cases of equality of triangles in the elliptic plane.

19.8.5. Consider any partition of the elliptic plane into ϕ triangles, and call the total number of vertices σ and the number of sides α. Show that $\sigma - \alpha + \phi = 1$.

19.8.6. Show that the only conformal maps of an elliptic space into itself are isometries.

* **19.8.7.** Find the fourth angle of a quadrilateral in the hyperbolic plane such that three of its angles are right and the lengths of the sides which join two right angles are a and b.

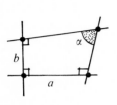

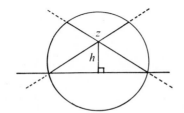
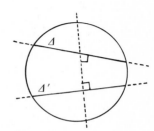

Figure 19.8.7   Figure 19.8.8   Figure 19.8.9

19.8.8. Show that, in the hyperbolic plane $\mathcal{B}$, the angles α between the extremal parallels to Δ through z (cf. 19.3.2) depends only on $d(z, \Delta) = h$. Find α as a function of h.

19.8.9. Give a geometric construction in $\mathcal{B}$, then in $\mathcal{C}$, for the common perpendicular Θ to two lines Δ, Δ' in the hyperbolic plane.

19.8.10. Prove 19.6.9 and the formulas in 19.7.1.

19.8.11. Given a line Δ in the hyperbolic plane (Klein model) and $h \in \mathbf{R}_+$, study the set

$$\{ m \in \mathcal{B} \mid d(m, \Delta) = h \}$$

. Show that such a set is bounded by segments of conics. Study the family of such sets as h ranges over $\mathbf{R}_+$ (make a drawing). Carry out the same study in $\mathcal{C}$, and compare the results.

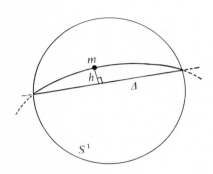

Figure 19.8.11

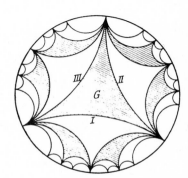

Figure 19.8.13 (Source: [KN2])

19.8.12. Given a point m in the hyperbolic plane (Klein model) and $h \in \mathbf{R}_+$, show that the set

$$\{ z \in \mathcal{B} \mid d(z, m) = r \}$$

is an ellipse in $\mathbf{R}^2$. Study the family of such sets as r ranges over $\mathbf{R}_+$ (make a drawing).

19.8.13. Explain figure 19.8.13.

19.8.14. Calculate explicitly the map Ξ in 19.6.1.

19.8.15. A *horosphere* in $\mathcal{C}$ is a sphere Σ interior to and tangent to S^{n-1} (cf. 19.6.8.3). Show that $\Sigma \cap \mathcal{C}$, considered as a submanifold of $\mathcal{C}$ with the induced Riemannian metric, is isometric to the Euclidean space $\mathbf{R}^{n-1}$. This means that $(n-1)$-dimensional creatures living in Σ would not be able to figure out that they were actually inside a hyperbolic space. If they lived in the whole of $\mathcal{C}$, how could they discover (even locally) that they did not live in a Euclidean space?

* **19.8.16.** Show that in n-dimensional hyperbolic space every set of $n + 2$ points z_i $(i = 1, \ldots, n + 2)$ obeys the relation

$$\det\bigl(\cosh[d(z_i, z_j)]\bigr) = 0.$$

19.8.17. Consider the set $\mathcal{B}$ with the new metric d_k $(k \in \mathbf{R}_+^*)$ given by $d_k(x, y) = kd(x, y)$. Show that, for $k \neq 1$, the metric space $(\mathcal{B}, d_k)$ is not isometric to $(\mathcal{B}, d)$. Define angles in $(\mathcal{B}, d_k)$, and write new formulas for triangles. What happens to these formulas as k approaches infinity?

19.8.18. Show that the length of a curve in the hyperbolic plane given, in the Klein model, by the polar coordinates $\bigl(\rho(t), \theta(t)\bigr)$, is equal to

$$\int_a^b \frac{\sqrt{\rho'^2 + \rho^2(1 - \rho^2)\theta'^2}}{1 - \rho^2}\, dt.$$

19.8.19. Study the analogues of 9.8.1 through 9.8.5 in elliptic space.

* **19.8.20.** Here n is an integer ≥ 3. We want to study n-sided polygons in the hyperbolic plane, all of whose sides are equal and all of whose angles have the value $2\pi/n$. Are there such polygons for any n? Are they unique up to isometries?

19.8.21. A THEOREM ABOUT OVERCROWDING. Prove that any set of nine points in the elliptic plain contains at least one triangle with perimeter less than or equal to π (cf. [BL, 262]).

19.8.22. THE FUBINI-STUDY METRIC ON $P^n(\mathbf{C})$. Consider $P^n(\mathbf{C})$ as the quotient of $S^{2n+1} \subset \mathbf{C}^{n+1} = \mathbf{R}^{2n+2}$ under the equivalence relation induced from the one in definition 4.1.1, and let $\pi : S^{2n+1} \to P^n(\mathbf{C})$ be the canonical projection. For $m, n \in P^n(\mathbf{C})$, set $d(m, n) = \bar{d}\bigl(\pi^{-1}(m), \pi^{-1}(n)\bigr)$, where $\bar{d}$ is the intrinsic metric on S^{2n+1}. Show that d makes $P^n(\mathbf{C})$ into a metric space. Study the shortest path between two points. When is it unique?

Extend your results to $P''(\mathbf{H})$.

19.8.23. What does figure 19.6.12.4 represent?

* **19.8.24.** ELLIPTIC EQUILATERAL SETS. An *equilateral set* in a metric space is any set $\{m_i\}_{m=1,\ldots,n}$ such that all the distances $d(m_i, m_j)$ $(i < j)$ are equal. Show that the elliptic plane P contains three-point equilateral sets with side lengths ranging from 0 to $\pi/2$; classify them under the action of $\mathrm{Is}(P)$. Show that P contains four-point equilateral sets with side length $\cos^{-1}(1/\sqrt{3})$ or $\cos^{-1}(1/\sqrt{5})$; study their behavior under the action of $\mathrm{Is}(P)$. Show that, up to isometries, P contains exactly one five-point and one six-point equilateral set, and their sides have length $\cos^{-1}(1/\sqrt{5})$. See [BL, 211–214].

19.8.25. Show that every hyperbolic triangle satisfies the inequalities

$$b^2 + c^2 - 2bc \cos \alpha \leq a^2 \leq b^2 + c^2 + 2bc \cos(\beta + \gamma).$$

Study the same question for spherical triangles.

19.8.26. Consider a fixed point p in three-dimensional hyperbolic space (Poincaré model), a regular polyhedron $P \subset \mathbf{R}^3$ with center a and vertices (b_i), and a real number $t > 0$. Construct a *regular polyhedron* $\Pi(P, t)$ in $\mathcal{C}$ as follows: start by drawing in $\mathcal{C}$ half-lines (D_i) originating at p and arranged in the same way that the half-lines $\overrightarrow{ab_i}$ are arranged around a. Mark along each D_i the point $q_i(t)$ whose distance to p is t. A *face* of $\Pi(P, t)$ is a plane in $\mathcal{C}$ containing the points $q_i(t)$ that correspond to the vertices of a face of P (check that such a plane exists and is well-defined). The polyhedron $\Pi(P, t)$ itself is the intersection of the appropriate half-spaces determined by the faces.

Find all regular hyperbolic polyhedra that tile $\mathcal{C}$.

Show that, as t increases, the dihedral angle between two adjacent faces of $\Pi(P, t)$ tends towards a limit which is a rational multiple of π.

Chapter 20
The space of spheres

In sections 18.10 and 19.2 we have seen the group $G(n)$, alias $\text{Möb}(n)$, act on the sphere S^n and on the open unit ball $B(0,1) = B$ of $\mathbf{R}^n$. The first action is on the isotropic cone of the quadratic form $q = -\sum_{i=1}^n x_i^2 + x_{n+1}^2$; the second, on the lines restricted to which q is positive definite. How about the lines where the restriction of q is negative definite?

The object of this chapter is to show that this set can be naturally identified with the set of spheres of a Euclidean affine space. Once the preliminaries have been laid out, everything follows from previous results on projective geometry and quadratic forms. The projective technique leads to an elegant theory of the space of spheres; in particular, it fulfils the need, felt since chapter 10, for a unified framework that permits the statement of results without reference to special cases. As usual, we get some bonus results, like the cyclids having six families of circles (section 20.7).

The fact that the set of lines where q is less than 0 is more complicated than the set where q is greater than 0 explains the difficulty of the geometry of spheres.

20.1. The space of generalized spheres

20.1.1. We wish to study the space of all spheres in a Euclidean affine space E. A geometric approach, similar in philosophy to 18.10.2, is illustrated in figure 20.1.1. Consider the stereographic projection f from a sphere S (minus its north pole) onto E, and associate to the sphere σ of E the vertex w of the cone tangent to S along $f^{-1}(\sigma)$. This correspondence between spheres in E and points in the space $\hat{E}$ where the projection is taking place can serve as the cornerstone for a theory of spheres, but it has two drawbacks: the choice of S and f is not canonical, and one must include the points at infinity of $\hat{E}$ and extend the relevant objects.

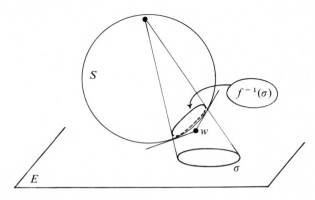

Figure 20.1.1

20.1.2. We shall avoid such problems by working algebraically, at least to get things started. The idea is to write spheres as affine quadrics, and work in the projective space of affine quadratic forms on E. What will make the whole thing work is that there exists on this projective space a canonical quadratic form (whose image corresponds to the points of the sphere S in 20.1.1). Our construction may appear artificial in the beginning, but we trust that the reader will eventually be pleased with the construction and the results obtained from it.

20.1.3. NOTATION. In this whole chapter E is a Euclidean affine space and $Q(E)$ is the vector space of affine quadratic forms on E. A form $q \in Q(E)$ has an associated *symbol* $\vec{q} \in Q(\vec{E})$. If we vectorialize E at a, we can write $q = q_2 + q_1 + q_0$, with $q_0 \in \mathbf{R}$, $q_1 \in E_a^*$ and q_2 identified with $\vec{q}$ (cf. 3.3).

We will now complete E by adding a single point ∞ at infinity. (Do not confuse this situation with the one in chapter 15, where the affine space X was completed by adding a hyperplane ∞_X.)

The study made in 10.7.6 shows that, in talking of spheres, it is natural to introduce the vector subspace of $Q(E)$ given by

20.1.4 $\check{S}(E) = \{\, q \in Q(E) \mid \vec{q} = k\| \cdot \|^2 \text{ for } k \in \mathbf{R} \,\}.$

Notice that $\check{S}(E)$ is an $(n+2)$-dimensional subspace of $Q(E)$. Our acquaintance with affine and projective quadrics leads us to introduce the projective space

20.1.5 $$p : \check{S}(E) \setminus 0 \rightarrow P\big(\check{S}(E)\big) = S(E),$$

called the *space of generalized spheres* of E. Thus $S(E)$ is an $(n+1)$-dimensional projective subspace of $\mathrm{PQ}(E)$ (cf. 14.1).

20.1.6. IMAGES. What are the images of quadrics $\lambda \in S(E)$? Let q be an equation for λ, and express q in E_a, as in 10.7.6:

20.1.7 $$q = k\| \cdot \|^2 + (\alpha \mid \cdot) + h,$$

where $\alpha \in \vec{E}$ and $h \in \mathbf{R}$. If $k = 0$ and $\alpha \neq 0$ the image is an affine hyperplane in E; if $k = 0$ and $\alpha = 0$ we find $\mathrm{im}(\lambda) = \emptyset$ because $h \neq 0$. If $k \neq 0$ we see from 10.7.6 that the image is a sphere, a point or the empty set, and $\mathrm{im}(\lambda)$, when non-empty, is enough to determine λ. Thus, if Θ is the set of affine hyperplanes in E and Σ is the set of spheres in E with positive radius, we have an injection

20.1.8 $$i : \Sigma \cup E \cup \Theta \rightarrow S(E),$$

where the injection on E is obtained by considering a point in E as a "sphere of radius zero" centered at that point. The complement of $i(\Sigma \cup E \cup \Theta)$ in $S(E)$ consists of two things: the point associated with the equation $q = 1$, *denoted by* $\infty \in S(E)$ and called the *point at infinity*, and the points associated with equations

$$q = k\| \cdot \|^2 + (\alpha \mid \cdot) + h \qquad \text{with} \quad \|\alpha\|^2 - 4kh < 0,$$

which can be considered as "spheres of non-zero imaginary radius". We generally identify Σ, E and Θ with the corresponding subsets of $S(E)$, and we set

20.1.9 $$\hat{E} = E \cup \infty \subset S(E), \qquad S(E) = \Sigma \cup \hat{E} \cup \Theta \subset S(E).$$

Recall the caveat in 20.1.3. The justification for all these objects will become clear as we go along, but the reader can also explain them in the context of 20.1.1.

20.2. The fundamental quadratic form

20.2.1. LEMMA. *For a form q on E_a having the expression 20.1.7, the quantity*

$$\rho(q) = \frac{\|\alpha\|^2 - 4kh}{4}$$

depends only on $q \in \check{S}(E)$ and not on a. The function ρ is a quadratic form on $\check{S}(E)$, with signature $(n+1, 1)$, and will be called the fundamental quadratic form on $\check{S}(E)$.

Proof. This is a reasonable result, since by 10.7.6 the formula for $\rho(q)$ bears a striking resemblance to the square of the radius of the sphere $q^{-1}(0)$. A calculation involving translations is all that is needed to establish invariance. The signature can be read off from the definition: $\|\alpha\|^2$ contributes n positive directions and $4kh = (k+h)^2 - (k-h)^2$ contributes a positive and a negative one (cf. 13.4.7). $\qquad\qquad\square$

The polar form of ρ will be *denoted* by $R(\cdot,\cdot)$; we have

20.2.2 $R\big(k\|\cdot\|^2 + (\alpha\,|\,\cdot\,) + h,\, k'\|\cdot\|^2 + (\alpha'\,|\,\cdot\,) + h'\big) = \dfrac{1}{4}\big((\alpha\,|\,\alpha') - 2(k'h + kh')\big).$

For example, if q, q' have as images the spheres a, a' of radii r, r', the distance between the centers is given by

20.2.3 $$d^2(a, a') = r^2 + r'^2 - \frac{2}{kk'}R(q, q').$$

20.2.4. The *fundamental quadric* in $S(E)$ is the projective quadric having ρ for equation; it is still *denoted* by ρ. The symbol $\perp$ will *stand* for polarity with respect to ρ. We know (cf. 14.3.3 or 18.10) that the image of ρ is homeomorphic to the sphere S^n; but the remarks above (cf. 10.7.6) show that $\mathrm{im}(\rho) = \hat{E}$. This means exactly that the embedding $i : E \to S(E)$ is a homeomorphism onto its image $i(E)$, and that $\hat{E}$ appears as the Alexandroff compactification of E, obtained by adding a point at infinity (here ∞). This is the tidying up promised in 10.8.4.2, and to be completed in 20.6.3.

20.2.5. Now 10.7.6 shows that points in $S(E)$ coming from forms $q \in \check{S}(E)$ such that $\rho(q) < 0$ are the spheres of imaginary radius, and those for which $\rho(q) > 0$ are the elements of $\Sigma \cup \Theta$, that is, true spheres and hyperplanes. If we multiply ρ by -1 we obtain a form with signature $(1, n+1)$, and 19.2 shows that the spheres of imaginary radius make up a hyperbolic space (of dimension $n + 1$); that is the piece of $S(E)$ that does not concern us here. The piece $\hat{E}$ has the geometry of the sphere S^n, under the action of Möb(n) (cf. 19.2.2 or 18.10); this allows us to talk about the conformal (or circular) group Conf(E) of E, which we shall do in 20.6. The last piece, $S(E)$, is the one that interests us most.

20.2.6. The geometry of $S(E)$ is not simple, as can be gleaned from 10.8.4.2, and this for two reasons: first, the topology of $S(E)$ is that of a projective space minus a point, which retracts to a real projective space of dimension one lower, and thus cannot be simply connected (already $P^2(\mathbf{R})$ minus a point is homeomorphic to a Möbius strip, cf. 4.3.9.1). Next, although the form ρ is everywhere positive in $S(E)$, it may or may not take negative values on a line joining two points in $S(E)$ (for an analysis of the possible cases, see 20.4.2).

20.2.7. From 20.2.2 and 14.5.2.1 it is clear that $\Theta \cup \infty$ is exactly the hyperplane tangent to $\hat{E}$ at ∞ (figure 20.2.7).

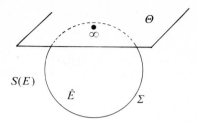

Figure 20.2.7

20.3. Orthogonality

A direct calculation involving 20.2.2, 20.2.3 and 10.7.10.2 gives the following

20.3.1. LEMMA. *Given $s, s' \in S(E)$, the condition $s \perp s'$ is equivalent to:*
— *if $s, s' \in \Sigma$, the two spheres are orthogonal in the sense of 10.7.10.2;*
— *if $s \in \Sigma$ and $s' \in \Theta$, the hyperplane s' goes through the center of s;*
— *is $s, s' \in \Theta$, the two hyperplane are orthogonal in the Euclidean sense;*
— *if $s \in \Sigma \cup \Theta$ and $s' \in E$, the point s' lies in s;*
— *if $s \in \Sigma \cup \Theta$ and $s' = \infty$, then s is a hyperplane.* $\square$

Thus orthogonality in $S(E)$ is synonymous with Euclidean orthogonality, if we gloss over a few special cases. If we want to take the special cases into account, we can *define* the image $\overline{\mathrm{im}(s)}$ of a conic $s \in S(E)$ in $\hat{E}$ as follows:

20.3.2
$$\begin{cases} \overline{\mathrm{im}(s)} = \mathrm{im}(s) & \text{if } s \in \Sigma; \\ \overline{\mathrm{im}(s)} = \mathrm{im}(s) \cup \infty & \text{if } s \in \Theta; \\ \overline{\mathrm{im}(s)} = s & \text{if } s \in E; \\ \overline{\mathrm{im}(s)} = \infty & \text{if } s = \infty; \\ \overline{\mathrm{im}(s)} = \emptyset & \text{if } \mathrm{im}(s) = \emptyset \text{ and } s \neq \infty. \end{cases}$$

Under this convention, 20.3.1 leads to the following interpretation for the points of $\mathrm{im}(s)$ in $S(E)$:

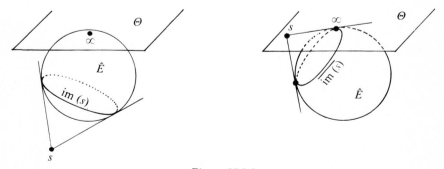

Figure 20.3.3

20.3.3. LEMMA. *For $s \in S(E)$ the set $\overline{\mathrm{im}(s)}$ coincides with $\hat{E} \cap s^{\perp}$, and with the set of points where the cone with vertex s circumscribed around $\hat{E}$ touches $\hat{E}$.* □

20.3.4. EXAMPLE. The center ω of s is the second point of $\hat{E}$ (distinct from ∞) where the line $\langle s, \infty \rangle$ intersects $\hat{E}$. In fact, ω is characterized by the fact that all hyperplanes through ω are orthogonal to s (in the Euclidean sense); by the lemma, the set of such hyperplanes is

$$\omega^{\perp} \cap \Theta = \omega^{\perp} \cap \infty^{\perp} = (\omega\infty)^{\perp}$$

(cf. 14.5.2.3). And saying that $s \perp (\omega\infty)^{\perp}$ is the same as saying that $s \in \omega\infty$.

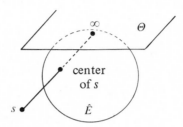

Figure 20.3.4

20.3.5. Thus we can say that hyperplanes in E are spheres with center at ∞, and ∞ is both a particular hyperplane ($\infty \in \Theta$) and a particular sphere of radius zero ($\infty \in \hat{E}$) and center everywhere (cf. [PC, §72]).

20.4. Intersection of two spheres

20.4.1. Consider $s, s' \in \Sigma \cup \Theta$, with equations q, q'. What is the quantity $\dfrac{R(q, q')}{\sqrt{\rho(q)\rho(q')}}$? Here we don't have an exact analogue for 19.2.5 or 8.6.3, and things depend on whether or not the projective line $\langle s, s' \rangle$ intersects $\hat{E}$. The situation is summarized in the next result, which is obtained by studying the discriminant of quadratic equation:

20.4.2. PROPOSITION AND DEFINITION. *If $s, s' \in \Sigma \cup \Theta$ have equations q, q', the following conditions are equivalent:*
- *the projective line ss' intersects $\hat{E}$ in at most one point;*
- $\dfrac{|R(q, q')|}{\sqrt{\rho(q)\rho(q')}} \leq 1$;
- *the restriction of ρ to ss' is non-negative;*
- $\overline{\mathrm{im}(s)} \cap \overline{\mathrm{im}(s')} \neq \emptyset$.

If these conditions are satisfied, we say that s, s' are secant, and define the angle between s and s', denoted by $[s, s']$, as

$$[s, s'] = \text{Arccos}\left(\frac{|R(q, q')|}{\sqrt{\rho(q)\rho(q')}}\right).$$

The angle is a real number in $[0, \pi/2]$, and depends only on s and s', not on q and q'.   □

20.4.3. The name "angle" is justified by the fact that, for $m \in \text{im}(s) \cap \text{im}(s')$, the angle between the tangent hyperplanes $T_m s$ and $T_m s'$ (where $T_m s = s$ if $s \in \Theta$) is given by $[s, s']$. This follows from 20.2.3 if $s, s' \in \Sigma$; the case where s or s' are hyperplanes is even easier.

For $[s, s'] = 0$, either s, s' are tangent at $m \in \text{im}(s) \cap \text{im}(s')$, or s, s' are parallel hyperplanes (the latter case happens when the images are disjoint, that is, $\overline{\text{im}(s)} \cap \overline{\text{im}(s')} = \infty$).

For $[s, s'] = \pi/2$ we are back in 20.3.

For $[s, s'] \in \,]0, \pi/2[$, this angle corresponds to two supplementary values of the angle defined in 10.7.7. This loss of information is the cost of incorporating hyperplanes into the picture and projectivizing. To make up for it, many problems that were tricky in elementary geometry (cf. 10.13.20 and 20.4.4) become very simple, since they can be solved by linear algebra.

20.4.4. ISOGONAL SPHERES. Given $s, s' \in \Sigma \cup \Theta$, the set

$$\{t \in \Sigma \cup \Theta \mid [t, s] = [t, s']\}$$

can be decomposed into two sets F, F', as follows: there exist $u, u' \in S(E)$ such that $t \in F$ (resp. $t \in F'$) implies $t \perp u$ (resp. $t \perp u'$). The spheres u and u' are called the *bisectors* of s and s'.

Proof: we take equations q, q' for s, s' such that $\rho(q) = \rho(q') = 1$, and look for the desired t with equation q'' such that $\rho(q'') = 1$. The condition $[t, s] = [t, s']$ implies $|R(q, q'')| = |R(q', q'')|$, which reduces to either $R(q + q', q'') = 0$ or $R(q - q', q'') = 0$. The spheres u, u' are given by the equations $q + q'$ and $q - q'$.

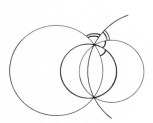   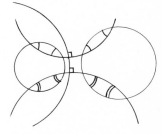

Figure 20.4.4

More generally, we can study the sets $\left\{ t \mid [t, s_i] = \alpha_i \text{ for all } i = 1, \ldots, k \right\}$, where $\alpha_i \in [0, \pi/2]$ and $s_i \in \Sigma \cup \Theta$ for all i. Compare with 10.7.8.

20.5. Pencils

We have already encountered in 20.4.1 projective lines in $S(E)$, which are a particular case of pencils of projective quadrics (cf. 14.2). If s, s' are secant and $[s, s'] \neq 0$, the two spheres intersect along an $(n-2)$-sphere (cf. 10.7.5), and the intersection depends only on the projective line ss'. This justifies the following definition:

20.5.1. DEFINITION. *A k-pencil of spheres in E is a k-dimensional projective subspace of $S(E)$. A (generalized) k-sphere in E is an $(n-k-1)$-pencil of spheres in E.*

For $n = 2$, zero-spheres are pairs of points (or the empty set). For $n = 3$, one-spheres are the usual (generalized) circles in the Euclidean space we live in. Proposition 20.4.2 can be generalized to give this:

20.5.2. LEMMA. *If K is a k-pencil of spheres (for k arbitrary), the intersection $\bigcap_{s \in K} \overline{\mathrm{im}(s)}$ is non-empty if and only if the restriction of ρ to K is non-negative.*

Proof. By 20.3.3, we have $\bigcap_{s \in K} \overline{\mathrm{im}(s)} = \hat{E} \cap K^{\perp}$; but ρ cannot be non-negative on both K and $K^{\perp}$, by 13.4.7. $\square$

20.5.3. DEFINITION. *The angle between the k-sphere K and the k'-sphere K' is defined as*

$$[K, K'] = \inf \left\{ [u, u'] \mid u \in K, u' \in K' \right\}.$$

This angle only makes sense in certain cases. For $k = 1$, $k' = 2$ and $n = 3$ we have, by 8.6.7 and 20.4.3, the usual angle between a circle and a sphere in our everyday Euclidean space.

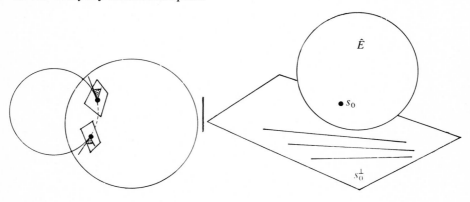

Figure 20.5.3   Figure 20.5.4

20.5.4. APPLICATION. The remarks above give a good explanation for 18.9. Consider a fixed sphere s_0 with imaginary radius; $s_0^\perp$ is a hyperplane in $S(E)$ on which ρ is positive definite. If $n = 3$ this $s_0^\perp$ is an elliptic space, and we can apply 19.1.4 to it, obtaining the properties stated in 10.12.3. Moreover, the group $PO(\rho)$ (cf. 13.7.1) acts transitively on the set of spheres of imaginary radius, so all pairs of circles that form a paratactic ring with angle α are conjugate under $\mathrm{Conf}(\hat{E})$. As an exercise, the reader can translate the results in 18.8.4 into this language.

20.5.5. LINKING OF TWO CIRCLES. Again for $n = 3$, we can read off the linking of two circles. Two circles K and K' will be linked if and only if the projective hyperplane $\langle K \cup K' \rangle$ spanned by them does not intersect $\hat{E}$.

20.5.6. CLASSIFICATION OF PENCILS. We study the positions that a line can take with respect to the elements that determine the geometry of E (namely, E, ∞ and Θ). There are six cases, shown in figure 20.5.6 and in the following table:

Type	Position	Nature	$(n-2)$-sphere
I	$K \subset \Theta$ and $K \ni \infty$	parallel hyperplanes	$\emptyset$
II	$K \subset \Theta$ and $K \not\ni \infty$	secant hyperplanes	$(n-2)$-dimensional projective subspace
III	$K \not\subset \Theta$ and $K \ni \infty$	concentric spheres	$\emptyset$
IV	$K \not\ni \infty$ and $K \cap E = 2$ points	pencil with limit points	$\emptyset$
V	$K \not\ni \infty$ and $K \cap E = 1$ point	tangent spheres	1 point
VI	$K \not\ni \infty$ and $K \cap E = \emptyset$	secant spheres	$(n-2)$-sphere

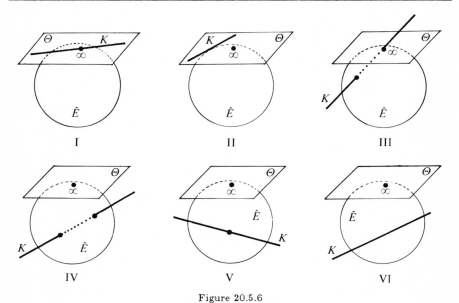

Figure 20.5.6

In the case of the plane we obtain the three figures in 10.10.1, corresponding to case IV, V and VI. Notice that the pencil of circles orthogonal to the pencil K is exactly $K^\perp$, by 20.3.1 and 10.10. Orthogonality between types is as follows, for $n = 2$: $\mathrm{I}^\perp = \mathrm{I}$, $\mathrm{II}^\perp = \mathrm{III}$, $\mathrm{IV}^\perp = \mathrm{VI}$ and $\mathrm{V}^\perp = \mathrm{V}$.

20.5.7. NOTE. By 14.3.7 there exists a natural (yes, natural) isomorphism between the circles in three-dimensional Euclidean space and the points of the complex projective quadric $C(4)$.

20.6. The circular group

20.6.1. We have already see in 20.2.5 that it is natural to introduce the projective group $\mathrm{PO}(\rho)$ of the quadratic form ρ. This group acts transitively on $\Sigma \cup \Theta$ by 13.7.1, and also on $\hat{E}$, by an action isomorphic to that of $\mathrm{M\ddot{o}b}(n)$. When considered as acting on $\hat{E}$ in this way, $\mathrm{PO}(\rho)$ is *written* $\mathrm{Conf}(\hat{E})$, and called the *circular group of* E, or the *conformal group of* $\hat{E}$ (notice that the conformal group of E is $\mathrm{Conf}(E) = \mathrm{Sim}(E)$—cf. 9.5.4.6). See 20.6.4.

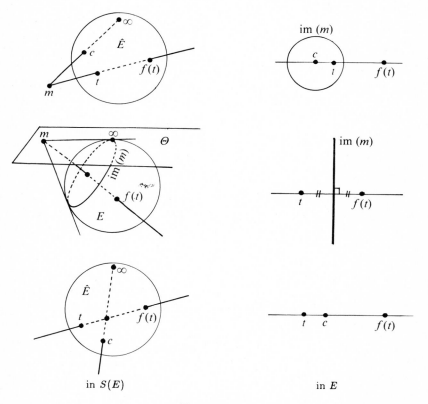

Figure 20.6.2

20.6.2. We now figure out what the elements of $\mathrm{Conf}(\hat{E})$ are. By 14.7.4, $\mathrm{Conf}(\hat{E})$ is generated by maps of the type shown in figure 14.7.4, where the point m lies in $S(E) \setminus \hat{E}$ and $f : t \mapsto f(t)$ is well-defined by the condition that t, $f(t)$ and m are collinear. In E, assuming that t, $f(t) \in E$ and $\mathrm{im}(m) \neq \emptyset$, this condition means that $\mathrm{im}(m)$ and the two spheres of radius zero form a pencil, or again that $f(t)$ is the image of t under the inversion whose pole is the center c of $\mathrm{im}(m)$ and whose fixed sphere is $\mathrm{im}(m)$. The extension to $\hat{E}$ is given by $f(c) = \infty$ and $f(\infty) = c$, by 20.3.4. If $\mathrm{im}(m)$ is a hyperplane, we obtain the reflection through $\mathrm{im}(m)$, extended by $f(\infty) = \infty$. If $\mathrm{im}(s) = \emptyset$, we obtain the inversion whose center is the center c of s and with power $\rho(q)$, where $q = \| \cdot \|^2 + (\alpha \mid \cdot) + h$ is a normalized equation for s. From 14.7.4 and 13.7.12 we deduce the following

20.6.3. THEOREM. *Every map in $\mathrm{Conf}(\hat{E})$ is the product of at most $n + 2$ maps f of the following types: either f is an inversion of E with pole c, extended to $\hat{E}$ by $f(c) = \infty$ and $f(\infty) = c$, or f is a hyperplane reflection extended to $\hat{E}$ by $f(\infty) = \infty$.* $\square$

20.6.4. NOTES. The name "circular group" comes from the fact that $\mathrm{Conf}(\hat{E})$ transforms spheres, and consequently circles, into spheres and circles. This is obvious by construction. All maps with this property lie in $\mathrm{Conf}(\hat{E})$, by 18.10.4. We now have completed what was left unfinished in 10.8.4.2 and 18.10.2.2.

There exists a theory of oriented spheres, due mostly to Laguerre. In the framework of $\check{S}(E)$, it consist in identifying points on each half-line (instead of each line, which gives $S(E)$). We have already mentioned this geometry and given some references in 10.11.6.

20.7. Polyspheric coordinates

20.7.1. The reader who leafs through books dating from a few decades back, like [KN2, 49 ff.], will encounter the expressions "tetracyclic coordinates" and "pentaspheric coordinates" in connection with the study of circles in a plane and spheres in three-dimensional space, respectively. In the picture above, these are just homogeneous coordinates in $S(E)$ that arise from an orthogonal basis for $\check{S}(E)$, in the sense of the fundamental form ρ. Such coordinates will be called here *polyspheric*. They can be used to study spheres in E, and also the points in E, which are but spheres satisfying $\rho = 0$. Here is an application for such coordinates.

20.7.2. CYCLIDS AND THEIR FAMILIES OF CIRCLES. We have encountered in 10.12 relatively simple (degree-four) surfaces that are connected and contain four one-parameter families of circles. Such surfaces, the tori of revolution, are a particular case of more general degree-four surfaces, which can contain up to six families of circles.

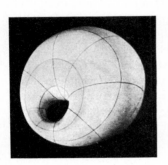
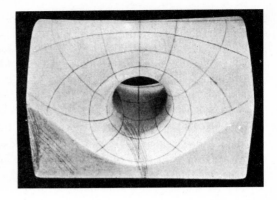

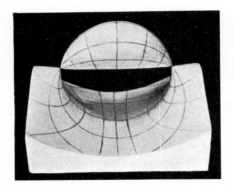

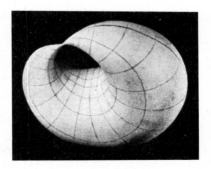

Figure 20.7.3 (Source: [H–C])

We define a *cyclid* in the three-dimensional Euclidean space E as a surface that can be written in the form $E \cap \mathrm{im}(\mu)$, where μ is a projective quadric in the projective space $S(E)$. We will not pursue a systematic study of cyclids, which the reader can find in [DX, 405–481], but we will consider one example. Suppose that μ and ρ can be simultaneously reduced (cf. 13.5); we can assume (cf. 13.4.2 and 20.2.2) that they have the form

$$\rho = z_1^2 + z_2^2 + z_3^2 + z_4^2 - z_5^2, \qquad \mu = \sum_{i=1}^{5} a_i z_i^2.$$

Looking for circles in $\mathrm{im}(\mu)$ is the same as looking for projective planes in $S(E)$ that lie in the pencil $k\rho + h\mu$ of quadrics in $S(E)$. The degenerate quadrics in this pencil are the following:

20.7.3 $a_1 \rho - \mu, \quad a_2 \rho - \mu, \quad a_3 \rho - \mu, \quad a_4 \rho - \mu, \quad a_5 \rho + \mu.$

By 14.1.7 their images are cones whose base is a quadric in a hyperplane of $S(E)$. For example, $a_5 \rho + \mu$ has equation $\sum_{i=1}^{4}(a_5 + a_i)z_i^2$, so its image is the cone with vertex $(0,0,0,0,1)$ and having for base the quadric $\sum_{i=1}^{4}(a_5 + a_i)z_i^2$ in the hyperplane $z_5 = 0$. If this latter quadric contains lines, $a_5\rho + \mu$ contains planes. We know that $\sum_{i=1}^{4}(a_5 + a_i)z_i^2$ contains two families of lines if among the $a_5 + a_1$ two have positive signs and two negative (cf. 14.4). To sum up, we find on the cyclid $E \cap \mathrm{im}(\mu)$ twice as many families of circles as there are equations in 20.7.3 having two positive and two negative coefficients. It is easy to see that at most three equations can satisfy this property, so the maximum number of families of circles is six, and this number is achieved, for example, by $2z_1^2 + 3z_2^2 - 4z_3^2 - 3z_4^2 - z_5^2$.

The torus is a particular case of the so-called *Dupin cyclid*, those for which at least two of the a_i are equal, resulting in the coalescence of some of the families of circles.

20.8. Exercises

* **20.8.1.** Construct geometrically the tangent hyperplane to $\mathrm{im}(s)$ in $S(E)$.

20.8.2. Justify the word "bisector" in 20.4.4.

20.8.3. Translate [HD, volume 2, 608–667] into the language of this chapter.

20.8.4. Translate [KN2, 49–58 and 100–105] into the language of this chapter.

* **20.8.5.** Show that the torus is a cyclid of Dupin (20.7.3).

20.8.6. Let C, D, C', D' be circles in E. Denote by r, s, r', s' their respective radii, and by d, d' the distance between the centers of C, D and C', D', respectively. Show that a necessary and sufficient condition for the existence of a map $f \in \mathrm{Conf}(\hat{E})$ taking $C \cup D$ into $C' \cup D'$ is that

$$\left| \frac{r^2 + s^2 - d^2}{2rs} \right| = \left| \frac{r'^2 + s'^2 - d'^2}{2r's'} \right|.$$

Compare your proof with the one given in [D–C1, 222].

∗ **20.8.7.** THEOREM OF DARBOUX. If three points of a line describe three spheres whose centers are collinear, then every point of that line also describes such a sphere, or possibly a plane for one exceptional point. Find a relation between four points of the line and the centers of the four spheres they describe.

Now consider two finite sets of lines on a one-sheet hyperboloid or hyperbolic paraboloid (see 15.3.3.3), each set contained in one family of generators (14.4). Think of the lines as metal rods, attached together at all their common points by means of joints that allow rotational freedom. Deduce from Darboux's theorem that the object thus obtained is deformable. Can it be completely flattened? If so, what is the final figure?

Figure 20.8.7 (Source: [KS])

Bibliography

[A–A] V. I. Arnol'd and A. Avez. *Ergodic Problems of Classical Mechanics.* Benjamin, New York, 1968.

[A–B] Annequin and Boutigny. *Cours d'Optique.* Vuibert, Paris.

[A–B–S] M. F. Atiyah, R. Bott, and A. Shapiro. Clifford modules. *Topology,* 3:3–38, 1964–1965, supplement.

[AE] Evry Schatzman. *Astronomie. Encyclopédie de la Pléiade, 13,* Gallimard, Paris, 1962.

[AN] Emil Artin. *Geometric Algebra.* Interscience, New York, 1957.

[AS] J.-M. Arnaudies. *Les Cinq Polyèdres de $\mathbf{R}^3$ et leurs Groupes.* Centre de Doc. Univ., Paris, 1969.

[AW1] A. D. Alexandrow. *Die innere Geometrie der konvexen Fläche.* Akademie-Verlag, Berlin, 1955.

[AW2] A. D. Alexandrow. *Konvexe Polyeder.* Akademie-Verlag, Berlin, 1958.

[BA] L. Bianchi. Sulle configurazioni mobili di Möbius nelle transformazione asintotiche delle curve e delle superficie. *Rend. del Circolo Mat. di Palermo,* 25:291–325, 1908.

[BAN] Thomas Banchoff. Non-rigidity theorems for tight polyhedra. *Archiv der Mathematik,* 21:416–423, 1970.

[BA–WH] Thomas Banchoff and James White. The behavior of the total twist and self-linking number of a closed space curve under inversions. *Mathematica Scandinavica,* 36:254–262, 1975.

[B–B] Edwin F. Beckenbach and Richard Bellman. *Inequalities. Ergebnisse der Mathematik, 30,* Springer-Verlag, Berlin, 1961.

[B- B- M] L. Bérard-Bergery, J.-P. Bourguignon, and E. Mazet. Séminaire de Géométrie Riemannienne 1970–1971: *Variétés à courbure négative*. *Publications Mathématiques de l'Université Paris VII, 8*.

B–B-N-W–Z] H. Brown, R. Bülow, J. Neubüser, H. Wondratschek, and H. Zassenhaus. *Crystallographic Groups of Four-Dimensional Space*. Wiley, New York, 1978.

[BD] Y. Brossard. *Rosaces, Frises et Pavages*. CEDIC, Paris, 1977.

[BE] Pierre Brousse. *Mécanique*. Armand Colin, Paris.

[BER1] Marcel Berger. *Lecture Notes on Geodesics in Riemannian Geometry*. Tata Institute, Bombay, 1965.

[BER2] Marcel Berger. Une caractérization purement métrique des variétés riemanniennes à courbure constante. In P. L. Butzer and F. Fehér, editors, *E. B. Christoffel*, pages 480–493, Birkhäuser, Basel, 1981.

[BES] Arthur Besse. *Manifolds all of whose geodesics are closed. Ergebnisse der Mathematik, 93*, Springer-Verlag, Berlin, 1978.

[B–F] O. Bonnesen and W. Fenchel. *Theorie der konvexen Körper*. Springer, Berlin, 1934.

[B–G] Marcer Berger and Bernard Gostiaux. *Differential Geometry*. Springer-Verlag, 1987.

[B–H] A. Borel and F. Hirzebruch. Characteristic classes of homogeneous spaces. *Americal Journal of Math.*, 80:458–538, 1958.

[B–H–H] W. L. Black, H. C. Howland, and B. Howland. A theorem about zigzags between two circles. *American Math. Monthly*, 81:754–757, 1974.

[BI0] Nicolas Bourbaki. *Theory of Sets. Elements of Mathematics, 1*, Addison-Wesley, Reading, Mass., 1968.

[BI1] Nicolas Bourbaki. *General Topology. Elements of Mathematics, 3*, Addison-Wesley, Reading, Mass., 1966.

[BI2] Nicolas Bourbaki. *Algèbre* (chapter IX). *Eléments de Mathématiques, 2*, Hermann, Paris, 1959.

[BI3] Nicolas Bourbaki. *Algebra* (chapters I–III). *Elements of Mathematics, 2*, Addison-Wesley, Reading, Mass., 1974.

[BI4] Nicolas Bourbaki. *Groupes et Algèbres de Lie* (chapters IV–VI). *Eléments de Mathématiques, 7*, Hermann, Paris, 1960.

[BI5] Nicolas Bourbaki. *Espaces Vectoriels Topologiques. Eléments de Mathématiques, 5*, Hermann, Paris, 1953.

[B- K] Herbert Busemann and Paul J. Kelly. *Projective Geometry and Projective Matrices*. Academic Press, New York, 1953.

[BL] Leonard M. Blumenthal. *Theory and Applications of Distance Geometry*. Chelsea, Bronx, N.Y., 1970.

[BLA1] Wilhelm Blaschke. *Kreis und Kugel*. de Gruyter, Berlin, second edition, 1956.

[BLA2] Wilhelm Blaschke. *Vorlesungen über Differentialgeometrie, II: Affine Differentialgeometrie*. *Grundlehren der Mathematischen Wissenschaften, 7*, Springer-Verlag, Berlin, second edition, 1923.

[BLA3] Wilhelm Blaschke. *Vorlesungen über Differentialgeometrie, III: Differentialgeometrie der Kreise und Kugeln*. *Grundlehren der Mathematischen Wissenschaften, 29*, Springer-Verlag, Berlin, 1929.

[BL-BO] W. Blaschke and G. Bol. *Geometrie der Gewebe*. Springer-Verlag, Berlin, 1938.

[B-M] Leonard M. Blumenthal and Karl Menger. *Studies in Geometry*. W. H. Freeman, San Francisco, 1970.

[BN] D. G. Bourgin. *Modern Algebraic Topology*. MacMillan, New York, 1963.

[BO1] Armand Borel. Sur l'homologie et la cohomologie des groupes de Lie compactes connexes. *Americal Journal of Math.*, 76:273–342, 1954.

[BO2] Armand Borel. Sur la cohomologie des espaces fibrés principaux et des espaces homogènes de groupes de Lie compactes. *Annals of Math.*, 57:115–207, 1953.

[BO3] Armand Borel. *Seminar on Transformation Groups*. *Annals of Mathematics Studies, 46*, Princeton University Press, Princeton, 1960.

[BOM] G. Bomford. *Geodesy*. Clarendon Press, Oxford, third edition, 1971.

[BOU1] H. Bouasse. *Optique Géométrique Elémentaire*. Delagrave, Paris, 1917.

[BOU2] H. Bouasse. *Appareils de Mesure*. Delagrave, Paris, 1917.

[BP] Paul Baudoin. *Les ovales de Descartes et le limaçon de Pascal*. Vuibert, Paris.

[B-P-B-S] M. Berger, P. Pansu, J.-P. Berry, and X. Saint-Raymond. *Problems in Geometry*. Springer-Verlag, New York, 1984.

[BR] R. Baer. *Linear Algebra and Projective Geometry*. Academic Press, New York, 1952.

[BR–TI] F. Bruhat and J. Tits. Groupes réductifs sur un corps local, I: Données radicielles valuées. *Publ. Math. I.H.E.S*, 41, 1972.

[B–S] Z. I. Borevich and I. R. Shafarevich. *Number Theory*. Academic Press, New York, 1966.

[BT] J. J. Burckhardt. *Die Bewegungsgruppen der Kristallographie*. Birkhäuser, Basel, 1947.

[BU1] Herbert Busemann. *Convex Surfaces*. Interscience, New York, 1958.

[BU2] Herbert Busemann. *Recent Synthetic Differential Geometry*. *Ergebnisse der Mathematik, 54*, Springer-Verlag, Berlin, 1970.

[BUR] C. D. Burnside. *Mapping from Aerial Photographs*. Wiley, New York, 1979.

[BU–ZA] Yu. D. Burago and V. A. Zalgaller. *Geometric Inequalities*. Springer-Verlag, to appear.

[B–W] Bart J. Bok and Frances W. Wright. *Basic Marine Navigation*. Houghton–Mifflin, Boston, 1944.

[BZ] Walter Benz. *Vorlesungen über Geometrie der Algebren. Grundlehren der Mathematischen Wissenschaften, 197*, Springer-Verlag, Berlin, 1973.

[CA] Stewart Scott Cairns. *Introductory Topology*. Ronald Press, New York, 1961.

[CAL] Francis P. Callahan. Morley polygons. *American Math. Monthly*, 84:325–337, 1977.

[CAR] Manfredo P. do Carmo. *Differential Geometry of Curves and Surfaces*. Prentice-Hall, Englewood Cliffs, N.J., 1976.

[C–B] P. Couderc and A. Ballicioni. *Premier Livre du Tétraèdre*. Gauthier-Villars, Paris.

[CD] C. Carathéodory. The most general transformation which transforms circles into circles. *Bull. Amer. Math. Soc.*, 43:573–579, 1937.

[C–D–L] Claude Cohen-Tannoudji, Bernard Diu, and Franck Laloë. *Quantum Mechanics*. Wiley, New York, 1977.

[CE1] Elie Cartan. Sur les propriétés topologiques des quadriques complexes. In *Oeuvres complètes*, CNRS, Paris, 1984.

[CE2] Elie Cartan. *The Theory of Spinors*. MIT Press, Cambridge, Mass., 1966.

[C–G] Jeff Cheeger and Detlef Gromoll. On the structure of complete manifolds of nonnegative curvature. *Ann. of Math.*, 96:413–443, 1972.

[CH1] Henri Cartan. *Differential Calculus*. Hermann, Paris, 1971.

[CH2] Henri Cartan. *Elementary theory of analytic functions of one or more complex variables.* Hermann, Paris, 1963.

[CH–GR] S. S. Chern and P. Griffiths. Abel's theorem and webs. *Jahresberichte der Deutschen Math. Vereinigung,* 80:13–110, 1978.

[CL1] Robert Connelly. A counterexample to the rigidity conjecture for polyhedra. *Publ. Math. I.H.E.S,* 47, 1978.

[CL2] Robert Connelly. A flexible sphere. *Mathematical Intelligencer,* 1:130–131, 1978.

[CL3] Robert Connelly. An attack on rigidity (I and II). Preprint, Cornell University.

[C–M] H. S. M. Coxeter and W. O. J. Moser. *Generators and Relations for Discrete Groups. Ergebnisse der Mathematik, 14,* Springer-Verlag, Berlin, fourth edition, 1980.

[CN] R. Cuénin. *Cartographie Générale.* Eyrolles, Paris.

[CN–GR] G. D. Chakerian and H. Groemer. Convex bodies of constant width. In P. M. Gruber and J. M. Wills, editors, *Convexity and Its Applications,* Birkhäuser, Basel, 1983.

[COO] Julian L. Coolidge. *A Treatise on the Circle and the Sphere.* Clarendon Press, Oxford, 1916.

[CP] Christophe. *L'idée fixe du savant Cosinus.* Armand Colin, Paris.

[CR1] H. S. M. Coxeter. *Introduction to Geometry.* Wiley, New York, second edition, 1969.

[CR2] H. S. M. Coxeter. *Regular Polytopes.* Dover, New York, third edition, 1973.

[CR3] H. S. M. Coxeter. *Non-Euclidean Geometry.* University of Toronto Press, Toronto, third edition, 1957.

[CR4] H. S. M. Coxeter. *Regular Complex Polytopes.* Cambridge University Press, London, 1974.

[CR5] H. S. M. Coxeter. The problem of Apollonius. *American Math. Monthly,* 75:5–15, 1968.

[CS] J. W. S. Cassels. *An Introduction to the Geometry of Numbers.* Springer-Verlag, Berlin, 1959.

[CT] Gustave Choquet. *Cours d'Analyse, II: Topologie.* Masson, Paris, 1964.

[CY] Claude Chevalley. *Theory of Lie Groups.* Princeton University Press, Princeton, 1946.

[CZ] Jean-Pierre Conze. *Le théorème d'isomorphisme d'Ornstein et la classification des systèmes dynamiques en théorie ergodique.* Lecture Notes in Mathematics, *383*, Springer-Verlag, Berlin, 1974 (Séminaire Bourbaki, November 1972).

[D–C1] Robert Deltheil and Daniel Caire. *Géométrie.* J. B. Baillière, Paris.

[D–C2] Robert Deltheil and Daniel Caire. *Compléments de Géométrie.* J. B. Baillière, Paris.

[DE1] Jean Dieudonné. *La Géométrie des Groupes Classiques. Ergebnisse der Mathematik, 5,* Springer-Verlag, Berlin, third edition, 1971.

[DE2] Jean Dieudonné. *Linear Algebra and Geometry.* Hermann and Houghton, Mifflin, Paris and Boston, 1969.

[DE3] Jean Dieudonné. *Infinitesimal Calculus.* Hermann, Paris, 1971.

[DE4] Jean Dieudonné. *Treatise on Analysis. Pure and Applied Mathematics, 10,* Academic Press, New York, 1969–.

[DE5] Jean Dieudonné. *History of Algebraic Geometry.* Wadsworth, Belmont, Ca., 1985.

[D–G–K] L. Danzer, B. Grünbaum, and V. Klee. Helly's theorem and its relatives. In *Convexity,* pages 101–180, AMS, Providence, R.I., 1963.

[DI] P. Dembowski. *Finite Geometries. Ergebnisse der Mathematik, 44,* Springer-Verlag, Berlin, 1968.

[DI–CA] Jean Dieudonné and James B. Carrell. *Invariant Theory, Old and New.* Academic Press, New York, 1971.

[DO] Heinrich Dorrie. *100 Great Problems of Elementary Mathematics.* Dover, New York, 1965.

[DP] Baron Charles Dupin. *Géométrie et Mécanique des Arts et Métiers et des Beaux-Arts.* Bachelier, Paris, 1825.

[DQ] Ernest Duporcq. *Premiers Principes de Géométrie Moderne.* Gauthier-Villars, Paris, third edition.

[DR] Jacques Dixmier. *Cours de Mathématiques du premier cycle, deuxième année.* Gauthier-Villars, Paris, 1968.

[DV] Aryeh Dvoretzky. Some results on convex bodies an Banach spaces. In *International Symposium on Linear Spaces,* pages 123–160, Jerusalem Academy Press, Jerusalem, 1961.

[DX] Gaston Darboux. *Principes de Géométrie Analytique.* Gauthier-Villars, Paris, 1917.

[DY] Adrien Douady. Le shaddock à six becs. *Bulletin A.P.M.E.P*, 281:699, 1971.

[EE] *Enzyklopädie der Mathematischen Wissenschaften*. Volume III.2.1, Teubner, Leipzig, 1903–1915.

[E–K] James Eells and Nicolas O. Kuiper. Manifolds which are like projective planes. *Publ. Math. I.H.E.S*, 14:5–46, 1962.

[EL] W. J. Ellison. Waring's problem. *American Math. Monthly*, 78:10–36, 1971.

[E–M–T] C. J. A. Evelyn, G. B. Money-Coutts, and J. A. Tyrell. *Le Théorème des Sept Cercles*. CEDIC, Paris.

[EN] H. G. Eggleston. *Convexity*. Cambridge University Press, Cambridge [Eng.], 1958.

[ER] M. C. Escher. *The graphic work of M. C. Escher*. Gramercy Publishing Co., New York, 1984.

[E–S] J. Eells and J. H. Sampson. Harmonic mappings of Riemannian manifolds. *American Journal of Math.*, 86:109–160, 1964.

[FA] K. J. Falconer. A characterization of plane curves of constant width. *J. London Math. Soc.*, 16:536–538, 1977.

[FI] Jay P. Fillmore. Symmetries of surfaces of constant width. *J. of Diff. Geometry*, 3:103–110, 1969.

[FL] Jean Frenkel. *Géométrie pour l'élève-professeur*. Hermann, Paris, 1973.

[FN] William Fulton. *Algebraic Curves*. Benjamin, New York, 1969.

[FR] Herbert Federer. *Geometric Measure Theory. Grundlehren der Mathematischen Wissenschaften, 153*, Springer-Verlag, Berlin, 1969.

[FT1] L. Fejes Tóth. *Lagerungen in der Ebene, auf der Kugel und im Raum. Grundlehren der Mathematischen Wissenschaften, 65*, Springer-Verlag, Berlin, second edition, 1972.

[FT2] L. Fejes Tóth. *Regular Figures*. Macmillan, New York, 1964.

[GA] Martin Gardner. Extraordinary non-periodic tiling that enriches the theory of tilings. *Scientific American*, January 1977.

[GE] F. Gonseth. Un théorème relatif à deux ellipsoïdes confocaux. *Bulletin des Sciences Mathématiques*, 42:177–180 and 193–194, 1918.

[GF] Phillip A. Griffiths. Variations on a theorem of Abel. *Inventiones Math.*, 35:321–390, 1976.

[GG1] Marvin J. Greenberg. *Euclidean and Non-Euclidean Goemetry, Development and History*. W. H. Freeman, San Francisco, 1980.

[GG2] Marvin J. Greenberg. *Lectures in Algebraic Topology.* W. A. Benjamin, Reading, Mass., 1967.

[G-G-V] I. M. Gel'fand, M. I. Graev, and N. Ya. Vilenkin. *Generalized Functions.* Volume 5: Integral Geometry and Representation Theory, Academic Press, New York, 1969.

[G-H] M. Gage and R. S. Hamilton. The heat equation shrinking convex plane curves. Preprint.

[GK] Heinrich Behnke et al. *Grundzüge der Mathematik.* Volume 4: Praktische Methoden und Anwendungen der Mathematik, Vandenhoek, Göttingen, 1966.

[G-K-M] D. Gromoll, W. Klingenberg, and W. Meyer. *Riemannsche Geometrie im Grossen. Lecture Notes in Mathematics, 55,* Springer-Verlag, Berlin, second edition, 1975.

[GL] Herman Gluck. Almost all simply connected closed surfaces are rigid. In *Geometric Topology*, pages 225–239, Springer-Verlag, Berlin, 1975.

[GM] André Gramain. *Topologie des Surfaces.* Presses Universitaires de France, Paris.

[GN] Paul Gérardain. Mathématiques élémentaires approfondies. Mimeographed notes, Université Paris VII, U.E.R de Mathématiques.

[G-O] Berdard D. Gelbaum and John M. H. Olmsted. *Counterexamples in Analysis.* Holden-Day, San Francisco, 1964.

[GR] Heinrich W. Guggenheimer. *Plane Geometry and its Groups.* Holden-Day, San Francisco, 1967.

[GR-HA1] P. Griffiths and S. Harris. On Cayley's explicit solution to Poncelet's porism. *L'Enseignement Mathématique (sér. 2),* 24:31–40, 1978.

[GR-HA2] P. Griffiths and S. Harris. *Principles of Algebraic Geometry.* Wiley, New York, 1978.

[GR-SH1] B. Grünbaum and G. C. Shephard. The eighty-one types of isohedral tilings in the plane. *Math. Proc. Cambridge Phil. Soc.,* 82, 1977.

[GR-SH2] B. Grünbaum and G. C. Shephard. The ninety-one types of isogonal tilings in the plane. *Trans. Amer. Math. Soc.,* 242, 1978.

[GS] O. Bottema et al. *Geometric Inequalities.* Wolters-Noordhoff, Groningen, 1969.

[GT] A. Guichardet. *Calcul intégral.* Armand Colin, Paris, 1969.

[GX] Lucien Godeaux. *Les géométries.* Armand Colin, Paris, 1937.

[G–W] J. C. Gibbons and C. Webb. *Circle-preserving maps of spheres.*
Preprint, Illinois Institute of Technology, Chicago.

[GZ] Heinz Götze. *Castel del Monte: Gestalt und Symbol der Architektur
Friedrichs II.* Prestel-Verlag, München, 1984.

[HA] M. Hall. *The Theory of Groups.* Macmillan, New York, 1959.

[HA–WR] G. H. Hardy and E. M. Wright. *An Introduction to the Theory of
Numbers.* Clarendon Press, Oxford, 1945.

[H–C] D. Hilbert and S. Cohn-Vossen. *Geometry and the Imagination.*
Chelsea, New York, 1952.

[HD] Jacques Hadamard. *Leçons de Géométrie Elémentaire.* Armand Colin,
Paris, fifth edition, 1911-15.

[HG] Hans Haug. *L'art en Alsace.* Arthaud, Grenoble, 1962.

[H–K] O. Haupt and H. Künneth. *Geometrische Ordnungen. Grundlehren
der Mathematischen Wissenschaften, 133,* Springer-Verlag, Berlin, 1967.

[HL] Alan Holden. *Shapes, space and symmetry.* Columbia Univ. Press,
New York, 1971.

[H–L–P] G. H. Hardy, J. E. Littlewood, and G. Pólya. *Inequalities.*
Cambridge University Press, Cambridge [Eng.], 1934.

[HM] P. Hartman. On isometries and a theorem of Liouville. *Mathematische
Zeitschrift,* 69:202–210, 1958.

[HN] Sigurdur Helgason. *Differential Geometry, Lie Groups and Symmetric
Spaces.* Academic Press, New York, 1978.

[HOL] Raymond d'Hollander. *Topologie Générale.* Volume 1, Eyrolles, Paris.

[HO–PE] W. V. D. Hodge and D. Pedoe. *Methods of Algebraic Geometry.*
Cambridge University Press, Cambridge [Eng.], 1947–1954.

[H–P] Daniel R. Hughes and Fred C. Piper. *Projective Planes. Graduate
Texts in Mathematics, 6,* Springer-Verlag, New York, 1973.

[HR] Ludwig Hadwiger. *Vorlesungen über Inhalt, Oberfläche und
Isoperimetrie. Grundlehren der Mathematischen Wissenschaften, 93,*
Springer-Verlag, Berlin, 1957.

[HS] Joseph Hersch. Quatre propriétés des membranes sphériques
homogènes. *Comptes Rendus Acad. Sci. Paris,* 270:1714–1716, 1970.

[HU] Dale Husemoller. *Fibre Bundles. Graduate Texts in Mathematics, 20,*
Springer-Verlag, New York, second edition, 1975.

[H–W] P. J. Hilton and S. Wylie. *Homology Theory.* Cambridge University
Press, Cambridge [Eng.], 1960.

[H-Y] John G. Hocking and Gail S. Young. *Topology*. Addison-Wesley, Reading, Mass., 1961.

[HZ] M. A. Hurwitz. Sur quelques applications géométriques des séries de Fourier. *Annales Ec. Norm.*, 19:357–408, 1902.

[I-B] I. M. Îaglom and V. G. Boltyanskii. *Convex Figures*. Holt, Rinehart and Winston, New York, 1961.

[I-R] G. Illiovici and P. Robert. *Géométrie*. Eyrolles, Paris.

[JE] Jürgen Joedicke. *Shell Architecture*. Reinhold, New York, 1963.

[KE] Paul Krée. *Introduction aux Mathématiques et à leurs applications fondamentales, M.P. 2*. Dunod, Paris, 1969.

[KF] Nicholas D. Kazarinoff. *Geometric Inequalities*. Random House, New York, 1961.

[KG1] Wilhelm Klingenberg. *A Course in Differential Geometry. Graduate Texts in Mathematics, 51*, Springer-Verlag, New York, 1978.

[KG2] Wilhelm Klingenberg. Paarsymmetrische alternierende Formen zweiten Grades. *Abhandl. Math. Sem. Hamburg*, 19:78–93, 1955.

[KH] A. G. Kurosh. *Lectures on General Algebra*. Pergamon, Oxford, 1965.

[KJ1] Marie-Thérèse Kohler-Jobin. Démonstration de l'inégalité isopérimétrique. *Comptes Rendus Acad. Sci. Paris*, 281:119–120, 1976.

[KJ2] Marie-Thérèse Kohler-Jobin. Une propriété de monotonie isopérimétrique qui contient plusieurs théorèmes classiques. *Comptes Rendus Acad. Sci. Paris*, 284:917–920, 1978.

[KM] Tilla Klotz-Milnor. Efimov's theorem about complete immersed surfaces of negative curvature. *Advances in Math.*, 8:474–543, 1972.

[KN1] Felix Klein. *Lectures on the Icosahedron*. Paul, London, 1913.

[KN2] Felix Klein. *Vorlesungen über Höhere Geometrie*. Chelsea, New York, third edition, 1949.

[KO-NO] Shoshichi Kobayashi and Katsumi Nomizu. *Foundations of Differential Geometry. Tracts in Mathematics, 15*, Interscience, New York, 1963–1969.

[KS] Gabriel Koenigs. *Leçons de Cinématique*. Hermann, Paris, 1897.

[KT] Herbert Knothe. Contributions to the theory of convex bodies. *Michigan Math. Journal*, 4:39–52, 1957.

[KY] George P. Kellaway. *Map Projections*. Methuen, London, 1970.

[LB1] Henri Lebesgue. *Leçons sur les Constructions Géométriques*. Gauthier-Villars, Paris, 1949.

[LB2] Henri Lebesgue. *Les coniques*. Gauthier-Villars, Paris, 1955.

[LB3] Henri Lebesgue. Octaèdres articulés de Bricard. *L'Enseignement Mathématique (ser. 2)*, 13:175–185, 1967.

[LE] Cornelis G. Lekkerkerker. *Geometry of Numbers*. Wolters-Noordhoff, Groningen, 1969.

[LEV] Paul Lévy. Le problème des isopérimètres et les polygones articulés. *Bull. Soc. Math. France*, 90:103–112, 1966.

[LF1] Jacqueline Lelong-Ferrand. *Géométrie Différentielle*. Masson, Paris, 1963.

[LF2] Jacqueline Lelong-Ferrand. Transformations conformes et quasi-conformes des variétés riemanniennes compactes. *Mémoires Acad. Royale Belg., Cl. Sci. Mém. Coll.*, 5, 1971.

[LF3] Jacqueline Lelong-Ferrand. Invariants conformes globaux sur les variétés riemanniennes. *J. of Diff. Geometry*, 8:487–510, 1973.

[LF–AR] J. Lelong-Ferrand and J.-M. Arnaudiès. *Géometrie et Cinématique. Cours de mathématiques, vol. 3*, Dunod, Paris, 1972.

[LG1] Serge Lang. *Elliptic Functions*. Addison-Wesley, Reading, Mass., 1973.

[LG2] Serge Lang. *Analysis II*. Addison-Wesley, Reading, Mass., 1969.

[LM1] J. Lemaire. *Hypocycloïdes et Epicycloïdes*. Blanchard, Paris, 1967.

[LM2] J. Lemaire. *L'hyperbole Equilatère*. Vuibert, Paris, 1927.

[LS] Jean-Jacques Levallois. *Géodésie Générale*. Volume 2, Eyrolles, Paris, 1969.

[LU] Lazar A. Lйusternik. *Convex Figures and Polyhedra*. Dover, New York, 1963.

[LW] K. Leichtweiß. *Konvexe Mengen*. Springer-Verlag, Berlin, 1980.

[LY] Harry Levy. *Projective and Related Geometries*. Macmillan, New York, 1964.

[LZ] V. F. Lazutkin. The existence of caustics for a billiard problem in a convex domain. *Math. USSR: Izvestia*, 7:185–214, 1973.

[MA] Paul Malliavin. *Géométrie Différentielle Intrinsèque*. Hermann, Paris, 1972.

[MB] Benoît B. Mandelbrot. *Fractals: Form, Chance and Dimension.* W. H. Freeman, San Francisco, 1977.

[MD] A. Marchaud. Les surfaces du second ordre en géométrie finie. *J. Math. pures et appl.*, 9–15:293–300, 1936.

[ME] Ricardo Mañé. *Ergodic Theory. Ergebnisse der Mathematik*, Springer-Verlag, Berlin, 1986.

[MG] Caroline H. MacGillavry. *Fantasy and Symmetry: The Periodic Drawings of M. C. Escher.* Harry N. Abrahams, New York, 1976.

[MI] John Milnor. A problem in cartography. *Amer. Math. Monthly*, 76:1101–1102, 1969.

[ML] Charles Michel. *Compléments de Géométrie Moderne.* Vuibert, Paris, 1926.

[M–P] P. S. Modenov and A. S. Parkhomenko. *Géométrie des Transformations.* Volume 1, Academic Press, New York, 1965.

[MR] John Mather. The nice dimensions. In *Liverpool Singularities Symposium*, pages 207–253. *Lecture Notes in Mathematics*, 192, Springer-Verlag, New York, 1971.

[M–T] André Martineau and François Trèves. *Eléments de la théorie des espaces vectoriels topologiques et des distributions.* Centre de Documentation Universitaire, Paris, 1962–1964.

[MW1] G. D. Mostow. *Strong Rigidity of Locally Symmetric Spaces. Annals of Mathematics Studies, 78*, Princeton University Press, Princeton, 1973.

[MW2] G. D. Mostow. Discrete subgroups of Lie groups. *Advances in Mathematics*, 15:112–123, 1975.

[MY] William S. Massey. *Algebraic Topology: an Introduction. Graduate Texts in Mathematics, 56*, Springer-Verlag, New York, 1977.

[NA] Rolf Nevanlinna. On differentiable mappings. In *Analytic Functions*, pages 3–9, Princeton University Press, Princeton, 1960.

[NU] Jason John Nassau. *Practical Astronomy.* McGraw-Hill, New York, second edition, 1948.

[NW] P. E. Newstead. Real classification of complex conics. *Matematika*, 28:36–53, 1981.

[OA] M. Obata. The conjecture on conformal transformations of riemannian manifolds. *J. of Diff. Geometry*, 6:247–258, 1972.

[OM] O. T. O'Meara. *Introduction to Quadratic Forms. Grundlehren der Mathematischen Wissenschaften, 117*, Springer-Verlag, Berlin, 1963.

[OS1] Robert Osserman. Bonnesen-style isoperimetric inequalities. *Amer. Math. Monthly*, 86:1–29, 1979.

[OS2] Robert Osserman. The isoperimetric inequality. *Bull. Amer. Math. Soc.*, 84:1182–1238, 1978.

[PA] Richard Palais. *The Classification of G-Spaces. AMS Memoirs, 36*, American Math. Society, Providence, R.I., 1960.

[PC] Blaise Pascal. *Pensées*. Dutton, New York, 1958.

[PE] D. Pedoe. *A Course of Geometry*. Cambridge University Press, London, 1970.

[PEN] Roger Penrose. The geometry of the universe. In Lynn Arthur Steen, editor, *Mathematics Today*, pages 83–125, Springer-Verlag, New York, 1978.

[PL] William F. Pohl. A theorem of géométrie finie. *J. of Diff. Geometry*, 10:435–466, 1975.

[PN] L. E. Payne. Isoperimetric inequalities and their applications. *SIAM Review*, 9:453–488, 1967.

[PO] Ian R. Porteous. *Topological Geometry*. Van Nostrand–Reinhold, London and New York, 1969.

[PR] T. I. Porter. A history of the classical isoperimetric problem. In Chicago University: *Contributions to the Calculus of Variations 1931-1932*, Chicago University Press, 1933.

[P–S] G. Pólya and G. Szegö. *Isoperimetric Inequalities in Mathematical Physics*. Princeton University Press, Princeton, 1951.

[PT] G. Pickert. *Projektive Ebenen. Grundlehren der Mathematischen Wissenschaften, 80*, Springer-Verlag, Berlin, 1955.

[PV1] A. V. Pogorelov. *Extrinsic Geometry of Convex Surfaces*. American Math. Society, Providence, R.I., 1973.

[PV2] A. V. Pogorelov. *Hilbert's Fourth Problem*. Washington, Winston, 1979.

[RA] Hans Rademacher. *Topics in Analytic Number Theory. Grundlehren der Mathematischen Wissenschaften, 169*, Springer-Verlag, Berlin, 1973.

[RB] Hugh S. Roblin. *Map Projections*. Edward Arnold, London, 1969.

[R–C] Eugène Rouché and Charles de Comberousse. *Traité de Géométrie*. Gauthier-Villars, Paris, 1922.

[RE] Robert Roussarie. Sur les feuilletages de variétés de dimension 3. *Annales de l'Institut Fourier*, 21(3):13–81, 1971.

[RM] Georges de Rham. Sur les polygones générateurs des groupes fuchsiens. *L'Enseignement Mathématique (ser. 2)*, 17:49–61, 1971.

[RN1] Raphael M. Robinson. Undecidability and nonperiodicity for tilings of the plane. *Inventiones Math.*, 12:177–209, 1971.

[RN2] Raphael M. Robinson. Undecidable tiling problems in the hyperbolic plane. *Inventiones Math.*, 44:259–264, 1978.

[RN3] Raphael M. Robinson. Comments on the Penrose tilings. Preprint, University of California, Berkeley.

[RO] V. G. Romanov. *Integral Geometry and Inverse Problems for Hyperbolic Equations.* Springer-Verlag, Berlin, 1974.

[RS] Claude A. Rogers. *Packing and Covering.* Cambridge University Press, Cambridge [Eng.], 1964.

[RU] Walter Rudin. *Real and Complex Analysis.* McGraw-Hill, New York, second edition, 1974.

[R–V] Arthur W. Roberts and Dale E. Varberg. *Convex Functions.* Academic Press, New York, 1973.

[RO–SA] Arthur H. Robinson and Randall D. Sale. *Elements of Cartography.* Wiley, New York, third edition, 1969.

[SA] Goro Shimura. *Introduction to the Arithmetic Theory of Automorphic Functions.* Princeton University Press, Princeton, 1971.

[SAL] G. Salmon. *A Treatise on the Analytic Geometry of Three Dimensions.* Hodges, Dublin, third edition, 1874.

[SB] Shlomo Sternberg. *Lectures on Differential Geometry.* Prentice-Hall, Englewood Cliffs, N.J., 1964.

[SC1] Doris Schattschneider. The plane symmetry groups: their recognition and notation. *American Mathematical Monthly*, 85:439–450, 1978.

[SC2] Doris Schattschneider. Tiling the plane with congruent pentagons. *Mathematics Magazine*, 51:29–44, 1978.

[SD1] Otto Staude. Fadenkonstruktionen des Ellipsoides. *Math. Annalen*, 20:147–184, 1882.

[SD2] Otto Staude. *Die Fokaleigenschaften der Flächen Zweiter Ordnung.* Teubner, Leipzig.

[SE1] Jean-Pierre Serre. *Local Fields. Graduate Texts in Mathematics, 67,* Springer-Verlag, New York, 1979.

[SE2] Jean-Pierre Serre. *A Course in Arithmetic. Graduate Texts in Mathematics, 7,* Springer-Verlag, New York, 1973.

[SE3] Jean-Pierre Serre. *Algèbres de Lie Semisimples Complexes*. W. A. Benjamin, New York, 1966.

[SE–TH] H. Seifert and W. Threlfall. *A Textbook of Topology*. Academic Press, New York, 1980.

[SF1] Hans Schwerdtfeger. Invariants of a class of transformation groups. *Aequationes Math.*, 14:105–110, 1976.

[SF2] Hans Schwerdtfeger. Invariants à cinq points dans the plan projectif. *Comptes Rendus Acad. Sci. Paris*, 285:127–128, 1977.

[SG] A. Seidenberg. *Lectures in Projective Geometry*. Van Nostrand, Princeton, 1962.

[SI] Ya. G. Sinai. *Introduction to Ergodic Theory*. Princeton University Press, Princeton, 1976.

[SK] Michael Spivak. *A Comprehensive Introduction to Differential Geometry*. Publish or Perish, Berkeley, Ca., second edition, 1979.

[SL1] G. T. Sallee. Maximal area of Reuleaux polygons. *Canadian Math. Bull.*, 13:175–179, 1970.

[SL2] G. T. Sallee. Reuleaux polytopes. *Mathematika*, 17:315–323, 1970.

[SO1] Luis Antonio Santaló Sors. *Introduction to Integral Geometry*. Hermann, Paris, 1953.

[SO2] Luis Antonio Santaló Sors. *Integral Geometry and Geometric Probability*. Addison-Wesley, New York, 1976.

[SR] Edwin H. Spanier. *Algebraic Topology*. McGraw-Hill, New York, 1966.

[S–T] Ernst Snapper and Robert J. Troyer. *Metric Affine Geometry*. Academic Press, New York, 1971.

[ST–RA] E. Steinitz and H. Rademacher. *Vorlesungen über die Theorie der Polyeder. Grundlehren der Mathematischen Wissenschaften, 41,* Springer-Verlag, Berlin, 1934.

[SU] Pierre Samuel. Unique factorization. *American Math. Monthly*, 75:945–952, 1968.

[SY] J. L. Synge and B. A. Griffith. *Principles of Mechanics*. McGraw-Hill, New York, 1942.

[SW] Ian Stewart. *Galois Theory*. Chapman and Hall, London, 1973.

[TM] René Thom. Sur la théorie des enveloppes. *J. de Math. Pures et Appl.*, 16:177–192, 1962.

[TS] Jacques Tits. *Buildings of spherical type and finite BN-pairs. Lecture Notes in Mathematics, 386,* Springer-Verlag, Berlin, 1974.

[VE] Frederik A. Valentine. *Convex Sets*. McGraw-Hill, New York, 1964.

[VG1] H. Voderberg. Zur Zerlegung eines ebenen Bereiches in Kongruente. *Jahresberichte d. Deutschen Math. Ver.*, 46:229–231, 1936.

[VG2] H. Voderberg. Zur Zerlegung der Ebene in kongruente Bereiche in Form einer Spirale. *Jahresberichte d. Deutschen Math. Ver.*, 47:159–160, 1937.

[VL] Patrick du Val. *Homographies, Quaternions and Rotations*. Clarendon Press, Oxford, 1964.

[VN] Georges Valiron. *The Geometric Theory of Ordinary Differential Equations and Algebraic Functions. Lie Groups: History, Frontiers and Applications, XIV*, Math Sci Press, Brookline, Mass., 1984.

[V-Y] O. Veblen and J. W. Young. *Projective Geometry*. Ginn and Co., Boston, 1910–1918.

[WF] Joseph A. Wolf. *Spaces of Constant Curvatures*. Boston, 1974.

[WK] R. J. Walker. *Algebraic Curves*. Princeton University Press, Princeton, 1950.

[WL] Hermann Weyl. *Symmetry*. Princeton University Press, Princeton, 1952.

[WN] Magnus J. Wenninger. *Polyhedron Models*. Cambridge University Press, Cambridge [Eng.], 1971.

[WO] Yung-Chow Wong. *Isoclinic n-planes in Euclidean 2n-space, Clifford parallels in elliptic $(2n-1)$-space and the Hurwitz matrix equations. AMS Memoirs, 41*, American Math. Society, Providence, R.I., 1961.

[WR] Frank Warner. *Foundations of Differentiable Manifolds and Lie Groups*. Scott, Foresman, Greenville, Ill., 1971.

[ZN] Michel Zisman. *Topologie Algébrique Elémentaire*. Armand Colin, Paris, 1972.

[ZR] C. Zwikker. *The Advanced Geometry of Plane Curves and their Applications*. Dover, New York, 1963.

Index of notations

Index

Entries in italics refer to definitions. Entries consisting of a roman numeral and a number refer to volume and page number.

H

Acknowledgements

We are pleased to acknowledge the permission of various publishers and institutions to reproduce some of the figures appearing in this book:

Figs. 9.12.7.2, 12.10.9.2.1, 18.1.1.3, 18.1.6.1: ©Armand Colin Editeur, Paris

Fig.   10.12.1.2: ©Librairie Arthaud, Paris

Figs. 1.7.4.8, 1.7.6.13, page 108 (vol. I), 19.6.12.5:
       ©BEELDRECHT, Amsterdam/BILD-KUNST, Bonn 1982

Figs. 1.7.4.1 - 1.7.4.5, 1.7.6.1 - 1.7.6.12: ©Cedic, Paris

Figs. 12.1.3.2.1, 12.1.3.2.2, 12.6.10.5.3, 12.6.10.5.4:
       ©Cambridge University Press, Cambridge

Figs. 4.3.9.11, 4.3.9.12, 15.3.3.3.2, 18.1.8.6, 20.7.3:
       ©Chelsea Publishing Company, New York

Figs. 12.1.1.5, 12.1.1.6: ©1971 Columbia University Press, New York

Fig.   18.9.4.2: ©Conference Board of the Mathematical Sciences, Washington

Figs. 12.5.6.1, 12.5.6.2.1 - 12.5.6.2.4: ©Dover Publications, Inc., New York

Fig.   18.1.8.7: ©Eyrolles, Paris

Figs. 4.7.4.1, 4.7.4.2: ©Institut Géographique National, Paris
       (Autorisation no. 99-0175)

Figs. 1.8.5, 19.6.12.4: ©John Wiley & Sons Inc., New York

Figs. 14.4.4.6.3, 14.4.6.4, 15.3.3.3.3, 15.3.3.3.4:
       ©Karl Krämer Verlag, Stuttgart

Fig.   page 22 (vol. II): ©Prestel-Verlag, München

Fig.   12.5.5.7: ©Princeton University Press, Princeton

Fig.   12.10.5.2: ©Vandenhoeck & Ruprecht, Göttingen